Contents

_____

Contributors

_____

Araus, J.L., Unit of Plant Physiology, Dept. of Plant Biology, University of Barcelona, Barcelona E-08028, Spain and International Maize and Wheat Improvement Center (CIMMYT) Km 45, Carretera México-Veracruz, Texcoco CP56130, Mexico. E-mail: j.araus@cgiar.org

Ascough, J.C. II, USDA-ARS, Agricultural Systems Research Unit, 2150 Centre Avenue, Building D, Suite 200, Fort Collins, CO 80526, USA. E-mail: jim.ascough@ars.usda.gov

Colalongo, C., Department of Agroenvironmental Sciences and Technology, University of Bologna, Viale Fanin 44, 40127, Bologna, Italy. E-mail: maria.colalongo2@unibo.it

Condon, A.G., CSIRO Plant Industry, PO Box 1600, Canberra, ACT 2601, Australia. E-mail: Tony.Condon@csiro.au

De Pauw, E., International Center for Agricultural Research in the Dry Areas (ICARDA), Aleppo, Syria. E-mail: e.de-pauw@cgiar.org

Ferrio, J.P., Dept. of Crop and Forest Sciences, University of Lleida, Lleida E-25198, Spain. E-mail: Pitter.Ferrio@pvcf.udl.cat

Foyer, C.H., Centre of Plant Science, Research Institute of Integrative and Comparative Biology, Faculty of Biological Science, University of Leeds, Leeds, LS2 9JT, UK. E-mail: C.Foyer@leeds.ac.uk

Göbel, W., International Center for Agricultural Research in the Dry Areas (ICARDA), Aleppo, Syria. E-mail: w.goebel@cgiar.org

Intrigliolo, D.S., Instituto Valenciano de Investigaciones Agrarias, Centro para el Desarrollo de la Agricultura Sostenible, Apartado Oficial 46113, Moncada (Valencia), Spain. E-mail: intrigliolo_die@ivia.gva.es

Lopes, M.S., International Maize and Wheat Improvement (CIMMYT), Km. 45, Carretera Mexico-Veracruz, El Batan, Texcoco, CP 56130 Mexico. E-mail: m.dasilva@cgiar.org

Maccaferri, M., Department of Agroenvironmental Sciences and Technology, University of Bologna, Viale Fanin 44, 40127, Bologna, Italy. E-mail: marco.maccaferri@unibo.it

McMaster, G.S., USDA-ARS, Agricultural Systems Research Unit, 2150 Center Avenue, Building D, Suite 200, Fort Collins, CO 80526, USA. E-mail: greg.mcmaster@ars.usda.gov

Ortiz, R., Centro Internacional de Mejoramiento de Maíz y Trigo (CIMMYT), Apdo, Postal 6-641, 06600 Mexico, D.F., Mexico. Current address: Swedish University of Agricultural Sciences, PO Box 101, SE-23053, Alnarp, Sweden. E-mail: rodomiroortiz@gmail.com

Peng, S., Crop and Environmental Sciences Division, International Rice Research Institute (IRRI), DAPO Box 7777, Metro Manila, Philippines. E-mail: s.peng@cgiar.org

Ramos, C., Instituto Valenciano de Investigaciones Agrarias, Centro para el Desarrollo de la Agricultura Sostenible, Apartado Oficial 46113, Moncada (Valencia), Spain. E-mail: ramos_carmon@ivia.gva.es

Rebetzke, G.J., CSIRO Plant Industry, PO Box 1600, Canberra, ACT 2601, Australia. E-mail: Greg.Rebetzke@csiro.au

Richards, R.A., CSIRO Plant Industry, PO Box 1600, Canberra, ACT 2601, Australia. E-mail: richard.richards@csiro.au

Salvi, S., Department of Agroenvironmental Sciences and Technology, University of Bologna, Viale Fanin 44, 40127, Bologna, Italy. E-mail: salvi@agrsci.unibo.it

Satorre, E.H., Cátedra de Cerealicultura, Department of Plant Production, Faculty of Agronomy, University of Buenos Aires and AACREA, Asociación Argentina de Consorcios Regionales de Experimentación Agrícola, Unidad de Investigación & Desarrollo, Argentina. E-mail: satorre@agro.uba.ar

Slafer, G.A., Department of Crop and Forest Sciences, University of Lleida, Av. Alcalde Rovira Roure 191, Lleida, 25198, Spain. E-mail: slafer@pvcf.udl.es

Thompson, R.B., Universidad de Almería, Depto Producción Vegetal, Almería, Spain. E-mail: rodney@ual.es

Tuberosa, R., Department of Agroenvironmental Sciences and Technology, University of Bologna, Viale Fanin 44, 40127, Bologna, Italy. E-mail: roberto.tuberosa@unibo.it

Voltas, J., Dept. of Crop and Forest Sciences, University of Lleida, Lleida E-25198, Spain. E-mail: jvoltas@pvcf.udl.cat

Watt, M., CSIRO Plant Industry, PO Box 1600, Canberra, ACT 2601, Australia. E-mail: michelle.watt@csiro.au

Yang, L., Key Lab of Crop Genetics & Physiology of Jiangsu Province, Yangzhou University, Yangzhou 225009, Jiangsu, China. E-mail: lxyang@yzu.edu.cn

Preface

This book addresses the challenges of the foreseen climate change for agriculture from a multidisciplinary point of view. Agriculture has shaped the world into its present form. It was only after the beginning of agriculture, and the relative food security and sufficiency thereby derived, that after 120,000 years populations have increased substantially, giving place to complex social structures created by civilization. Interestingly, the 'Neolithic revolution' (i.e. the agricultural revolution determining the transition from hunting-gathering to settlement facilitated by the beginning of agriculture) took place somewhat simultaneously in several different parts of the world. It should have been a global force determining such a change in culture. A remarkable feature is that this 'revolution' was temporally coincident with the occurrence of sudden climate changes that, unlike those expected nowadays, determined an increase in humidity which, in turn, determined a climate amelioration for crop growth in the early Holocene. Therefore, in this book the initial chapter is devoted to discussing the global changes that, some 10,000 years ago, gave rise to the beginning of agriculture. The rest of the book is subdivided in two major parts: first, towards an understanding of the present and future challenges imposed by climate change on several different agricultural systems; and, secondly, to reviewing research avenues to cope with the environmental conditions expected in the near future from climate change.

The current climate change, even if global in nature, affects the diverse world agro-ecosystems in different ways. Nevertheless, it is in the dryland systems of the Mediterranean basin where many global change models predict the most severe consequences, due to increases in temperature together with decreased precipitation, which would be less evenly distributed than at present. Concurrent social and demographic changes in the region may complicate this scenario even further. Chapter 2 discusses the predictions for this fragile region, while Chapter 3 analyses the situation in the highly productive agricultural systems of irrigated rice in southern Asia, where in the absence of water stress the increase in atmospheric concentration of CO_2 may represent a positive factor for species with a C_3 metabolism, such as rice, providing that high temperatures are prevented (advancing sowings, changing phenological patterns) and soil fertility is maintained. Chapter 4 deals with Pampean agriculture, another of the World food baskets, also challenged by increases in temperature and changes in levels and patterns of precipitation. Chapter 5 addresses the challenges expected in the already highly technological and added-value horticultural systems, where the possibilities for controlling the environment, particularly temperature and efficiency in the use of water, must be further improved.

In the following part, this book deals with different scientific and technical avenues that are being envisaged to mitigate the expected environmental constraints. First, Chapter 6 offers a detailed discussion on physiological plant responses to an increase in carbon dioxide and to the interaction of this factor with the occurrence of abiotic stresses, such as drought. Chapter 7 illustrates the practical experience in crop breeding of the Australian CSIRO, one of the institutions most credited worldwide concerning breeding for drought adaptation. Undoubtedly molecular techniques have, and will have even more in the future, a key role in breeding efforts to produce crops better suited to global change challenges. However, only through a multidisciplinary approach, combining molecular techniques, field breeding and adequate phenotyping, will advances in breeding be ensured (Chapter 8). Crop management is the other pillar in the amelioration of crop adaptation to the expected environments of the future. Information technologies will have a paramount role in the coming years, helping, for example, to define target environments for crop improvement or to process the flux of information associated with precision agriculture (Chapter 9).

Chapter 10 highlights the need for a global effort, from science to policy, to cover the challenges involved in improving agriculture in a changing environment, particularly in the developing world where political structures are weak but social networks may be of assistance.

1 Global Change and the Origins of Agriculture

J.P. Ferrio, J. Voltas and J.L. Araus

1.1 Introduction

According to Gordon Childe's concept of 'Neolithic Revolution', the shift from hunting and gathering to food production (i.e. agriculture and husbandry) transformed human lifestyles radically and was the most important event since the discovery of fire (Childe, 1952). Indeed, labour-intensive agriculture led to an unprecedented rise in population together with social changes. It was the basic cause of the transformation of villages into cities and the onset of modern civilization (Harlan, 1992).

Agriculture was adopted independently in various parts of the world during the Holocene, a period of global warming that followed the end of the last Ice Age, at about 11,600 cal. BP (see references in Balter, 2007; Purugganan and Fuller, 2009). Consequently, the particular climatic conditions for this period could have played an important role in the success of these early farming communities. It is also after the adoption and spread of agriculture that the first unequivocal examples of large-scale, human-induced impacts on natural ecosystems are detected (Yasuda et al., 2000; Hill, 2004). Comprehensive information on climate factors prevailing during the early stages of agriculture would help explain the context in which agriculture developed and evolved, providing clues on its long-term effects that have contributed to shaping today's landscapes. The aim of this chapter is to provide an overview of the various approaches to investigating the role of climate in the origins of agriculture, as well as the effects of ancient agriculture on the environment.

1.2 Where and When? Agriculture at its Origins

The first irrefutable evidence of domesticated plants (wheat and barley) is found in the Near East about 10,500 cal. BP (Hillman and Davies, 1990; Tanno and Willcox, 2006), but pre-domestic cultivation in this area may go back to 11,500 cal. BP (Willcox et al., 2008) or even 13,000 cal. BP (Hillman et al., 2001; Byrd, 2005). On the other hand, the appearance of sickles dating back to 15,000 cal. BP has been proposed as evidence for cereal harvesting (e.g. Unger-Hamilton, 1989), but it is likely that these tools were used for other purposes well before cereal domestication (Fuller, 2007). The Near East is the most frequently studied area for the origins of agriculture, but at least ten regions are known where agriculture developed independently during the Holocene (Balter, 2007; Purugganan and Fuller, 2009). In fact, increasing evidence suggests that agriculture may have begun in several regions as early as in the Near East (Fuller, 2007; Ranere et al., 2009).

Figure 1.1 summarizes current knowledge about the timing of agriculture's emergence around the world. In East Asia, two early independent centres of the origin of farming have been located, the loess terraces of the Yellow river for millet and the lowlands of the Yangtze river for rice. First domesticated forms date back to at least 9000 cal. BP in both cases (Fuller, 2007; Purugganan and Fuller, 2009), whereas indirect evidence suggests that initial steps towards settled agricultural villages began around 12,000–11,000 cal. BP (Underhill, 1997; Pechenkina et al., 2005). In the New World, the poor preservation of plant macroremains in wet

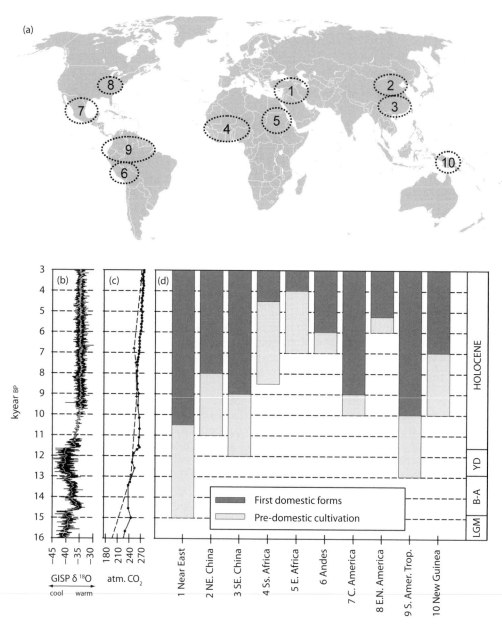

Fig. 1.1. (a) Main core areas where agriculture developed independently (Diamond, 2002; Balter, 2007). (b) Oxygen isotope composition ($\delta^{18}O$, in ‰) from Greenland ice core GISP-2 (Grootes *et al.*, 1993; Stuiver *et al.*, 1995; Stuiver and Grootes, 2000). Black, high-resolution data; grey, mid-resolution. (c) Atmospheric CO_2 levels (μmol/mol), from Taylor Dome (solid line) and Vostok (broken line) ice cores (Barnola *et al.*, 1987; Indermühle *et al.*, 1999). (d) Approximate chronology for cultivation and domestication in areas where agriculture emerged independently (see text for details). LGM, Late Glacial Maximum; B-A, Bölling-Allerød interstadial; YD, Younger Dryas.

tropical areas hinders proper establishment of the beginnings of agriculture in the Central and South American tropics, but there is growing evidence of independent domestication of squash dating back to about 10,000 cal. BP in coastal Peru and Mexico (Smith, 1997; Dillehay *et al.*, 2007); and the study of phytoliths suggests that maize domestication may go back to at least 9000 cal. BP (Ranere *et al.*, 2009). Other independent agricultural systems emerged in eastern North America (sunflower, squash) and in the Andes region (potato and other tuber crops), dating back to about 5000 and 7000 cal. BP, respectively (Smith, 1995; Balter, 2007). Although the process here is not so well understood, two regions for the origin of agriculture have been proposed in Africa: sub-Saharan West Africa, with pearl millet as the main crop, and North-East Africa, in which sorghum was predominant (Fuller, 2007; Purugganan and Fuller, 2009). Unlike in other areas, African agriculture appeared well after the adoption of herding and settled life (9000–7000 cal. BP). Domesticated forms were already cultivated in a wide range of sites by 4500 and 4000 cal. BP for sub-Saharan and North-East Africa, respectively (Fuller, 2007; Purugganan and Fuller, 2009), but the timing for the actual transition to agriculture remains unclear. In the New Guinean Highlands, a controversial interpretation of drainage channels and other soil structures suggests that agriculture based on banana, yam and taro could have evolved independently as early as 10,000 cal. BP (Smith, 1995; Neumann, 2003). Nevertheless, as is the case for the South American tropics, plant macroremains are not well preserved in this area, and only recently have plant phytoliths confirmed the existence of plant domesticates in the region dating back to 7000 cal. BP (Neumann, 2003).

1.3 Views on the Origins of Agriculture: Did Climate Change Play a Key Role?

The reasons why hunter-gatherers started to cultivate are still subject to intense debate,

mostly focusing on the particular case of the Near East. The earliest model for the origins of agriculture is Childe's 'oasis theory' (Childe, 1952). Childe suggested that, after the last Ice Age, South-west Asia became drier and humans began to aggregate in areas where water was available. However, evidence so far indicates that landscape in the Fertile Crescent shifted from a cool, dry steppe during the Younger Dryas to a warmer and perhaps moister open forest when agriculture started (Bottema and Woldring, 1984; Willcox, 1996; Harlan, 1998). Although Childe's theory is not supported by current palaeo-environmental data, it was a crucial stimulus for archaeological research aimed at determining the origin and spread of ancient strategies of food production. Later on, other authors also emphasized the role of climate in the adoption of agriculture in several areas, mainly through its effect on the distribution of the wild ancestors of cultivated plants (MacNeish, 1992; Hillman, 1996).

To test Childe's theory, Braidwood (1958) pioneered a major interdisciplinary approach that became standard practice in field archaeology. Braidwood proposed that the development of agriculture was due to a gradual cultural process, 'as the culmination of the ever increasing cultural differentiation and specialization of human communities'. He rejected the causal role of climate, based on the assumption that comparable changes occurred in previous interglacial periods that did not lead to cultivation routines. However, Braidwood based this assumption on field observations that turned out to be erroneous (Wright, 1993). We now know that none of the previous interglacial periods were as long or as stable as the Holocene (IPCC, 2001; COACC, 2002). Alternatively, Binford (1968) and Flannery (1969), in what has been defined as the 'marginal theory', proposed that a food crisis occurred among semi-sedentary communities of highly developed foragers (e.g. Natufians in the Near East), which forced them to move to marginal regions where cultivation soon became advantageous. Most archaeological evidence, however, indicates that in most cases a substantial decline in nutritional health occurred among early farmers (Cohen and Amelagos, 1984; Larsen, 1995), and that population increase was a consequence, rather than a cause, of the adoption of agriculture (Harris and Hillman, 1989; Harlan, 1992).

More recently, Richerson *et al.* (2001) argued that cultural and demographic hypotheses cannot alone explain why the onset of agriculture took so long to appear in areas like the Near East, where wild progenitors were gathered as early as 23,000 years ago, during the Pleistocene (e.g. Ohalo II in the Levant, Kislev *et al.*, 2004; Weiss *et al.*, 2006). These authors attributed the apparent delay in cultural and/or demographic changes to climate constraints that made agriculture an impossible task until the Holocene. In particular, high-definition data from ice cores suggest that previous interglacial periods were not only shorter and cooler than the Holocene, but also much more variable (Ditlevsen *et al.*, 1996; see also Fig. 1.1a). According to Sage (1995) and Richerson *et al.* (2001), the onset of agriculture might also have been limited by the low CO_2 concentration preceding the Holocene (Fig. 1.1c), which would have reduced photosynthesis and, therefore, harvests by up to 50%. Climate might also have played a role by further delaying agriculture adoption in areas with high climate instability, e.g. in the African Sahel or in the high Andes (De Menocal, 2001). In summary, while climate factors probably assisted in triggering the onset of agriculture, it is obvious that demographic, social and environmental factors other than climate (e.g. CO_2 concentration, soil fertility, orography, vegetation) restricted the number of areas where this transition became effective (Diamond, 2002).

1.4 Environmental Background

After the cold–dry period known as Younger Dryas (ca. 12,800–11,600 cal. BP), the longest warm and stable period in the last 400,000 years, the Holocene, began (IPCC, 2001; COACC, 2002). This period coincided with accelerating human expansion, and environmental factors were probably crucial

for the onset and further spread of agriculture – a development that was at the core of creating modern civilization (see above, Fig. 1.1b, c). Although climate fluctuations were also identified during the Holocene, they are lower in magnitude than those characterizing the earlier Pleistocene. Between 11,500 and 10,500 cal. BP, the climate was still cooler than present, but showed steady improvement towards wetter and warmer conditions until 8200 cal. BP (IPCC, 2001; COACC, 2002). This phase was followed by a short, cold period of about 200 years, around 8200 cal. BP (Alley *et al.*, 1997). Between 8000 and 4500 cal. BP, the climate was somewhat warmer and wetter than today, reaching since then similar features to those of today, except for some cold phases (2900–2300 cal. BP, Iron Age Cold Epoch; 16th–19th centuries, Little Ice Age) (Gribbin and Lamb, 1978; Van-Geel *et al.*, 1998) and warm episodes (9th–14th centuries, Medieval Warm Period) (Gribbin and Lamb, 1978; Bradley *et al.*, 2001).

The above-mentioned climate fluctuations occurred on a global scale, but their actual characteristics were highly variable at the regional or local levels. In particular, precipitation regimes show considerable geographical heterogeneity in response to climate forcing, as shown by historical records (see e.g. Rodó *et al.*, 1997; Cullen *et al.*, 2002). Temperature fluctuations, on the other hand, occur mainly on a global scale, but are associated with changes in atmospheric circulation, which actually leads to a contrasting spatial distribution of precipitation (Rodó *et al.*, 1997; Cullen *et al.*, 2002; Magny *et al.*, 2003). Beyond circulation of air masses, the varying response of precipitation is also due to a multiplicity of factors, including orography and ocean temperature. This heterogeneity is evident not only from current meteorological data, but also from palaeo-environmental records. Thus, for example, the 8.2 ky BP cold period was characterized by a humid climate in Central Europe and the North of the Mediterranean Basin, but a dry one in the South of the Mediterranean Basin (Pérez-Obiol and Julià, 1994; Magny *et al.*, 2003). A similar contrasting pattern among coastal and continental areas of the eastern Iberian Peninsula has been described for the Iron Age Cold Epoch (Aguilera *et al.*, 2009). Therefore, conclusions based only on global palaeo-records on the role of climate as the element triggering the emergence of agriculture should be treated with caution. Instead, the particular conditions in each area of origin need to be examined. As an example, we will focus on the Near East region, the most extensively studied area to date.

1.5 Climate and Origins of Agriculture in the Near East

Mediterranean agriculture, based on winter cereals (mainly wheat and barley) and pulses (pea and lentil), appeared in the 'Fertile Crescent', an area of the Near East comprising the steppes of Syria and southern Levant, as well as some mountain areas of Anatolia. The origins of agriculture in the Fertile Crescent, the species adopted for cropping and the selection pressure experienced by these species as a result of cultivation and their further spread into the Mediterranean Basin are intimately related to moisture availability. Thus, proper characterization of past changes in available water is crucial to the understanding of the origins of agriculture. The climate experienced during the Younger Dryas has been put forward as forcing the transition from gathering to cultivation in the Near East, and it has been suggested that pre-domestic cultivation began at the start of this period at Abu Hureyra (Hillman *et al.*, 2001). However, other authors suggest that agriculture did not appear until the beginning of the Holocene in northern Syria (Willcox *et al.*, 2008) and in the southern Levant (Lev-Yadun *et al.*, 2000; Weiss *et al.*, 2006). Further north, in Anatolia and the Upper Euphrates, contemporary sites show no evidence of cultivation (Savard *et al.*, 2006). The first unequivocal signs of cereal domestication appear at least one millennium after the start of the Holocene and pre-domestic cultivation, between 10,500 and 10,000 cal. BP, in a range of sites in both the southern and northern Levant (see Fuller, 2007 for an

extensive review). The shift from gathering to cultivation is poorly understood. The two modes of food procurement probably coexisted for millennia (gathering continues today in some areas). If agriculture started during the Younger Dryas, as suggested for Abu Hureyra, it probably did not become the main food procurement until the onset of more stable climate conditions during the Holocene (see Fig. 1.2). However, is there any evidence that climate changes affected the availability of food plants in the Near East and thus triggered the beginning of cultivation?

According to global trends in temperature (Fig. 1.2a), the late glacial interstadial (Bølling-Allerød), which preceded the Younger Dryas, was characterized by relatively warm conditions throughout. Pollen records from lake sediments indicate a steady replacement of cold steppe species (e.g. *Artemisia*, chenopods) by arboreal and grass pollen in the Near East, particularly at low altitudes (see Bottema and Woldring, 1984; Yasuda *et al.*, 2000; Roberts *et al.*, 2001; Stevens *et al.*, 2001; Wick *et al.*, 2003). Oxygen isotopes in carbonates from Soreq Cave (Israel) demonstrate higher freshwater inputs (Bar-Matthews *et al.*, 1999; Fig. 1.2b), suggesting higher water availability than at present. Furthermore, charcoal and seed analyses from the Epipalaeolithic Abu Hureyra site (Middle Euphrates, Syria) point to forest–steppe vegetation similar to that found today in wetter and cooler areas (Willcox, 2002b). This favourable climate is followed by the Younger Dryas, a well-defined synchronous cold period, particularly in the northern hemisphere (Berger, 1990; Fig. 1.2a). In the Near East, cool and dry conditions are shown by lake level changes, pollen and diatom analyses and stable isotopes from lake sediments (e.g. Yasuda *et al.*, 2000; Roberts *et al.*, 2001; Hajar *et al.*, 2009), stable isotopes in cave carbonates (Bar-Matthews *et al.*, 1999; Fig. 1.2b) and archaeo-botanical data (Willcox *et al.*, 2009). This is also confirmed by carbon isotope records of cereal grains and wood charcoal from Abu Hureyra (the authors, unpublished), indicating high levels of water stress during this period. Contrasting with

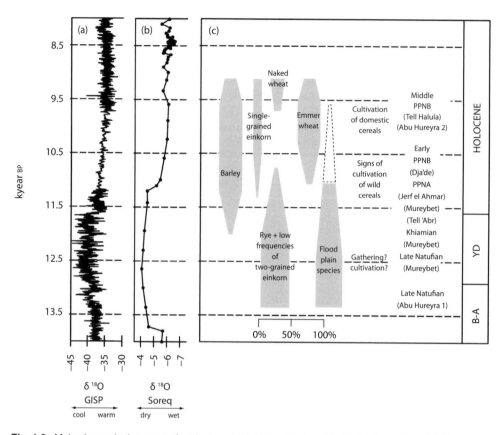

Fig. 1.2. Main phases in the onset of agriculture in the Near East and their climate context. (a) Oxygen isotope composition ($\delta^{18}O$, in ‰) from Greenland ice core GISP-2 (Grootes *et al.*, 1993; Stuiver *et al.*, 1995; Stuiver and Grootes, 2000). (b) $\delta^{18}O$ in carbonates from Soreq cave in Israel (Bar-Matthews *et al.*, 1999). (c) Relative frequencies of selected food plants from Middle Euphrates sites (Willcox *et al.*, 2009). PPNA and PPNB, Pre-Pottery Neolithic A and B, respectively; B-A, Bölling-Allerød interstadial; YD, Younger Dryas.

the harsh conditions during the Younger Dryas, the early Holocene showed an increase in global temperature, which in most regions was accompanied by higher rainfall (IPCC, 2001). Additionally, the Holocene was a much more stable period than the Younger Dryas or the Bölling–Allerød phases (Fig. 1.2a). In the Near East, palaeo-environmental records indicate a reduction in steppe species and forest expansion, particularly in lowlands (e.g. Roussignol-Strick, 1999; Roberts *et al.*, 2001; Willcox *et al.*, 2009), and there are several indications that water availability was indeed much higher than in present times (Araus *et al.*, 1999, 2007; Bar-Matthews *et al.*, 1999; Willcox, 2002a).

Bearing in mind this climate background, Willcox *et al.* (2009) report clear parallels between climate changes over time and the steady introduction of agriculture in the Near East. According to the frequency of seed species found at archaeological sites (Fig. 1.2c), wild rye was present at late Pleistocene sites such as Abu Hureyra 1 and Mureybet 1, but was less common than *Polygonum/Rumex*. The latter species complex, presumably gathered from the flood plains, would have been a more reliable source of food, as it is less affected by drought episodes than rye (see Fig. 1.2c). After the Younger Dryas, rye diminished and wild emmer and barley, which are adapted to warmer conditions, became predominant as warming increased during the early Holocene (Fig. 1.2c). Also, the gathering of *Polygonum/Rumex* and other small-seeded food plants probably diminished. Willcox *et al.* (2009) correlated more stable climate conditions at the beginning of the Holocene with the appearance of cultivation, and concluded that it was these climatic conditions that allowed cultivation to develop into a reliable subsistence economy. Indeed, the advent of stable climate conditions and further establishment of agriculture explain the spectacular cultural developments starting in the Near East between ca. 13,000 and 11,000 cal. BP (see e.g. Molist, 1998; Schmidt, 2006).

1.6 Human-induced Environmental Impact and Sustainability of Early Mediterranean Agriculture

The adoption of agriculture would have been the starting point for large-scale human impact on the environment. The Near East is particularly susceptible to landscape degradation, since deforestation has brought about drastic landscape changes (Zohary and Hopf, 1973). Cultivation, wood exploitation but particularly grazing have led to the destruction of the steppe forests in northern Syria (see Fig. 1.3a). Only a few relic stands or isolated trees are preserved in less accessible areas (Willcox, 2002a). Charcoal records from the early Holocene suggest the presence of mixed open forests, dominated by wild oriental terebinth and wild almond in the south of the area, and deciduous oaks as dominant formations in the north (Willcox, 2002a). The former community is now restricted to a few remaining relic formations (such as those found on the Jebel Abdul Aziz in north-east Syria; Fig. 1.3b), while some isolated open stands of deciduous oaks still exist just north of the Syrian border in Turkey (Fig. 1.3c) (Willcox, 2002a). The more humid conditions during the early Holocene would have broadened the distribution of oaks to the south, compared with their present potential distribution (Willcox, 2002a). Nevertheless,

their actual distribution is mainly defined by the expansion of cultivated land in valleys and on hill flanks, and overgrazing in more abrupt areas.

The *Epic of Gilgamesh*, a text from Ancient Mesopotamia describing the fates of Gilgamesh, 5th king of the 1st Dynasty of Uruk (ca. 4700 cal. BP), depicts the first historical evidence of intensive forest exploitation (http://www.ancienttexts.org/library/mesopotamian/gilgamesh/, tablets II–V). Gilgamesh, against the advice of his friend Enkidu and the elders of the city, decides to cut down all the cedar trees in the Great Cedar Forest, including the largest tree, which will be used to build a new gate for the city. The last part of the epic, far from celebrating this event, could be interpreted as a symbolic alert to the risks of overexploitation, which causes the premature death of Enkidu cursed by the dying Humbaba, guardian-demon of the forest. Could this part of the ancient epic have sought to emphasize the importance of preserving a natural resource? Indeed, deforestation may have started in the Near East much earlier than the 5th millennium BP. For example, Yasuda *et al.* (2000) suggest evidence of forest clearance that may go back to 10,000 cal. BP in the Ghab Valley (north-west Syria). By examining pollen in the sediments of an ancient lake, they found an abrupt rise in charcoal frequency, together with a drop in deciduous oak pollen, and followed by different steps of a secondary forest succession (ferns, pines; Fig. 1.4a). The authors attributed these findings to forest fires associated with extensive forest clearance, probably contemporary with the onset of cultivation by Pre-Pottery Neolithic people. Nevertheless, dating of carbonate-rich lakes can be overestimated by several millennia due to a 'memory effect' (Meadows, 2005), and thus the forest fire event shown in Fig. 1.4 might be linked to a later phase of agriculture expansion during the Bronze Age, rather than to the beginning of cultivation. Alternative evidence, however, suggests that increased anthropogenic disturbance associated with the expansion of agriculture and husbandry may have already started

Fig. 1.3. (a) The site of Tell Halula in northern Syria, showing the denuded landscape prevailing in the region. (b) Relic stand of wild oriental terebinth on the Jebel Abdul Aziz, Syria. (c) Relic deciduous oak stand in south-east Turkey.

between 10,000 and 9000 cal. BP. Human impact is likely to have increased during this period, since the number of sites in the Near East increased dramatically after the Pre-Pottery Neolithic A (ca. 11,000 cal. BP; Fig. 1.4b, c). Indeed, during the same period, a significant number of sites increased their population (Rollefson and Köhler-Rollefson, 1992; Moore *et al.*, 2000), and around 10,000 so-called megasites appeared, which can be considered as the first urban populations (Simmons, 2007). However, the increase in the number of sites and population may not have been linear. Some studies of early Holocene sites carried out at the local scale suggest periods of expansion,

but also periods of abandonment and possibly relocation of settlements (e.g. Ur, 2002; Hill, 2004). For example, pollen work by Van Zeist and Bottema (1991) found no evidence of large-scale forest clearance until 5000–4000 cal. BP in the north-western part of the Ghab Valley, not far from the study outlined above (Fig. 1.4a). In the latter case, olive pollen increased just before forest clearance occurred, suggesting that such activity was a consequence of olive cultivation. Hajar *et al.* (2009) found evidence of anthropogenic disturbance and deforestation in Mount Lebanon from ca. 8000 cal. BP, whereas in the nearby Antilebanon range major disturbance started much later,

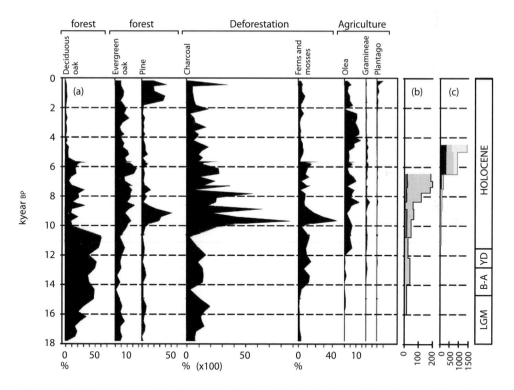

Fig. 1.4. (a) Pollen diagram of the Ghab Valley. Selected taxa indicate vegetation changes that can be related to deforestation and agriculture (Yasuda *et al.*, 2000). The time axis has been rescaled to a calibrated temporal scale (using Cal Pal 2007-online; Danzeglocke *et al.*, 2009), although variable residence time in lake carbonates may have caused an overestimation of pollen age (Meadows, 2005). (b) Number of sites per period (ca. 16,000–4500 cal. BP) in ASPRO Database (http://www.mom.fr/Atlas-des-Sites-du-Proche-Orient.html), including data from Syria, Lebanon, Iraq, Turkey, Israel and Palestine. Dark grey, pre-desert; light grey, steppe. (c) Number of sites per period in TAY Database (http://www.tayproject.org/enghome.html). Age scale based on average dating of cultural periods (from Pre-Pottery Neolithic to Early Bronze Age). Black, dark grey and light grey: eastern, south-eastern and central Anatolia, respectively. LGM, Late Glacial Maximum; B-A, Bölling–Allerød interstadial; YD, Younger Dryas.

during the Roman period (ca. 2000 cal. BP). Regional variability in landscape patterns across the Near East, together with dating problems in palaeo-environmental archives, makes it difficult to establish a general chronology. Nevertheless, the overall trend during the Holocene is of a steady increase in population pressure and deforestation, but locally disrupted by abandonment episodes (Hill, 2004; Hajar *et al.*, 2009).

On the other side of the Mediterranean, Carrión and Van-Geel (1999) provide an enlightening example of how human impact during the Neolithic could have overcome changes in the biosphere driven by natural climate-forcing mechanisms. The palynological record of Navarre (Valencia, Spain) indicates that the primary glacial pine forest (probably *Pinus nigra*) persisted for ca. 30,000 years, starting in the last Glaciation (Würm) and lasting until the mid-Holocene. Neither the climate optima of the Bölling–Allerød warm phase and the early Holocene, nor the drought episodes of the Younger Dryas affected the composition of the original forest. However, coinciding with the onset of a Neolithic settlement in the area around 7000 cal. BP (Dupré *et al.*, 1998), and in less than one century, a sharp increase in charcoal in the sediment record indicates the occurrence of massive forest fires, leading to the replacement of the pine forest by fire-adapted evergreen oaks.

Current evidence therefore points to the beginning of agriculture and husbandry as a crucial step in the anthropization of landscapes. How did this impact affect the agricultural practices themselves? Were they sustainable in the long term, or rather a self-defeating economic activity? By characterizing the evolution over time of stable isotopes and crop distribution at the Tell Halula site (north-west Syria), Ferrio *et al.* (2007) reported a trend towards agriculture intensification that might have threatened settlement subsistence. In the early phases of the settlement (Pre-Pottery Neolithic B, ca. 9800–9200 cal. BP), cereal grains accounted for 56% of crop remains, followed by legumes (27%), flax (13%) and fruit trees (4%). In the Pre-Halaf phase (ca. 8900–8300 cal. BP), the pattern changed to cereal monoculture (92% of the remainder), resembling current conditions in the area, where 91% of the land is dedicated to cereal crops. Interestingly, this change was accompanied by a reduction in cereal yield, as derived from carbon isotopes (Araus *et al.*, 2001, 2007), particularly for wheat, which dropped to ca. 30% in estimated grain yield. The reduction in wheat productivity was accompanied by increased cultivation of barley (better adapted to harsher growing conditions), suggesting steady land degradation over time. These observations were paralleled by a significant reduction in settlement size (i.e. occupied area) and, presumably, in population, along with some sort of cultural decadence, judging by the quality of built structures, e.g. replacement of *cimento*-coated basements by bare soils (Molist, 1996). After two millennia of continuous occupation, the settlement was finally abandoned and the region remained unsettled till the second half of the 19th century, being used only by Bedouins for grazing.

The case of Tell Halula may exemplify how agriculture intensification in the past could have forced the shift towards less intensive land use, once a certain threshold of pressure on land resources was reached. A complementary study in the Western Mediterranean that also involved a long-time occupied site (Los Castillejos de Montefrío, Granada, south-east Spain) suggests a similar, non-sustainable pattern of land exploitation over time (Aguilera *et al.*, 2008). The combination of stable isotopes in cereals and grain dimensions (Aguilera *et al.*, 2008) pointed to a reduction in grain yield of 35% for wheat and 30% for barley from ca. 6000 to 4500 cal. BP, along with significant decreases in grain size (33–38%) and nitrogen content (33–56%). These changes were attributed to a loss of soil fertility, since carbon isotope values indicated good plant water status throughout the whole period. This interpretation was supported by concomitant changes in weed assemblages, suggesting an increase over time in the area devoted to cereals at the expense of pastures and unploughed land. Similar patterns of overexploitation of agricultural land, alternating with periods of land abandonment, might have been common throughout the Mediterranean for the last ten millennia (Rollefson and Köhler-Rollefson, 1992; Wilkinson, 1994; Ur, 2002; Hill, 2004; Butzer, 2005; Riehl and Marinova, 2008). As quoted by Butzer (2005), Mediterranean agriculture is defined by cyclic patterns of intensification and de-intensification, driven by environmental changes, demographic pressure or warfare.

1.7 Concluding Remarks

Although the onset of agriculture in different areas of the world cannot be unequivocally and exclusively attributed to climatic factors, evidence so far suggests that climate stability during the Holocene could have favoured cultivation, allowing it to develop into a reliable production economy. On the other hand, there is no recognizable evidence of large-scale human impacts on landscape before the onset of agriculture, and such impacts were likely to modify the consequences of climate changes for landscape and land use. Alternating patterns of land use intensification and abandonment observed during the Holocene (Hill, 2004; Butzer, 2005; Riehl and Marinova, 2008) may correlate with growth and collapse observed in ancient settlements (e.g. Cullen et al., 2000; De Menocal, 2001; Riehl and Marinova, 2008). In the Mediterranean region, for example, such patterns of land management may have allowed relatively sustainable development in the long term, but with drastic consequences for human communities (e.g. forced migration, famine, war, land abandonment) at the centennial or even millennial scales (Hill, 2004; Butzer, 2005). Nowadays, technological advances and globalization may allow human communities to overcome the negative consequences of climate change, overexploitation and soil degradation, thus preventing collapse (Adger, 2000; Folke, 2006). However, local food crises and their consequences still persist, being comparable to those found in former times, and are likely to increase according to climate predictions (OECD-FAO, 2009). In this context, a look at the past may help to keep in mind the major role of agriculture in human development, as well as its dependence on environmental constraints.

1.8 Acknowledgements

This work has been supported by the projects *PALEOISOMED* (CGL2009-13079-C02, Spanish Ministry for Science and Innovation) and *AGRIWESTMED* (ERC-Advanced Grants, Proposal number 230561). J.P. Ferrio has a grant from the *Ramón y Cajal* programme from the Spanish Ministry. We thank George Willcox for useful comments and corrections.

References

Adger, W.N. (2000) Social and ecological resilience: are they related? *Progress in Human Geography* 24, 347–364.

Aguilera, M., Araus, J.L., Voltas, J., Rodríguez-Ariza, M.O., Molina, F., Rovira, N. *et al.* (2008) Stable carbon and nitrogen isotopes and quality traits of fossil cereal grains provide clues on sustainability at the beginnings of Mediterranean agriculture. *Rapid Communications in Mass Spectrometry* 22, 1653–1663.

Aguilera, M., Espinar, C., Ferrio, J.P., Pérez, G. and Voltas, J. (2009) A map of autumn precipitation for the third millennium BP in the eastern Iberian Peninsula from charcoal carbon isotopes. *Journal of Geochemical Exploration* 102, 157–165.

Alley, R.B., Mayewski, P.A., Sowers, T., Stuiver, M., Taylor, K.C. and Clark, P.U. (1997) Holocene climatic instability: A prominent, widespread event 8200 yr ago. *Geology* 25, 483–486.

Araus, J.L., Febrero, A., Catala, M., Molist, M., Voltas, J. and Romagosa, I. (1999) Crop water availability in early agriculture: evidence from carbon isotope discrimination of seeds from a tenth millennium BP site on the Euphrates. *Global Change Biology* 5, 201–212.

Araus, J.L., Slafer, G.A., Romagosa, I. and Molist, M. (2001) FOCUS: Estimated wheat yields during the emergence of agriculture based on the carbon isotope discrimination of grains: evidence from a 10th millennium BP site on the Euphrates. *Journal of Archaeological Science* 28, 341–350.

Araus, J.L., Ferrio, J.P., Buxó, R. and Voltas, J. (2007) The historical perspective of dryland agriculture: Lessons learned from 10000 years of wheat cultivation. *Journal of Exploratory Botany* 58, 131–145.

Balter, M. (2007) Seeking agriculture's ancient roots. *Science* 316, 1830–1835.

Bar-Matthews, M., Ayalon, A., Kaufman, A. and Wasserburg, G.J. (1999) The Eastern Mediterranean paleoclimate as a reflection of regional events: Soreq cave, Israel. *Earth and Planetary Science Letters* 166, 85–95.

Barnola, J.M., Raynaud, D., Korotkevich, Y.S. and Lonius, C. (1987) Vostok ice core provides 160,000-year record of atmospheric CO_2. *Nature* 329, 408–414.

Berger, W.H. (1990) The Younger Dryas cold spell – a quest for causes. *Palaeogeography Palaeoclimatology Palaeoecology* 89, 219–237.

Binford, L.R. (1968) Post pleistocene adaptations. In: Binford, S.R. and Binford, L.R. (eds) *New Perspectives in Archaeology.* Aldine, Chicago, Illinois.

Bottema, S. and Woldring, H. (1984) Late Quaternary vegetation and climate of south-western Turkey. Part II. *Palaeohistoria* 26, 123–149.

Bradley, R.S., Briffa, K.R., Crowley, T.J., Hughes, M.K., Jones, P.D. and Mann, M.E. (2001) The scope of medieval warming. *Science* 292, 2011–2012.

Braidwood, R.J. (1958) Near Eastern prehistory. *Science* 127, 1419–1430.

Butzer, K.W. (2005) Environmental history in the Mediterranean world: cross-disciplinary investigation of cause-and-effect for degradation and soil erosion. *Journal of Archaeological Science* 32, 1773–1800.

Byrd, B.F. (2005) Reassessing the emergence of village life in the Near East. *Journal of Archaeological Research* 13, 231–290.

Carrión, J.S. and Van-Geel, B. (1999) Fine-resolution Upper Weichselian and Holocene palynological record from Navarres (Valencia, Spain) and a discussion about factors of Mediterranean forest succession. *Review of Palaeobotany and Palynology* 106, 209–236.

Childe, V.G. (1952) *New Light on the Most Ancient East.* Routledge and Paul, London.

COACC (2002) *Abrupt Climate Change: Inevitable Surprises.* National Academy of Sciences, Washington, DC.

Cohen, M.N. and Amelagos, G.J. (1984) *Paleopathology at the Origins of Agriculture.* Academic Press, London.

Cullen, H.M., De Menocal, P.B., Hemming, S., Hemming, G., Brown, F.H., Guilderson, T. *et al.* (2000) Climate change and the collapse of the Akkadian empire: Evidence from the deep sea. *Geology* 28, 379–382.

Cullen, H.M., Kaplan, A., Arkin, P.A. and De Menocal, P.B. (2002) Impact of the North Atlantic Oscillation on Middle Eastern climate and streamflow. *Climatic Change* 55, 315–338.

Danzeglocke, U., Jöris, O. and Weninger, B. (2009) *CalPal-2007online.* http://www.calpal-online.de/

De Menocal, P.B. (2001) Cultural responses to climate change during the Late Holocene. *Science* 292, 667–673.

Diamond, J. (2002) Evolution, consequences and future of plant and animal domestication. *Nature* 418, 700–707.

Dillehay, T.D., Rossen, J., Andres, T.C. and Williams, D.E. (2007) Preceramic adoption of peanut, squash and cotton in northern Peru. *Science* 316, 1890–1893.

Ditlevsen, P.D., Svensmark, H. and Johnsen, S. (1996) Contrasting atmospheric and climate dynamics of the Late Glacial and Holocene periods. *Nature* 379, 810–812.

Dupré, M., Carrión, J.S., Fumanal, M.P., La Roca, N., Martínez, J. and Usera, J. (1998) Evolution and palaeoenvironmental conditions of an interfan area in eastern Spain (Navarrés, Valencia). *Italian Journal of Quaternary Sciences* 11, 97–105.

Ferrio, J.P., Arab, G., Bort, J., Buxó, R., Molist, M., Voltas, J. *et al.* (2007) Land use changes and crop productivity in early agriculture: comparison with current conditions in the Mid-Euphrates Valley. *Options Méditerranéennes* 59B, 167–174.

Flannery, K. (1969) Origins and ecological effects of early domestication in Iran and the Near East. In: Ucko, P.J. and Dimbledy, G.W. (eds) *The Domestication and Exploitation of Plants and Animals.* Aldine, Chicago, Illinois.

Folke, C. (2006) Resilience: The emergence of a perspective for social-ecological systems analyses. *Global Environmental Change* 16, 253–267.

Fuller, D.Q. (2007) Contrasting patterns in crop domestication and domestication rates: Recent archaeobotanical insights from the Old World. *Annals of Botany* 100, 903–924.

Gribbin, J. and Lamb, H.H. (1978) Climatic change in historical times. In: Gribbin, J. (ed.) *Climatic Change.* Cambridge University Press, Cambridge, UK, pp. 68–82.

Grootes, P.M., Stuiver, M., White, J.W.C., Johnsen, S. and Jouzel, J. (1993) Comparison of oxygen-isotope records from the GISP2 and GRIP Greenland ice cores. *Nature* 366, 552–554.

Hajar, L., Haydar-Boustani, M., Khater, C. and Cheddadi, R. (2009) Environmental changes in

Lebanon during the Holocene: Man vs. climate impacts. *Journal of Arid Environments* (in press).

Harlan, J.R. (1992) *Crops and Man*. American Society of Agronomy: Crop Science Society of America, Madison, Wisconsin.

Harlan, J.R. (1998) *The Living Fields: Our Agricultural Heritage*. Cambridge University Press, Cambridge, UK.

Harris, D.R. and Hillman, G.C. (1989) *Foraging and Farming: the Evolution of Plant Exploitation*. Unwin Hyman, London.

Hill, J.B. (2004) Land use and an archaeological perspective on socio-natural studies in the Wadi Al-Hasa, West-Central Jordan. *American Antiquity* 69, 389–412.

Hillman, G.C. (1996) Late Pleistocene changes in wild plant-foods available to hunter-gatherers of the northern Fertile Crescent: possible preludes to cereal cultivation. In: Harris, D.R. (ed.) *The Origins and Spread of Pastoralism in Eurasia*. University College London Press, London, pp. 159–203.

Hillman, G.C. and Davies, M.S. (1990) Measured domestication rates of wild wheats and barley under primitive cultivation, and their archaeological implications. *Journal of World Prehistory* 4, 157–222.

Hillman, G.C., Hedges, R., Moore, A., Colledge, S. and Pettitt, P. (2001) New evidence of late glacial cereal cultivation at Abu Hureyra on the Euphrates. *The Holocene* 11, 383–393.

Indermühle, A., Stocker, T.F., Joos, F., Fischer, H., Smith, H.J., Wahlen, M. *et al.* (1999) Holocene carbon-cycle dynamics based on CO_2 trapped in ice at Taylor Dome, Antarctica. *Nature* 398, 121–126.

IPCC (2001) *Climate Change 2001: the Scientific Basis*. Cambridge University Press, Cambridge, UK (http://www.grida.no/climate/ipcc_tar/).

Kislev, M.E., Weiss, E., and Hartmann, A. (2004) Impetus for sowing and the beginning of agriculture: ground collecting of wild cereals. *Proceedings of the National Academy of Sciences of the United States of America* 101, 2692–2695.

Larsen, C.S. (1995) Biological changes in human populations with agriculture. *Annual Review of Anthropology* 24, 185–213.

Lev-Yadun, S., Gopher, A. and Abbo, S. (2000) Archaeology – The cradle of agriculture. *Science* 288, 1602–1603.

MacNeish, R.S. (1992) *The Origins of Agriculture and Settled Life*. University of Oklahoma Press, Norman, Oklahoma.

Magny, M., Begeot, C., Guiot, J. and Peyron, O. (2003) Contrasting patterns of hydrological changes in Europe in response to Holocene climate cooling phases. *Quaternary Science Reviews* 22, 1589–1596.

Meadows, J. (2005) The Younger Dryas episode and the radiocarbon chronologies of the Lake Huleh and Ghab Valley pollen diagrams, Israel and Syria. *The Holocene* 15, 631–636.

Molist, M. (1996) *Tell Halula (Siria) Un yacimiento neolítico del Valle Medio del Eufrates. Campañas de 1991–1992*. Ediciones del Ministerio de Educación y Cultura, Madrid.

Molist, M. (1998) Des représentations humaines peintes au IXe millénaire B.P. sur le site de Tell Halula (Vallée de l'Euphrate, Syrie). *Paléorient* 24, 81–87.

Moore, A.M.T., Hillman, G.C. and Legge, A.J. (2000) *Village on the Euphrates: From Foraging to Farming at Abu Hureyra*. Oxford University Press, Oxford, UK.

Neumann, K. (2003) New Guinea: a cradle of agriculture. *Science* 301, 180–181.

OECD-FAO (2009) *Agricultural Outlook 2009-2018*. OECD Publishing, Paris.

Pechenkina, E.A., Ambrose, S.H., Xiaolin, M. and Benfer, R.A., Jr (2005) Reconstructing northern Chinese Neolithic subsistence practices by isotopic analysis. *Journal of Archaeological Science* 32, 1176–1189.

Pérez-Obiol, R. and Julià, R. (1994) Climatic-change on the Iberian Peninsula recorded in a 30,000-yr pollen record from lake Banyoles. *Quaternary Research* 41, 91–98.

Purugganan, M.D. and Fuller, D.Q. (2009) The nature of selection during plant domestication. *Nature* 457, 843–848.

Ranere, A., Piperno, D.R., Holst, I., Dickau, R. and Iriarte, J. (2009) The cultural and chronological context of early Holocene maize and squash domestication in the Central Balsas River Valley, Mexico. *Proceedings of the National Academy of Sciences* 106, 5014–5018.

Richerson, P.J., Boyd, R. and Bettinger, R.L. (2001) Was agriculture impossible during the Pleistocene but mandatory during the Holocene? A climate change hypothesis. *American Antiquity* 66, 387–411.

Riehl, S. and Marinova, E. (2008) Mid-Holocene vegetation change in the Troad (W Anatolia): Man-made or natural? *Vegetation History and Archaeobotany* 17, 297–312.

Roberts, N., Reed, J.M., Leng, M.J., Kuzucuoglu, C., Fontugne, M., Bertaux, J. *et al.* (2001) The tempo of Holocene climatic change in the eastern Mediterranean region: new high-resolution crater-lake sediment data from central Turkey. *Holocene* 11, 721–736.

Rodó, X., Baert, E. and Comin, F.A. (1997) Variations in seasonal rainfall in southern

Europe during the present century: Relationships with the North Atlantic Oscillation and the El Nino Southern Oscillation. *Climate Dynamics* 13, 275–284.

Rollefson, G.O. and Köhler-Rollefson, I. (1992) Early Neolithic exploitation patterns in the Levant: Cultural impact on the environment. *Population and Environment: a Journal of Interdisciplinary Studies* 13, 243–254.

Roussignol-Strick, M. (1999) The Holocene climatic optimum and pollen records of sapropel 1 in the eastern Mediterranean, 9000–6000 BP. *Quaternary Science Reviews* 18, 515–530.

Sage, R.F. (1995) Was low atmospheric CO_2 during the Pleistocene a limiting factor for the origin of agriculture? *Global Change Biology* 1, 93–106.

Savard, M., Nesbitt, M. and Jones, M.K. (2006) The role of wild grasses in subsistence and sedentism: new evidence from the northern Fertile Crescent. *World Archaeology* 38, 179–196.

Schmidt, K. (2006) *Sie bauten den ersten Tempel. Das rätselhafte Heiligtum der Steinzeitjäger. Die archäologische Entdeckung am Göbekli Tepe.* C.H. Beck, Munich, Germany.

Simmons, A.H. (2007) *The Neolithic Revolution in the Near East.* University of Arizona Press, Tucson, Arizona.

Smith, B.D. (1995) *The Emergence of Agriculture.* Scientific American Library, New York.

Smith, B.D. (1997) The initial domestication of *Cucurbita pepo* in the Americas 10,000 years ago. *Science* 276, 932–934.

Stevens, L.R., Wright, H.E. and Ito, E. (2001) Proposed changes in seasonality of climate during the Lateglacial and Holocene at Lake Zeribar, Iran. *Holocene* 11, 747–755.

Stuiver, M. and Grootes, P.M. (2000) GISP2 oxygen isotope ratios. *Quaternary Research* 53, 277–283.

Stuiver, M., Grootes, P.M. and Braziunas, T.F. (1995) The GISP2 d18O climate record of the past 16,500 years and the role of the sun, ocean and volcanoes. *Quaternary Research* 44, 341–354.

Tanno, K. and Willcox, G. (2006) How fast was wild wheat domesticated? *Science* 311, 1886.

Underhill, A.P. (1997) Current issues in Chinese neolithic archaeology. *Journal of World Prehistory* 11, 103–160.

Unger-Hamilton, R. (1989) The epi-Palaeolithic southern Levant and the origins of cultivation. *Current Anthropology* 30, 88–103.

Ur, J.A. (2002) Settlement and landscape in Northern Mesopotamia: The Tell Hamoukar Survey 2000–2001. *Akkadica* 123, 57–88.

Van-Geel, B., Buurman, J. and Waterbolk, H.T. (1998) Archaeological and palaeoecological indications of an abrupt climate change in The Netherlands, and evidence for climatological teleconnections around 2650 BP. *Journal of Quaternary Science* 11, 451–460.

Van Zeist, W. and Bottema, S. (1991) *Late Quaternary Vegetation of the Near East.* Dr. Ludwig Reichert, Wiesbaden, Germany.

Weiss, E., Kislev, M.E. and Hartmann, A. (2006) ANTHROPOLOGY: Autonomous cultivation before domestication. *Science* 312, 1608–1610.

Wick, L., Lemcke, G. and Sturm, M. (2003) Evidence of Late glacial and Holocene climatic change and human impact in eastern Anatolia: high-resolution pollen, charcoal, isotopic and geochemical records from the laminated sediments of Lake Van, Turkey. *Holocene* 13, 665–675.

Wilkinson, T.J. (1994) The structure and dynamics of dry-farming states in Upper Mesopotamia. *Current Anthropology* 35, 483–520.

Willcox, G. (1996) Evidence for plant exploitation and vegetation history from three Early Neolithic pre-pottery sites on the Euphrates (Syria). *Vegetation History and Archaeobotany* 5, 143–152.

Willcox, G. (2002a) Evidence for ancient forest cover and deforestation from charcoal analysis of ten archaeological sites on the Euphrates. In: Thiebault, S. (ed.) *Charcoal Analysis. Methodological Approaches, Palaeoecological Results and Wood Uses.* BAR Int. Series 1063, pp. 141–145.

Willcox, G. (2002b) Geographical variation in major cereal components and evidence for independent domestication events in Western Asia. In: Cappers, R.T.J. and Bottema, S. (eds) *The Dawn of Farming in the Near East.* Oriente, Berlin, pp. 133–140.

Willcox, G., Fornite, S. and Herveux, L. (2008) Early Holocene cultivation before domestication in northern Syria. *Vegetation History and Archaeobotany* 17, 313–325.

Willcox, G., Buxó, R. and Herveux, L. (2009) Late Pleistocene and early Holocene climate and the beginnings of cultivation in northern Syria. *The Holocene* 19, 151–158.

Wright, H.E., Jr (1993) Environmental determinism in Near Eastern prehistory. *Current Anthropology* 34, 458–469.

Yasuda, Y., Kitagawa, H. and Nakagawa, T. (2000) The earliest record of major anthropogenic deforestation in the Ghab Valley, northwest Syria: a palynological study. *Quaternary International* 73/74, 127–136.

Zohary, D. and Hopf, M. (1973) Domestication of pulses in the Old World. *Science, USA* 182, 887–894.

2 Climate Change in Drylands: From Assessment Methods to Adaptation Strategies

E. De Pauw and W. Göbel

2.1 Projections of Global Climate Change

Since the publication of the 4th Assessment Report (AR4) of the Intergovernmental Panel on Climate Change (IPCC), a broad consensus has developed among the scientific community that climate change is real, has started to show in the current weather and that it has a discernible human signature (IPCC, 2007).

The maps in Figs 2.1 and 2.2 visualize what climate change actually means in terms of precipitation and temperature changes in different parts of the globe over a 100-year period (from 1980/89 to 2080/99) under the greenhouse gas emission scenario A1b (see further). The areas in different temperature and precipitation classes are cross-tabulated in Tables 2.1 and 2.2. For convenience, as this chapter will further concentrate on the drylands, the change classes refer to different dryland[1] categories.

The temperature increases expected for the drylands (Table 2.1) are in the range 2–4°C, with a tendency in the tropical drylands towards the lower part of the range, and in the non-tropical drylands towards the higher part of the range. For those parts of the globe that have higher precipitation (non-drylands), the range in expected temperature increases is wider (2–5°C).

The changes in precipitation totals (Table 2.2) show some extremes, with a clear tendency towards increase in the non-drylands and tropical drylands, and a mixed pattern of increase or decrease in the non-tropical drylands.

Global circulation models (GCMs), complex models that simulate the interactions between the atmosphere, land and ocean surfaces, geosphere, biosphere and human interventions, have been at the forefront in drawing the main conclusions of AR4, as summarized in Figs 2.1 and 2.2 and Tables 2.1 and 2.2. In making these predictions, or rather projections, about climate in a specified future (e.g. 2010–2040, 2070–2100), there are several sources of uncertainty.

The first of these is that the future itself is only one possibility out of many that materializes. Given the strong linkage between greenhouse gas (GHG) emissions and global warming, the practice is therefore to run the climate models under specific GHG emission assumptions.

The three most commonly used scenarios for assessing the impact of climate change are the SRES[2] scenarios A1b, A2 and B1 (IPCC, 2007). The following description of these scenarios is taken from the IPCC summary report.

A1. The A1 storyline and scenario family describes a future world of very rapid economic growth, global population that peaks in mid-century and declines thereafter, and the rapid introduction of new and more efficient technologies. Major underlying themes are convergence among regions, capacity building and increased cultural and social interactions, with a substantial reduction in regional differences in per capita income. The A1 scenario family develops into three groups that describe

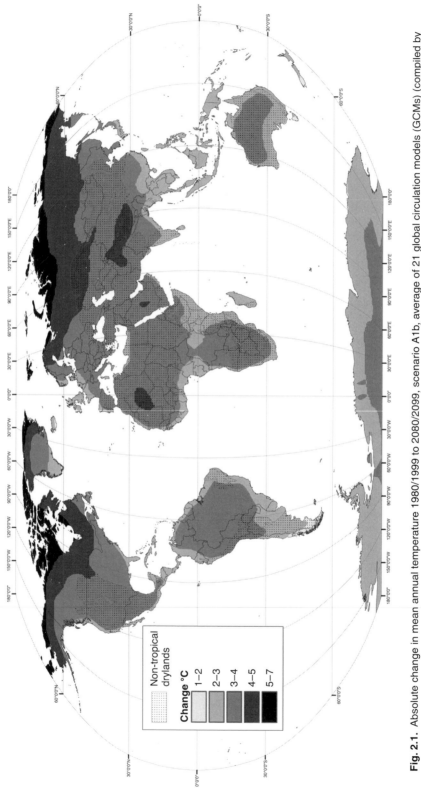

Fig. 2.1. Absolute change in mean annual temperature 1980/1999 to 2080/2099, scenario A1b, average of 21 global circulation models (GCMs) (compiled by GIS Unit ICARDA, based on partial maps in Christensen *et al.*, 2007).

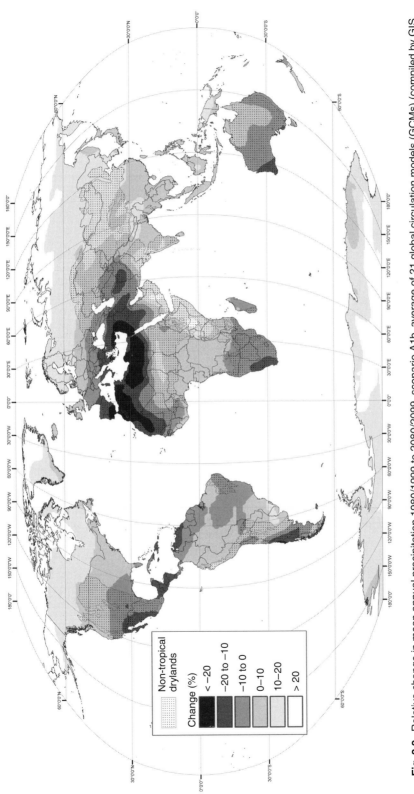

Fig. 2.2. Relative change in mean annual precipitation 1980/1999 to 2080/2099, scenario A1b, average of 21 global circulation models (GCMs) (compiled by GIS Unit ICARDA, based on partial maps in Christensen *et al.*, 2007).

Table 2.1. Changes (°C) in annual temperature for different dryland categories (from 1980/1989 to 2080/2099, scenario A1b).

Dryland category	0.5–1.0	1.0–1.5	1.5–2.0	2.0–2.5	2.5–3.0	3.0–3.5	3.5–4.0	4.0–5.0	5.0–7.0
Non-tropical drylands	0	1	4	10	27	48	10	0	0
Tropical drylands	0	0	1	34	49	16	0	0	0
True deserts	0	0	0	4	36	51	9	0	0
Non-drylands	0	1	3	11	22	17	31	15	0

Table 2.2. Changes (%) in annual precipitation for different dryland categories (from 1980/1989 to 2080/2099, scenario A1b).

Dryland category	−50 to −30	−30 to −20	−20 to −10	−10 to 0	0 to 10	10 to 20	20 to 30	30 to 50	Loss	Gain
Non-tropical drylands	0	4	12	32	32	19	0	0	48	52
Tropical drylands	0	0	6	21	51	19	2	0	27	73
True deserts	0	31	23	18	17	10	1	0	72	28
Non-drylands	0	0	2	10	27	39	17	4	13	87

alternative directions of technological change in the energy system.

A1b. The A1b scenario assumes a balance between fossil-intensive and non-fossil energy sources, where balance is defined as not relying too heavily on one particular energy source, on the assumption that similar improvement rates apply to all energy supply and end-use technologies.

A2. The A2 storyline and scenario family describes a very heterogeneous world. The underlying theme is self-reliance and preservation of local identities. Fertility patterns across regions converge very slowly, resulting in continuously increasing population. Economic development is primarily regionally oriented and per capita economic growth and technological change are more fragmented and slower than in other storylines.

B1. The B1 storyline and scenario family describes a convergent world with the same global population, which peaks in mid-century and declines thereafter, as in the A1 storyline, but with rapid change in economic structures towards a service and information economy, with reductions in material intensity and the introduction of clean and resource-efficient technologies. The emphasis is on global solutions to economic, social and environmental sustainability, including improved equity, but without additional climate initiatives.

A1b is the middle-of-the-road GHG emission scenario, A2 the more pessimistic one and B1 the more optimistic one. With no progress on reducing GHG emissions, the A2 scenario is now being considered more realistic, whereas A1b is slowly becoming the 'optimistic' scenario and B1 a kind of 'pie-in-the-sky' scenario.

The second uncertainty is that the IPCC's 4th Assessment Report is based on simulations of 21 GCM models. Since the IPCC published its first Assessment Report in 1990, these three-dimensional mathematical representations of the processes responsible for climate have grown in complexity, and are now able to model the complex interactions between atmosphere, land surface, oceans and sea ice (Fig. 2.3), and to simulate global distributions of temperature, winds, cloudiness and precipitation.

Despite increasing sophistication, there are still considerable differences between the predictions of different models originating from different research groups. For this reason it is important to select those models that are considered the most appropriate for developing adaptation strategies, or, alternatively, to apply a kind

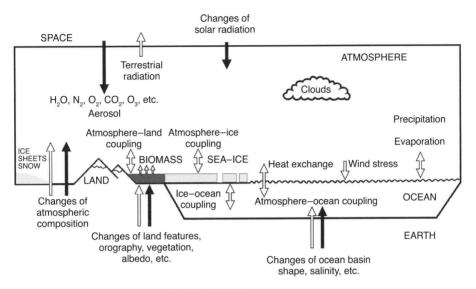

Fig. 2.3. Schematic representation of a typical GCM.

of averaging process of the output from different GCMs to obtain a 'middle of the road' prediction.

Typical for GCM models is that parameter estimation is at a relatively coarse spatial resolution (typically 2 to 3 degrees, corresponding to a grid cell of 10,000–36,000 km² depending on the model and geographical latitude). This scale is too coarse to include small-scale processes, the ones responsible for local weather patterns, and particularly in hilly to mountainous terrain these can be very important. Apart from these possible distortions, the coarse resolution of GCMs is perhaps the main bottleneck for planning of adaptation to climate change, as it prevents linkage to features with variability at much finer spatial variability, such as arable land, water resources, human settlements, agricultural production systems, poverty hot spots, etc. Downscaling the output of GCMs is therefore an extremely important step for mainstreaming climate change projections into development planning, and will be discussed further in this chapter.

From a planning perspective, climate predictions 50 to 100 years into the future are difficult to grapple with, especially given all the uncertainties associated with those projections. Hence agents involved in

development and food security tend now to focus on a shorter-term time horizon to formulate response and adaptation strategies. Besides the ability to ignore the uncertainties of the far futures, this has the advantage that the adaptation strategies can be rooted more into those that are already in place to cope with the challenges of the current climates.

2.2 Climate and Agricultural Systems in the Drylands of the Mediterranean Zone

Figure 2.2 and its enlargement (Fig. 2.4) indicate that, in terms of precipitation decline, the drylands around the Mediterranean are projected to be one of the most severely affected by climate change. In this chapter we focus on this particular dryland region as a very relevant case for illustrating how the continuity between current climate trends and the projections for the near future allow existing as well as new land, water and crop management practices to serve as models for coping with climate change.

The Mediterranean zone is mostly characterized by Mediterranean-type climates, which typically have warm and dry

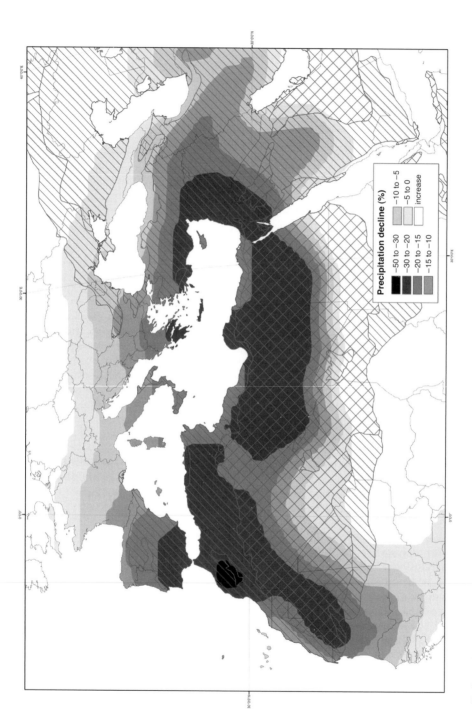

Fig. 2.4. Projected precipitation decline in the area around the Mediterranean (1980–1999 to 2070–2099) (source: Christensen *et al.*, 2007). Hatched, non-tropical drylands; cross-hatched, deserts.

summers and mild and rainy winters. Within this overall Mediterranean distinctiveness, the climates of the region show great diversity. In particular, moisture and temperature conditions can differ markedly as a result of differences in local topography, nearness to regions with either temperate climates (especially in the north) or arid climates (especially in the south and east), and exposure to either maritime or continental influence.

In response to the agroecological diversity, land use and agricultural systems are very diversified in the Mediterranean region, and a wide variety of crops are being grown under rainfed and irrigated conditions. Rainfed agriculture is the dominant form of crop production, with wheat, barley and food legumes the dominant crops. Irrigation is practised on only a small proportion of the land, usually 10% or less. Although the area under irrigation is still expanding, supply constraints are likely to increase for a variety of reasons, as summarized by Margat and Vallée (2000). Livestock plays a key role in this region; in most cases, it is characteristically interrelated with other land uses, through residue and stubble grazing or use of marginal lands, in particular rocky shrubs or woodlands, and the more arid rangelands, most of which are overgrazed (Ryan et al., 2006).

In accordance with the terminology developed by Dixon et al. (2001), the following 'model' types of agricultural systems occur in the Mediterranean zone (De Pauw, 2004a).

- Rainfed mixed: highly diversified systems with a wide range of rainfed crops, including tree crops (olives, fruits and nuts) and field crops (mainly wheat, barley, lentils, chickpeas, potatoes, sugarbeet and faba beans). Terracing is common in hilly areas. Seasonal interaction with livestock, mainly sheep and goats, and use of crop residues and other fodder are common features.
- Dryland mixed: less diverse than the rainfed mixed systems, with barley and wheat as main crops grown in alternation with single or double-season fallows or

with legumes (lentil, chickpea). Interactions with small livestock systems mainly take the form of barley and stubble grazing and are stronger than in the previous system.
- Highland mixed: dualistic land-use systems at higher altitude (1500–3000 m) with cropping pattern dominated by wheat and barley on arable land, and communal grazing on marginal land; mostly monoculture with occasional fallow, terracing common, sometimes supplemental irrigation.
- Irrigated: traditionally along major river systems downstream from dams, but more recently also based on groundwater extraction. Systems can be either large-scale or small-scale and include a wide variety of crops and cropping patterns depending on temperature regime.
- Pastoral: systems based on the mobility of flocks and herds moving between more humid and drier areas, with the availability of grazing and water. Resources under a wide precipitation range (typically 100–400 mm) are accessed.
- Sparse: too dry for productive land use, remaining limited to opportunistic grazing following rainstorms.

For an in-depth treatment of the climatic characteristics, land use patterns, agricultural systems and soils of the Mediterranean zone, the reader is referred to Ryan et al. (2006).

Already under current climatic conditions the main challenge for the agricultural systems of the region is coping with moisture deficits and drought. According to Ward et al. (1999), the Mediterranean zone is characterized by some of the most variable climates in the world, in which drought is endemic. Water availability is the main constraint for agriculture in the Mediterranean zone. Agricultural production of major grain crops is strongly affected by precipitation fluctuations (Keatinge et al., 1986), and crop and livestock losses due to drought can have very severe repercussions on both the countries' balance of payments and the livelihoods of individual producers. For this reason, irrigation plays a critical role in the widening of agricultural options and has a stabilizing influence on rainfed agriculture and farmer livelihoods that goes far beyond its areal extent.

2.3 Precipitation Trends in the 20th Century

The precipitation trends for the area around the Mediterranean zone have been derived by linear regression from the Full Data Reanalysis Product Version 4 of the Global Precipitation Climatology Centre (GPCC) at resolution 0.5°. This quality-controlled data set (Schneider *et al.*, 2008) maps monthly precipitation across the entire world, with the exception of Antarctica, between 1901 and 2007, based on between around 8000 and 45,000 stations, depending on the year.

Simple linear regression models were fitted to the 107-year time series of annual precipitation of each 0.5° × 0.5° grid cell by the least-squares method. From these models, the following trend surfaces have been derived and mapped:

- average absolute change in annual precipitation (mm/year) (Fig. 2.5a);
- average relative change in annual precipitation (%/10 years) (Fig. 2.5b);
- correlation between annual precipitation and time (Fig. 2.5c); and
- t-significance level of linear time trend of precipitation (two-sided t-test) (Fig. 2.5d).

Figure 2.5a indicates that most of the area around the Mediterranean (with the exception of Tunisia) has witnessed between 1901 and 2007 a trend of absolute decline of 0–3 mm/year in annual precipitation. The relative changes (Fig. 2.5b) in most of the area are in the range 0–5% per decade, except in deserts, where a 2 mm absolute decline per year obviously represents a much higher relative decline than in higher-rainfall areas. The correlation coefficients (Fig. 2.5c) in most of the area are negative but low, but the significance levels of the linear trends (Fig. 2.5d) are generally high (<0.1).

Simple linear regression proves to be an adequate model to demonstrate the trend of precipitation and drought in the region. In spite of the high year-to-year variability, which is the reason for the generally low correlation coefficients (Fig. 2.6), there is a clear and often highly significant trend as evidenced by the highly significant t-probabilities. Only in some of the extremely dry parts of the region does this simple linear model lead to an obvious overestimation of change (areas mapped as having 15 to >30% relative change of annual precipitation per decade).

Comparing Fig. 2.5 with Fig. 2.4, it can be concluded that the negative precipitation trend during the past century is of a similar magnitude to that predicted by most of the GCMs for the Mediterranean region in the coming decades. This suggests that forces of climate change have been active in the region for at least a century and that human activity is only exacerbating an already existing trend.

2.4 Downscaling the Potential Impact of Climate Change in the Eastern Mediterranean Region

Worldwide, many development agents are already developing strategies for adaptation to climate change. As mentioned earlier, one major bottleneck they face is the coarse resolution of the GCM predictions. For this reason there is an increasing interest in downscaled projections in the form of high-resolution maps. In this chapter we illustrate, through a case study from the eastern Mediterranean region, how GCM information can be downscaled and how this new information can be used to make projections of possible impact and plan adaptation strategies.

As presented earlier in this chapter, the eastern Mediterranean region is, according to AR4 and several follow-up studies, likely to be one of the most severely affected by climate change in the world. Predictions from GCMs are for lower precipitation, increase in precipitation variability leading to more extreme events and more droughts,

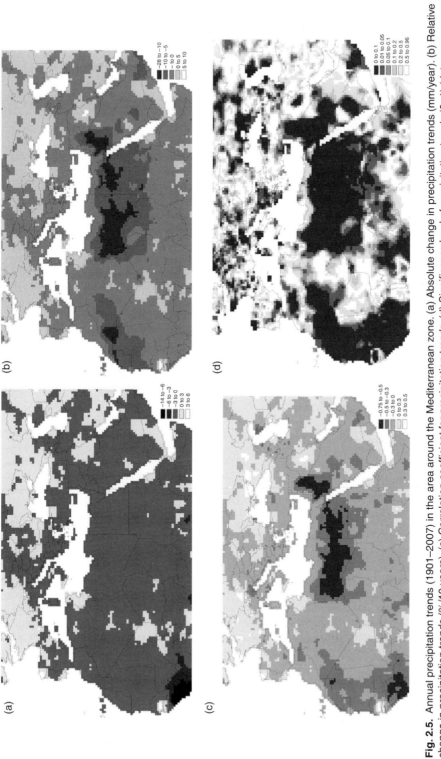

Fig. 2.5. Annual precipitation trends (1901–2007) in the area around the Mediterranean zone. (a) Absolute change in precipitation trends (mm/year). (b) Relative change in precipitation trends (%/10 years). (c) Correlation coefficient for precipitation trends. (d) Significance level of precipitation trends (0–1) (data source: Schneider *et al.*, 2008).

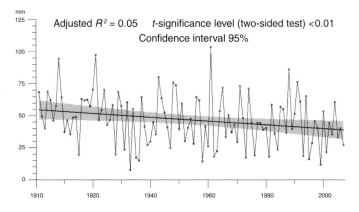

Fig. 2.6. Time trend of annual precipitation totals for a site in the Nile delta north of Cairo, Egypt (30.75°N, 31.25°E). In spite of a low R^2 caused by the high inter-annual variability of precipitation, the negative trend stands out clearly and is highly significant.

and of course higher temperatures, resulting in severe stress on already scarce water resources.

2.4.1 General approaches for climate change downscaling

Generally speaking, three methods are available for downscaling GCM output to higher spatial resolutions:

- calibration of current climate surfaces with GCM output;
- statistical downscaling with or without weather typing; and
- dynamical downscaling with regional climate models (RCM).

Statistical downscaling yields good results, in terms of reproducing current climates from GCMs, and can be applied to output of different GCMs. On the down side, statistical relationships have to be established individually for each station and GCM, requiring quality data. Surfaces have to be created from point data, a problem in data-scarce regions such as the eastern Mediterranean. Moreover, this method is computationally challenging.

Dynamical downscaling using an RCM yields the best results, even in areas with complex topography, and directly generates climate surfaces. It is the only technique capable of modelling complex changes of topographical forcing. A disadvantage is that different methods of dynamical downscaling are linked to specific GCMs, thus transferring inherent flaws in particular models from a lower to a higher resolution. They are also methodologically and computationally challenging.

In the absence of any downscaled data obtained from an RCM, in this study we used the calibration method of GCM downscaling, which essentially involves the superimposition of a low-resolution future climate change field on a high-resolution current climate surface. This method leads to fast results, which are directly applicable to individual models or averages of multi-model ensembles and lead directly to climate surfaces. One drawback of this approach is that only linear changes to current topographical forcing are assumed.

2.4.2 Methodology for downscaling

Four climatic variables were considered: precipitation, minimum, maximum and mean temperatures. Climate change as represented by these variables was assessed for the 2010–2040 time horizon, a 'near' future of interest to planners.

The transformation of GCM data into high-resolution climate maps required the following steps:

- selection of GCMs and emission scenarios;
- data extraction procedures;
- change mapping at coarse resolution;
- re-sampling;
- generating downscaled climate surfaces; and
- calculating averages.

Selection of GCMs and emission scenarios

A first screening was based on data availability. For only 17 GCM models out of the 23 on which the IPCC report is based, the necessary climatic variables were available online. A final selection of seven GCMs was based on age of the model, the spatial resolution of the GCM grid cells and how well they represented a particular modelling approach. The selected models are listed in Table 2.3. Two GHG emission scenarios were selected – the fairly optimistic A1b and the more pessimistic A2.

Data extraction procedures

Data sets for each GCM were retrieved from the sources mentioned above in a NetCDF format (.nc), a self-describing format for weather and climate data files developed by UCAR[3], using the program GrADS[4] (for Grid

Analysis and Display System), which runs under Linux platforms.

Change mapping at coarse resolution

After computing all monthly averages for each climatic variable, GHG scenario and time horizon, the averages were subtracted by the grid of the 1961–1990 time period (also a GCM output) in the case of temperature data. In the case of precipitation data, the ratio was computed.

For mean, minimum and maximum temperature (°C),

$$\Delta T = T_{LR,21} - T_{LR,20} \quad (2.1)$$

for precipitation (dimensionless),

$$r_{prec} = P_{LR,21} / P_{LR,20} \quad (2.2)$$

with LR, low-resolution; 20, 20th-century data; 21, 21st-century data.

One GCM model (no. 10) contained too many missing data in the area of interest during the months June–August; for this model no grids were generated during these months. For other GCMs missing data in particular pixels were replaced by using the mean of the surrounding pixels using the Neighbourhood function in the GIS software ArcGIS[5].

Table 2.3. GCM models used in the study.

No.	Name	Country	Year	Resolution (°) and (levels)	Source
01	BCCR-BCM2.0	Norway	2005	2.8 x 2.8 (31)	http://www.ipcc-data.org/ https://esg.llnl.gov:8443/home/ publicHomePage.do
02	CSIRO-MK3.0	Australia	2001	1.9 x 1.9 (18)	http://www.ipcc-data.org/ https://esg.llnl.gov:8443/home/ publicHomePage.do
04	MIROC3.2	Japan	2004	2.8 x 2.8 (20)	http://www.ipcc-data.org/
08	CGCM3.1(T63)	Canada	2005	2.8 x 2.8 (31)	http://www.ipcc-data.org/ http://www.cccma.ec.gc.ca/data/cgcm3/ cgcm3.shtml
09	CNRM-CM3	France	2005	2.8 x 2.8 (45)	http://www.ipcc-data.org/ https ://esg.llnl.gov:8443/home/ publicHomePage.do http://www.mad.zmaw.de/projects-at-md/ ensembles/experiment-list-for-stream-1/ cnrm-cm3/
10	ECHAM5/MPI-OM	Germany	2003	1.9 x 1.9 (31)	http://www.ipcc-data.org/
12	GFDL-CM2.0	USA	2005	2.0 x 2.5 (24)	http://www.ipcc-data.org/

Re-sampling

In order to refine the coarse climate change raster maps, a re-sampling was carried out in ArcGIS down to a resolution of 0.008333 decimal degrees (about 1 km), which corresponds to that of the reference climate maps of the study area. The re-sampling was done using the cubic convolution method, which computes new pixel values on the basis of a weighted average of the 16 nearest pixels of the original map (4 × 4 window). This method is relatively time consuming, but it offers a smoother appearance than other available methods (nearest neighbour or bilinear interpolation). Possible edge effects (where the 16 pixel values are not all available) were avoided by selecting an area of interest larger than the study area. In our case the re-sampling of the climate change maps was carried out in ArcGIS over the rectangle 0–55°N × 3–64°E.

In the case of precipitation, the cubic convolution method sometimes produces negative values of relative change when the original values are close to 0 mm. To avoid this, re-sampling for precipitation was done on the logarithm of the original values, and the final change grids were obtained by exponential transformation of the latter layers.

To avoid indeterminate expressions of the precipitation ratio in cases where an average of 0 mm of precipitation occurred for both the reference period 1961–1990 and the future period under consideration, precipitation totals lower than 0.0167 mm (or $6.43*10^{-8}$ $kg/m^2/s$), corresponding to a total rainfall of 1 mm in 60 years, were reset to this amount.

Generating downscaled climate surfaces of precipitation and temperature

Downscaled high-resolution (1 km) monthly climate surfaces were obtained for each GCM by adding the re-sampled monthly change maps to monthly high-resolution reference climate surfaces (De Pauw, 2008) for temperature variables, and by multiplying for precipitation.

The calculations were performed in ArcGIS using simple raster algebra according to the following formulae.

For mean, minimum and maximum temperature (°C),

$$T_{HR,21} = T_{HR,20} + \Delta T_{re\text{-}sampled} \qquad (2.3)$$

and for precipitation (mm),

$$P_{HR,21} = P_{HR,20} * r_{re\text{-}sampled} \qquad (2.4)$$

with HR, high resolution.

Calculating averages

Given the vast amount of data generated and the divergence between the results from the selected GCMs, averages were computed from the seven GCMs for the re-sampled high-resolution change maps of precipitation, mean, maximum and minimum temperature during each month and over the year, as well as the winter, spring, summer and autumn seasons under the two emission scenarios A1b and A2.

Derived climatic variables

Using the high-resolution monthly precipitation and temperature rasters, the following derived climate grid surfaces were produced for the time frame 2010–2040 under the two GHG scenarios:

- climatic zones according to the Köppen classification system;
- potential evapotranspiration (mm) on monthly and annual bases;
- aridity index on annual basis; and
- growing period.

The Köppen climate classification system (Köppen and Geiger, 1928) is based on the annual and monthly averages of precipitation and temperature and is, despite its venerable age, still the most widely used to date.

The potential evapotranspiration is the rate of evapotranspiration from an extensive surface of an 8–15 cm tall, green grass cover of uniform height, actively growing, completely shading the ground and not short of water (Doorenbos and Pruitt, 1984).

The aridity index is the ratio of annual precipitation to annual potential evapotranspiration (UNESCO, 1979).

The climatic growing period is a concept developed by the Food and Agriculture Organization of the United Nations (FAO, 1978) to estimate the duration of the period during the year in which neither moisture nor temperature is limiting to plants.

A detailed account of the methods used in deriving these climatic variables from the primary climatic variables is provided by Göbel and De Pauw (2010).

2.4.3 Climate change projections for the near future

The time slice chosen for this study, 2010 to 2040 with the mid-point 2025, lies in the near future. The climate change between now and then may be less impressive than long-term predictions until the end of the century would be, but the results are more useful for the analysis of impending vulnerabilities of populations and directly applicable to the planning of appropriate adaptation strategies in ensuring continuing food security.

Precipitation

Excepting the extremely arid desert areas of Egypt, where the decrease in precipitation is very high in relative terms but the absolute loss is very small and therefore of little consequence, annual precipitation during the period 2010 to 2040 will decline in the region studied by around 10% in comparison with the recent past (Fig. 2.7a). This decrease is more pronounced under scenario A1B than under A2, but overall the difference between the two scenarios is small (typically <10%). This is to be expected as the modelled greenhouse-gas emissions diverge significantly only from 2030 onwards. The precipitation decrease is least pronounced in Iraq (5–10% for both scenarios), while elsewhere it typically varies between 5 and 10% for scenario A2 and between 10 and 20% for scenario A1B. Most of the decline takes place during winter and spring, when decreases of up to 20% are expected to occur everywhere, while during summer and autumn the picture is less uniform and some areas show-

ing slight precipitation increases, particularly in Iraq and eastern parts of Jordan and Syria, but also in Cyprus, occur alongside areas of decreasing precipitation. The loss of precipitation in winter and spring during the growth cycle of most of the major field crops has potentially serious consequences for agriculture.

Temperatures

Temperatures are going to rise everywhere and all year round, with very small differences between the two scenarios. The mean annual temperature is expected to increase by 0.5–1.0°C (Fig. 2.7b), with most of the change occurring in summer and least in winter. In winter, the expected increase will be in the range of 0.5–1.0°C everywhere except in the Egyptian desert, where it may reach up to 1.5°C. In summer, the increase will range between approximately 1.0 and 2.0°C, with the largest increases occurring in Iraq, Syria and parts of Jordan. In spring and autumn, increases will be intermediate between those expected for winter and summer.

Climatic zones

Ninety per cent of the lands in the study area will remain in the same climatic zone according to the classification of Köppen, the other 10% changing to another climatic zone. Most affected will be Syria and Jordan, where 30% of the land will change from a steppe (BS) to a desert (BW) climate (Fig. 2.7d). Lebanon and the West Bank will also witness substantial changes due to shifts in climate zones. The predicted changes in climate zones are very similar for scenarios A1B and A2.

Annual potential evapotranspiration (PET)

An increase of 2–4% is expected, slightly more under scenario A2 than under scenario A1B, particularly in Syria. Overall, this represents a very modest increase only in evaporative demand for both rainfed winter crops and irrigated crops.

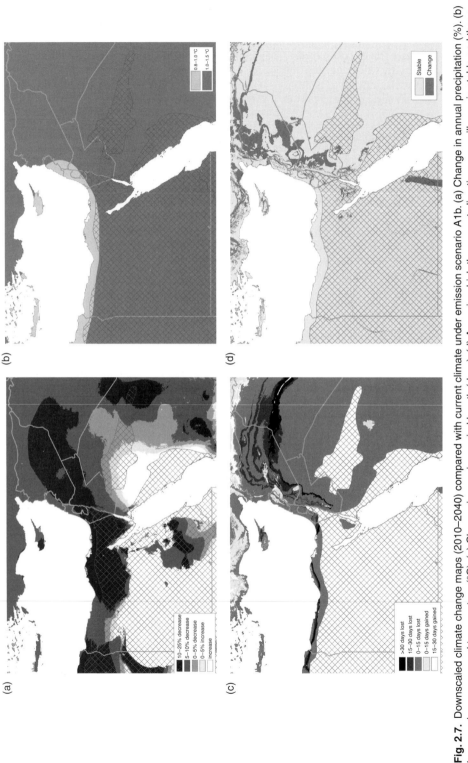

Fig. 2.7. Downscaled climate change maps (2010–2040) compared with current climate under emission scenario A1b. (a) Change in mean annual temperature (°C). (b) Increase in mean annual temperature (°C). (c) Change in growing period length (days). (d) Areas in which the current climatic zones will remain stable and those in which it will change; cross-hatching, hyper-arid (desert) areas.

Aridity

Declining precipitation and increased PET both work towards more arid conditions, as witnessed by the change in the aridity index. The trend is similar under scenarios A1B and A2. It is most pronounced in Lebanon, the West Bank and Cyprus.

Growing periods

The changes in length of the growing periods (Fig. 2.7c), too, are very similar for both scenarios examined. Most of the study area is likely to experience moderate reductions of up to 15 days, but in parts of Syria, the West Bank and Cyprus the decline will be more pronounced and in the range of 15–30 days. Only in some of the high mountain areas of Lebanon will the length of the growing period actually increase due to the rise in temperature, as this will reduce the number of days when low temperatures limit growth.

Drought

On the basis of consistency with current precipitation trends, the IPCC AR4 study globally projects an increase in droughts, especially in the tropics and subtropics (Christensen *et al.*, 2007). In order to assess how these coarse projections would materialize on a small scale, a downscaling was undertaken of the Standardized Precipitation Index (SPI) for the eastern Mediterranean zone. The SPI is a tool, based on precipitation data, developed for measuring how 'normal' or exceptional a drought or wetness event is, and has a uniform meaning across different places. Its calculation procedure is well explained by Edwards and McKee (1997). The SPI is used for periods between one month and several years. Göbel and De Pauw (2010) mapped the annual SPI for each year from 1901 to 2007. Interpretation of the SPI values was in accordance with the theoretical class limits shown in Table 2.4. In practice, extremely dry or wet periods with SPI values below −2 or above +2 tend to occur more frequently than expected, according to the theory.

After calculating the SPI for each year of the reference period, a time trend analysis of the SPI was undertaken, similar to that for precipitation (see Section 2.3) and downscaled using the re-sampling method of Section 2.4.2. The spatial trend of the SPI during the period 1901–2007 is shown in Fig. 2.8.

Across the more important agricultural areas of the countries studied, the SPI has dropped over slightly more than 100 years by about 0.5 to 1.0 points. This means that what would have been regarded as a normal year 100 years ago would now be considered as *moderately wet*, and many situations that are considered normal now would early last century have been regarded as a *moderate drought*. Similarly, a *moderately dry* year now would a century ago have been classified as *very dry*, and a *very dry* one now would then have been seen as an *extremely dry* year. Except for parts of Iraq (Fig. 2.8), the annual SPI and precipitation totals have been

Table 2.4. Expected frequencies of SPI values.

SPI value	Theoretical frequency from standard normal distribution	Event expected to happen approximately every ... years	Description
Over +4.0	$3.1671243*10^{-5}$	31,574	
Over +3.0	0.001349898	741	Extremely wet
Over +2.0	0.022750132	44	
Over +1.5	0.0668072	15	Very wet
Over +1.0	0.15865526	6	Moderately wet
+1.0 to −1.0	0.6826895	2 out of 3	Near normal
Under −1.0	0.15865526	6	Moderately dry
Under −1.5	0.0668072	15	Very dry
Under −2.0	0.022750132	44	
Under −3.0	0.001349898	741	Extremely dry
Under −4.0	$3.1671243*10^{-5}$	31,574	

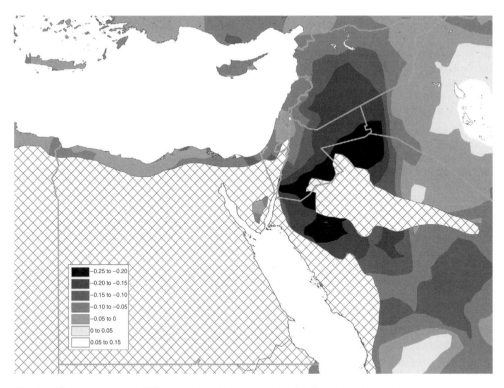

Fig. 2.8. Change in annual SPI, 1901–2007 (points per decade). Cross-hatching, desert areas.

decreasing everywhere. This negative trend is highly significant apart from the wettest areas along the Mediterranean coast. Desert areas have been masked in Fig. 2.8 as the SPI values in extremely arid areas, with an average annual precipitation of about 10 mm or below, are without real meaning.

Taking as a guide the continuity of the precipitation trend of the recent past and the projections of the near future, together with the concurrence of the precipitation and drought trends of the recent past, more frequent and severe droughts can be expected in the near future.

2.5 From Impact Assessment to Adaptation Strategies

Since time immemorial the agricultural systems of the drylands have been coping with the key ecological constraints of their environment: aridity, pronounced precipit-

ation variability, drought and restricted water resources for irrigation. As indicated by Table 2.1, with climate change some systems will become better off, at least in terms of precipitation totals, but those in the Mediterranean zone will be hit twice – by higher temperatures, raising the risk of heat stress to the traditional crops of the region, and by lower precipitation and increased risk of drought.

The key to adaptation will be in reviewing how these agricultural systems have been coping in the past, revisiting the recommended practices established after decades of dryland agricultural research and fine-tuning these dryland management principles in order to deal with the additional challenges imposed by climate change.

Of paramount importance is to recognize that the drylands and the agricultural systems that developed within them are extremely diverse and that, hence, adaptation measures cannot be of the 'one-size-fits-all' kind and that planning for climate change needs to take into consideration the site-specific constraints and potential for adaptation.

2.5.1 Geographical shifts in agricultural systems

A good starting point for assessing how the projected changes of the climatic parameters are likely to affect the existing agricultural systems of the drylands, as defined by Dixon *et al.* (2001), is by mapping their current position in an agroecological niche. As Fig. 2.9 indicates, each system occupies a specific segment of the aridity spectrum, which can be wide or narrow, although overlap is considerable. The irrigated systems constitute the only notable exception, since they occur under all aridity regimes.

With the expected increase of aridity in the Mediterranean zone, shifts are likely in the geographical location of the agricultural systems: those that currently occur within a particular aridity class will tend to occupy the agroecological niche of those systems

currently in a more humid zone, and will themselves be substituted by systems currently in a more arid climate. In this scenario of shifting systems, given the diversity of crops they can support, the *rainfed mixed systems* are more likely to maintain themselves, albeit in a modified form, than the *dryland mixed systems*. In fact, with declining precipitation and increased risk of drought, it is likely that parts of the dryland mixed systems will no longer be able to support wheat, which will be replaced by the more hardy barley crop, itself coming under threat at the low-rainfall margins of the dryland mixed systems. As barley-based systems have to abandon these low-rainfall areas, pastoral systems can occupy them, but they too will be forced to leave behind previously productive steppe areas that are likely to become too dry to produce biomass for animals. Possibly the only systems to benefit from climate change could be the *highland mixed systems* due to an extension in the thermal growing period, a reduction in the number of frost days and a topography conducive for water harvesting and diversion.

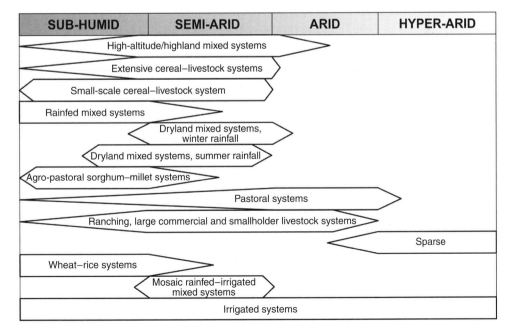

Fig. 2.9. Dryland agricultural systems and aridity (source: De Pauw, 2004b).

A second option is to look at the response mechanisms available in each agricultural system. In a general way, the agricultural systems of the drylands can be subdivided into three response groups: (i) rainfall-based systems; (ii) irrigated systems; and (iii) intermediate systems. The last of these rely on spatially and temporally variable mixes of rain and irrigation water.

2.5.2 Climate-proofing rainfall-based systems

The rainfall-based systems are those most likely to come under pressure from climate change and, in order to retain their productivity, will need to draw inspiration from the established principles for successful dryland crop management: retain the precipitation on the land, reduce evaporation and use crops with drought tolerance and that fit the rainfall pattern (Stewart, 1988).

Climate change is likely to be accompanied by an increase in high-intensity precipitation events. Hence, the retention of precipitation on the land will require more control of runoff, which can be small or substantial depending on the particular rainfall events, slope characteristics of the land and type and state of the soil surface. Tillage, covering the surface, land shaping and use of ponds are the main practices for the control of runoff. Some of the particular land-shaping practices used for water harvesting (see further) and retention may be useful for the control of runoff in the higher-rainfall areas.

Improved water conservation leading to higher soil moisture is a key dryland principle for rainfed systems. Reduction of evaporation can be achieved through weed control – manual, mechanical or chemical, depending on the system – and surface mulches. Weed control has proved to be one of the most dependable methods of conserving water for crops. The use of mulches is more contentious, given the dependence of some agricultural systems on stubble grazing and the complexities of soil surface management for evaporation control, which require a careful evaluation of the site-specific interactions between surface cover,

tillage, soil texture, rainfall patterns and evaporation losses (Papendick and Campbell, 1988).

Under climate change the traditional practice of fallow periods to conserve water may again need to be revived in areas with marginal precipitation. The principle is that, if a crop cannot get enough moisture from precipitation for its transpiration needs, part of its water requirement could be met by moisture retained in the soil profile from a previous season's rainfall when no crop was grown. With the expected decrease in precipitation, a price has to be paid in terms of lost productivity, as the efficiency of fallow systems generally decreases with increases in length of the fallow period.

The growing period in drylands is short and limited by soil moisture availability, and in the colder areas also by temperature. Due to pronounced rainfall variability the 'dependable' growing period may in fact be significantly smaller than the average growing period (De Pauw, 1982). In dryland agricultural research, drought avoidance by the development of high-yielding, short-maturing varieties for the main crops, in combination with statistically based recommendations for optimum sowing periods, has been the preferred strategy to date. Probability analysis to determine drought risk and periods of particular cold or heat stress is recommended, but is often hampered by lack of meteorological data in marginal environments. To deal with this problem, spatial decision-support tools based on weather generators calibrated by a limited meteorological data set can be very useful (Mauget and De Pauw, 2010).

However, as climate change is likely to be accompanied by more severe intra-seasonal drought, more salvation may come in the future from incorporating drought tolerance in traditional crops through breeding or genetic manipulation, trading off productivity against security. There is also scope, if markets can be created, for introducing drought-tolerant crops that are part of local farming systems or have shown high potential under research conditions, in other places where they are currently not known. Hinman and Hinman (1992) mention a number of little-known crops in this category, such as grain amaranth (*Amaranthus* spp.), quinoa (*Chenopodium quinoa*), the prickly pear cactus (*Opuntia* spp.), the bambara groundnut (*Voandzeia subterranea*), the marama bean (*Tylosema esculentum*), the tepary bean (*Phaseolus acutifolius*), the buffalo gourd (*Cucurbita foetidissima*) and halophytes, such as the pickle weed (*Salicornia* spp.) and salt bushes (*Atriplex* spp.). Their successful introduction in other environments is obviously subject to similarity in agroecological conditions with the 'home' location, social acceptance, economic attractiveness, ability to fit in the new farming environment and availability of markets for the new crops.

2.5.3 Making irrigated systems more efficient

As a result of limits on the water supply and increasing water demand, irrigated systems in the Mediterranean zone are already under considerable pressure to become more efficient. While there are major differences between countries, periodic water shortages are already common in some, particularly in the south and east, and are exacerbated during drought years. The ability to tap additional supplies is limited, as the scope for building more major dams is restricted by lack of suitable dam sites: there are few places left where the ratio of storage volume to flooded area is still acceptable.

It is now also recognized that, in the medium to long term, water storage in small and medium reservoirs is unsustainable owing to heavy sediment loads in floods and active siltation and, with climate change, more frequent droughts and higher evaporation rates. As a result, there has been a growing tendency to adopt other supply situations that are basically unsustainable, such as the increasing use and over-exploitation of groundwater. In coastal areas this has in most cases resulted in salt water intrusion, which is often irreversible because the balance between saline and fresh water has been disrupted. In other places there has been an exhaustion of shallow aquifers,

leading to exploitation of deeper ones with fossil water.

While the supply is reaching a ceiling, water demand doubled from the beginning of the 20th century to 2000, and has increased by 60% since 1975 (Margat and Vallée, 2000). Whereas in the countries north of the Mediterranean demand is growing slowly and tends to stabilize, in the south demand is still growing strongly owing to population growth, increasing urbanization, the predominance of agriculture in water use and inadequate controls on water use.

Although there are differences between countries, on average 80% of the water in the countries south and east of the Mediterranean is used in agriculture, nearly all for irrigation (Margat and Vallée, 2000). The fundamental reason for the high proportion of agriculture in water demand is that irrigation water is mostly needed in summer when temperatures and atmospheric demand are high. To grow a crop under such conditions much water is needed. Obviously, under climate change this situation will not improve. The second reason is considerable inefficiency in water application for irrigation. Average losses in irrigation projects suggest that only about 45% of water diverted or extracted for irrigation is actually used by the crop; the remaining 55% is lost to inefficiencies in the distribution and application of the irrigation water.

Within these trends of ceilings on the total water supply and growing competition from other water users (urban users, industry, tourism, etc.), climate change is an additional driver in forcing agricultural systems relying on irrigation water to become more efficient. To continue its paramount role in stabilizing crop production and improving livelihoods for the farmers of the region, irrigation will need to produce more with less water. This process is already happening: irrigation systems are adjusting, in the first place by reducing distribution losses through the modernization of existing schemes. The conversion to drip or sprinkler systems of gravity irrigation schemes can lead to field application efficiency rates of

70%, which would constitute gains that compensate for more than the eventual losses to be expected under climate change.

Irrigation systems are also adjusting to changes in the amount of available irrigation water, by shifting emphasis from more water-demanding systems based on relatively low-value staple crops, such as cereals and cotton, to vegetables, fruits and other niche crops serving the growing needs of nearby urban agglomerations or even global markets. Climate change will necessitate changes in crop calendars to avoid extreme heat and evapotranspiration losses. Dryland regions with the financial resources for wastewater treatment may use treated sewage effluent as an important source of irrigation water. In others, more use can be made of irrigation return flow runoff, agricultural subsurface drainage water and saline groundwater aquifers for salt-tolerant crops.

2.5.4 Expanding the role of intermediate rainfed–irrigated systems

The increased stress on rainfed dryland systems expected under climate change, the need for stabilization of production and diversified farm income as well as the pressure on irrigated systems to become more efficient in response to growing water shortages are likely to promote the growth of so-called 'hybrid' systems, which are neither fully irrigated nor fully rainfed.

The alternating use of rainfall and irrigation water is a potentially valuable management principle under conditions of water scarcity. Supplemental irrigation is the addition of small amounts of water to essentially rainfed crops during times of serious rainfall deficits, but the regulatory role of irrigation can become more substantial, especially if intra-seasonal droughts tend to become more prolonged. The aim is to reduce risk of crop failure, where rainfall is normally sufficient but vulnerability to drought high, and thus to stabilize yields. The water use efficiency of supplemental irrigation can be very high. Obviously, supplemental irrigation is only practical

where rainfall is high enough to count as a significant water source.

A spatial variant of supplemental irrigation is water harvesting. This practice covers various techniques to collect rainwater from natural terrains or modified areas and concentrating it for use on smaller sites or cultivated fields to ensure economic crop yields. Collected runoff is stored in the soil, behind dams or terraces, cisterns, gullies or recharged to aquifers. Water harvesting systems come in a variety of implementations, but the common components are invariably a catchment or source area, a storage facility and a use area. In micro-catchment systems the source and target areas are essentially so close together that they cannot be separated at scales larger than the field level, and the storage facility is either the soil's root zone for immediate use or a small reservoir for later use. In macro-catchment systems, runoff water is collected from a relatively large catchment outside a relatively small target area, with storage provided by surface structures such as small farm reservoirs, subsurface structures such as cisterns, or the soil in the target area itself.

Water-harvesting systems are relevant in moisture-deficit areas and constitute a compromise: a choice is made to sacrifice part of the land on which (in theory) a crop could be grown, yielding poorly in most years, in order to concentrate water on a smaller fraction of the land where a higher soil moisture supply would allow for better yields in most years. Water-harvesting systems remain dependent on precipitation and therefore offer no panacea for prolonged droughts. Nevertheless, they certainly offer a useful dryland land management practice that may gain in relevance under the climate change future envisaged for the Mediterranean zone.

As for all alternatives to traditional practices, the feasibility of water harvesting needs to be assessed not only from a technological perspective (suitable catchment areas, soils, storage sites, etc.) but also an economic (is it attractive in comparison with other land use options?) and cultural one (is it acceptable?).

2.6 Conclusions

In a context of global warming and precipitation increase, the drylands around the Mediterranean are, due to a projected precipitation decline, likely to become one of the regions most severely affected by climate change. In its current setting this region, characterized by one of the most variable climates in the world and with precipitation and drought trends in the recent past that are already very similar to the projections for the near future, has a great diversity of agricultural systems adapted to their agroecological niches and coping with their climatic constraints. For this reason it is well suited to assessment of the coping mechanisms of these systems to climatic variability, and to the drawing of general lessons where established dryland management principles may require fine-tuning in order to climate-proof agricultural systems.

At least for the near future, there appears no particular reason why the agricultural systems of the region would be unable to cope with climate change. The irrigated systems are already forced into greater efficiency due to increasing pressure on water supplies and demand, in which the projected precipitation decline is an additional, but not dominant, stress factor. The rainfed systems may have some adaptive capacity by moving their agroecological niche with the shift in precipitation zones. However, it is unclear whether the rigorous implementation of the tested dryland management principles, as explained in Section 2.5.2, will be sufficient to compensate for the additional stress of higher aridity and more severe droughts. It is therefore likely that innovative hybrid systems, partially irrigated, partially rainfed, will become more prominent, as they offer a high water productivity in an age of increasing water shortage and the prospect of greater yield stability than under rainfed conditions.

Notes

[1] Under the term drylands, we mean those areas with aridity index (AI = ratio of annual precipitation to annual potential evapotranspiration) <0.75, covering hyper-arid (AI <0.03), arid (AI 0.03–0.20), semi-arid (AI 0.2–0.5) and sub-humid (AI 0.5–0.75) zones in accordance with the criteria of UNESCO (1979).
[2] SRES: Special Report on Emission Scenarios.
[3] http://www.unidata.ucar.edu/software/netcdf/
[4] http://www.iges.org/grads/
[5] http://www.esri.com/software/arcgis/index.html

References

Christensen, J.H., Hewitson, B., Busuioc, A., Chen, A., Gao, X., Held, I. et al. (eds) (2007) *Regional Climate Projections. Climate Change 2007: The Physical Science Basis. Contribution of Working Group I to the Fourth Assessment Report of the Intergovernmental Panel on Climate Change.*

De Pauw, E. (1982) The concept of dependable growing period and its modelling as a tool for land evaluation and agricultural planning in the wet and dry tropics. *Pédologie* 3, 329–348.

De Pauw, E. (2004a) Technologies integration in Mediterranean rainfed conditions: a spatial perspective on the results from the MEDRATE project. In: Cantero-Martinez, C. and Gabina, D. (eds) *Mediterranean Rainfed Agriculture: Strategies for Sustainability. Options méditerranéennes.* CIHEAM, Saragossa, Spain, pp. 269–281.

De Pauw, E. (2004b) Management of dryland and desert areas. In: Encyclopedia: Land Use, Land Cover and Soil Sciences Vol. 4, Encyclopedia for Life Support Systems. UNESCO–EOLSS Publishers, Oxford, UK. Available at: http://www.eolss.net (accessed 23 February 2011).

De Pauw, E. (2008) Climatic and soil datasets for the ICARDA Wheat Genetic Resource Collections of the Eurasia Region. Explanatory Notes. ICARDA GIS Unit, Aleppo, Syria. 68 pages. Available at: http://geonet.icarda.cgiar.org/geonetwork/data/regional/GRU_NetBlotch/Doc/Report_NetBlotch.pdf (accessed 16 March 2011).

Dixon, J., Gulliver, A. and Gibbon, D. (2001) *Farming Systems and Poverty.* FAO/World Bank, Rome and Washington, DC, 412 pp.

Doorenbos, J. and Pruitt, W.O. (1984) *Guidelines for Predicting Crop Water Requirements.* FAO Irrigation and Drainage Paper 24, FAO, Rome.

Edwards, D.C. and McKee, T.B. (1997) *Characteristics of 20th Century Drought in the United States at Multiple Time Scales.* Climo Report 97-2, Dept. of Atmospheric Science, CSU, Fort Collins, Colorado, May, 155 pp.

FAO (1978) *Report on the Agro-ecological Zones*

Project, Vol. 1. *Methodology and results for Africa*. World Soil Resources Report 48, FAO, Rome.

Göbel, W. and De Pauw, E. (2010) *Climate and Drought Atlas for Parts of the Near East: a Baseline Dataset for Planning Adaptation Strategies to Climate Change*. Final report. GIS Unit, ICARDA, Syria.

Hinman, C.W. and Hinman, J.W. (1992) *The Plight and Promise of Arid Land Agriculture*. Columbia University Press, New York, 253 pp.

IPCC (2007) Summary for policymakers. In: Solomon, S., Qin, D., Manning, M., Chen, Z., Marquis, M., Averyt, K.B. *et al.* (eds) *Climate Change 2007: the Physical Science Basis*. Contribution of Working Group I to the Fourth Assessment Report of the Intergovernmental Panel on Climate Change. Cambridge University Press, Cambridge, UK and New York.

Keatinge, J.D.H., Dennett, M.D. and Rogers, J. (1986) The influence of precipitation regime on the crop management in dry areas in northern Syria. *Field Crops Research* 12, 239–249.

Köppen, W. and Geiger, H. (1928) *Handbuch der Klimatkunde*. Berlin.

Margat, J. and Vallée, D. (2000). Mediterranean vision on water, population and the environment for the XXIst century. PNUE. PAM. Plan Bleu, Valbonne, France. Online. Available at http://www.gwpmed.org/files/Water%20Vision%20 Mediterranean.pdf (accessed 16 March 2011).

Mauget, S. and De Pauw, E. (2010) The ICARDA Agro-climate Tool. *Meteorological Applications* 17, 105–116.

Papendick, R.I. and Campbell, G.S. (1988) Concepts and management strategies for water

conservation in dryland farming. In: Unger, P.W., Jordan, W.R., Sneed, T.V. and Jensen, R.W. (eds) *Proceedings of the International Conference on Dryland Farming*, 15–19 August 1988, Amarillo/Bushland, Texas, pp. 119–127.

Ryan, J., De Pauw, E., Gomez, H. and Mrabet, R. (2006) Drylands of the Mediterranean Zone: biophysical resources and cropping systems. In: Peterson, G.A., Unger, P.W. and Payne, W.A. (eds) *Dryland Agriculture*, 2nd edn. ASA, CSSA and SSSA. Agronomy Monograph 23, pp. 577–624.

Schneider, U., Fuchs, T., Meyer-Christoffer, A. and Rudolf, B. (2008) Global Precipitation Analysis Products of the GPCC. Global Precipitation Climatology Centre (GPCC), DWD, Internet publication, 1–12. Data and description can be downloaded from http://gpcc.dwd.de (accessed 16 March 2011).

Stewart, B.A. (1988) Dryland farming: the North American experience. In: Unger, P.W., Jordan, W.R., Sneed, T.V. and Jensen, R.W. (eds) *Proceedings of the International Conference on Dryland Farming*, 15–19 August 1988, Amarillo/Bushland, Texas, pp. 54–59.

UNESCO (1979) *Map of the World Distribution of Arid Regions (Scale 1:25,000,000 with Explanatory Note)*. United Nations Educational, Scientific and Cultural Organization, Paris, 54 pp.

Ward, M.N., Lamb, P.J., Portis, D.H., El Hamly, M. and Sebbari, R. (1999) Climate variability in Northern Africa: understanding droughts in the Sahel and the Maghreb. In: Navarra, A. (ed.) *Beyond El Niño – Decadal Variability in the Climate System*. Springer Verlag, Berlin.

3 Agronomic Avenues to Maximize the Benefits of Rising Atmospheric Carbon Dioxide Concentrations in Asian Irrigated Rice Systems

L. Yang and S. Peng

3.1 Introduction

Rice (*Oryza sativa* L.) is produced in at least 95 countries across the globe and it provides a staple food for more than half of the world's current population (Maclean *et al.*, 2002; Coats, 2003). With an annual production of about 650 million t, rice is and will continue to be the most important food crop for the world's population (Maclean *et al.*, 2002). Given a current population of more than six billion with a projected increase of an additional one billion every 12 years, the demand for rice will grow to an estimated 880 million t by 2030 (FAO, 2002). Meeting this 35% increase in demand will be a significant challenge to rice production.

In the meantime, global climate change will provide an additional challenge by significantly altering many elements of the future rice production environment (Bouman *et al.*, 2007); foremost among these changes is the rising atmospheric $[CO_2]$, which is projected to reach at least 550 ppm and 730–1020 ppm by 2050 and 2100, respectively (IPCC, 2007). In contrast to other aspects of global climate change (i.e. high temperatures, rising tropospheric ozone concentration and drought stress), rising $[CO_2]$ is unique in being globally almost uniform and it has been shown to stimulate photosynthesis, growth and yield of the world's major food crops (Kimball *et al.*, 2002). As a result, the response of crop yields to elevated $[CO_2]$ is one of the major sources of uncertainty when assessing the potential impacts of climate change on food security, as well as the effectiveness of different adaptation strategies (Parry *et al.*, 2004; Long *et al.*, 2005, 2006). For example, in a recent assessment of climate change impacts, removal of the beneficial effect of elevated $[CO_2]$ in crop models was found to increase the projected number of malnourished people in 2080 by as much as 500 million (Parry *et al.*, 2004). Therefore, accurate quantification of rice production and the development of adaptation strategies under rising $[CO_2]$ are increasingly important for the future food supply.

To understand how rice crops would respond to CO_2, earlier CO_2 studies were conducted in greenhouses and soil–plant–atmosphere research (SPAR) units (e.g. Baker *et al.*, 1996), temperature gradient tunnels (e.g. Kim *et al.*, 1996) and open-top chambers (e.g. Moya *et al.*, 1998), whereas more recently free-air (chamberless) CO_2 enrichment (FACE) facilities have also been established (Liu *et al.*, 2002; Kobayashi *et al.*, 2006a). The relevance of results from chamber experiments has been widely questioned because of potential interactions of CO_2 with biotic and abiotic factors. These factors are generally different within chambers compared with farmers' fields. It has been suggested that rice yield increases under elevated $[CO_2]$ in FACE experiments are lower than in chamber studies (Long *et al.*, 2005, 2006; Ainsworth, 2008), although comparisons are often difficult because of differences in $[CO_2]$, cultivars, nutrient input rates and other factors between experiments (Tubiello *et al.*, 2007; Ziska and Bunce, 2007). FACE technology provides a means to examine the impact of CO_2 with minimal alteration of microclimate and the soil–plant–atmosphere continuum (McLeod and Long, 1999; Long *et al.*, 2004). Moreover, it provides sufficient treatment area to meet the standard of agronomic trials and allows sufficient area for the destructive harvests that are needed for growth analysis during the growing season of the crop (Long *et al.*, 2004). Asia contains around 60% of the global human population, and rice provides an average of 21% of people's total caloric intake globally (Maclean *et al.*, 2002). The importance of rice in Asia justified the commencement of the rice FACE project in Japan in a cool temperate climate (1998–2000, 2003–2004) and in China in a warm subtropical climate (2001–2010; Fig. 3.1). Two large-scale and fully replicated FACE facilities provide agronomists and breeders with the best opportunity to test adaptation measures and, by taking agronomic adaptations into account, we will have a better capability to predict future climate change impacts on crop production.

In this chapter, we summarized the adaptation strategies of rice crops under the predicted future elevated [CO_2] conditions by examining the use of different cultivars and crop management practices, and present priority areas for further research.

3.2 Plant Density

Adjusting plant density would be helpful for maximizing the CO_2 fertilization effect on rice crops. Recent chamber experiments showed that competition for resources with increased plant density limited the CO_2 enhancement of biomass and especially grain yield (Reid and Fiscus, 2008). Similar results were found in greenhouse studies on wheat, where elevated [CO_2] increased leaf area index and shoot biomass for plants grown at low density but not at high density (Du Cloux *et al.*, 1987). Liu *et al.* (2008) and Yang *et al.* (2009) also reached the same conclusion in their FACE rice experiments.

Fig. 3.1. A view of free-air CO_2 enrichment (FACE) octagonal plots in China. The target [CO_2] in the FACE plots was controlled to 200 ppm above that of ambient levels by a computer system.

These findings suggest that increased competition at high plant density limits the potential beneficial effect of [CO$_2$]. Therefore, an appropriately reduced planting density should be advantageous, but further study is required before recommendations on optimal density can be made.

3.3 Fertilizer Application

Nutrient availability is one of the most important edaphic factors related to rice productivity. The interactive effects of elevated [CO$_2$] and nitrogen (N) supply on grain yield of rice were investigated only under open-air field conditions (Kim *et al.*, 2001, 2003a, b; Yang *et al.*, 2006, 2009; Liu *et al.*, 2008; Shimono *et al.*, 2008). These FACE experiments demonstrated that N fertilizer management strategies for rice

production systems should be adjusted in order to maximize rice productivity in a future high-[CO$_2$] world.

First, in regions with low N application (e.g. Japan; Table 3.1), future management practices must include the application of higher N. Low N fertilization limited N uptake during vegetative growth, which constrained any increase in spikelet number, and therefore limited the yield response to elevated [CO$_2$] (Kim *et al.*, 2001, 2003a, b; Weerakoon *et al.*, 2005). Low N may also cause more pronounced acclimatization of photosynthesis to elevated [CO$_2$], which can limit total dry matter and leaf area increases at elevated [CO$_2$] (Ziska *et al.*, 1996b; Ainsworth *et al.*, 2003). However, for regions with high N application (e.g. China; Table 3.1), there is little room for increase in the total N application rate, as indicated by the significant yield reduction that occurred

Table 3.1. Summary of experimental conditions and percentage responses of rice yield and its components to free-air CO$_2$ enrichment (FACE) in Japanese and Chinese study locations.

	Japan		China	
Experimental condition				
Years	1998–2000	2003–2004	2001–2003	2004–2006
Location	Iwate (39° 38' N, 140° 57' E)	Iwate	Wuxi (31° 37' N, 120° 28' E)	Yangzhou (32°35.5'N, 119°42'E)
Soil type	Typical andosol	Typical andosol	Stagnic anthrosol	Shajiang aquic cambosol
Cultivar	Akitakomachi	Akitakomachi	Wuxiangjing 14	Shanyou 63, Liangyoupeijiu
Cultivar type	Japonica inbred	Japonica inbred	Japonica inbred	Hybrid
N application rate (g/m^2)	4, 8, 9, 12, 15	8, 9	15, 25, 35	12.5, 25
Response in grain yield (%)	+12.8**	+11.8**	+12.8**	+34.1**, +30.1**
Response in yield components (%)				
Panicles/m^2	+8.6*	+14.5ns	+18.8**	+10.3**, +7.8*
Spikelets/panicle	+1.9**	−0.7ns	−7.6**	+10.3**, +9.6**
Spikelets/m^2	+11.5**	+13.3	+9.1**	+ 21.7**, +19.0**
Filled spikelets	+0.8ns	−0.7ns	+4.9*	+4.9**, +5.3**
Grain weight	+1.3*	−1.0ns	+1.3*	+4.3**, +4.4**
Reference(s)	Kim *et al.* (2003a)	Shimono *et al.* (2008)	Yang *et al.* (2006)	Liu *et al.* (2008); Yang *et al.* (2009)

ns, no significance; *, $P < 0.05$; **, $P < 0.01$

with a further increase in N supply and the lack of a synergistic effect between [CO_2] and N (Yang et al., 2006, 2009; Liu et al., 2008). However, the optimal N rate needed with increasing [CO_2] to maximize rice yield remains unaddressed.

Secondly, and perhaps more importantly, the proportion of N fertilizer applied after panicle initiation (PI) should be increased, especially in regions with high N application rates (e.g. China; Table 3.1), in order to: (i) suppress the CO_2-induced enhancement in unproductive tiller production and the subsequent reduction in productive tiller percentage and harvest index (Kim et al., 2001, 2003a; Yang et al., 2006); (ii) counteract the CO_2-induced inability of the root system to supply sufficient N to the plant after PI (Yang et al., 2008) and thus maintain canopy C assimilation capacity after PI (Weerakoon et al., 2000); and (iii) enhance the reproductive sink capacity by increasing the number of differentiated spikelets per panicle, while reducing the number of degenerated spikelets per panicle during the reproductive growth stage (Yang et al., 2006). A heavy N top dressing at this time may enhance the elongation of basal internodes, and consequently lead to a higher probability of lodging. However, the Japanese FACE study found that elevated [CO_2] could significantly decrease lodging under high N fertilization with shortened and thickened lower internodes, thereby increasing ripening percentage and grain yield (Shimono et al., 2007). This reduced risk of lodging later in the season under FACE suggests that the optimal rate of N application after PI can be increased for plants growing under high-atmospheric [CO_2].

Thirdly, controlled-release N fertilizer (CRN hereafter) is a feasible option under future high-[CO_2] environments (Yang et al., 2007a; Shimono et al., 2008). CRN might promote crop response to elevated [CO_2] because growth enhancement due to elevated [CO_2] often decreases as a crop ages (Kim et al., 2003b), possibly because of diminishing N supply and uptake (Kim et al., 2003b; Sakai et al., 2006; Yang et al., 2007a). A slow but continuous N supply

over a long time might alleviate this problem, and therefore increase the yield enhancement by elevated [CO_2] (Shimono et al., 2008). In the future, other new technological developments are anticipated to facilitate the adaptation of rice crops to elevated [CO_2].

In addition to N management, what of the supply of other nutrients under future high-[CO_2] conditions, particularly phosphorus (P) and potassium (K)? To date, only one enclosure experiment has investigated the interaction of CO_2 with P supply (Seneweera et al., 1996; Seneweera and Conroy, 1997), but no study has considered the interaction of CO_2 with K. As with N fertilizer, varying the supply of P influenced the magnitude of the CO_2 response, with the greatest responses occurring at medium- rather than luxury- or low-P supplies. However, yield enhancement by high CO_2 was observed even when P supply was a severely growth-limiting factor, possibly because P can be recycled more efficiently within the rice plant than other nutrients (Seneweera et al., 1996). However, Chinese FACE experiments indicated a remarkable increase in shoot P uptake across the growing season under CO_2 enrichment (Yang et al., 2007b). If this is confirmed for a wide range of rice varieties, it could have important implications for P cycling in rice–soil systems. Rice crops exhausted soil P to a great extent, leading to very low available soil P content when they were harvested, especially in soil receiving no or small amounts of P fertilizer (Yang et al., 2007b). This situation must be avoided, as P deficiency has been identified as a major factor limiting rice yields on many soils throughout the world (Hedley et al., 1994).

3.4 Irrigation Management

How does water management affect yield response to elevated [CO_2]? Given the importance of water to rice production, it is surprising that no information in the literature is available to answer this question. The rice FACE experiments with different N treatments indicated that elevated [CO$_2$] increased root growth rate during the early growth period (EGP), while a slight inhibition of root growth rate occurred during the middle growth period (MGP) (Yang *et al.*, 2008). In addition, specific root activity declined with elevated [CO$_2$] during MGP and the late growth period (LGP), the decline being larger during MGP than during LGP. An implication of this result is the importance of developing a robust root system at MGP and LGP to achieve maximum rice productivity under future higher [CO$_2$]. Management for irrigation during MGP would need redesigning in a way that the mid-season drainage period may have to be made earlier and longer, which is supposed to be beneficial for increasing root activity (Yang *et al.*, 2008). Such an adaptation strategy is also supported by reduced total water use and increased water-use efficiency by the rice plant under high [CO$_2$] (Yoshimoto *et al.*, 2005).

3.5 Selection of Genotypes

Breeding for varietal improvement is one of the key methods of increasing rice yield and improving crop quality, as well as adapting to high [CO$_2$] environments. Given that there are over 100,000 rice varieties, analysis of varietal differences in CO$_2$ responsiveness and exploitation of such differences could significantly improve global food security as [CO$_2$] increases. Unfortunately, we are unaware of any such systematic evaluation at the government, university or corporate level (Ziska and Bunce, 2007). Only a limited number of enclosure-based studies have examined the genotypic differences among rice cultivars in response to increasing [CO$_2$], and they found substantial genotypic variation (Ziska *et al.*, 1996a; Moya *et al.*, 1998), which implies possibilities for enhancing rice production by selecting varieties with higher [CO$_2$] responsiveness.

Many studies indicated that CO$_2$ enrichment increased grain yield largely through an increase in tillering and panicle number (Imai *et al.*, 1985; Baker and Allen,

1993; Kim *et al.*, 1996, 2001, 2003a; Ziska *et al.*, 1996a) and/or the number of spikelets per panicle (Baker, 2004; Liu *et al.*, 2008; Yang *et al.*, 2009). Moya *et al.* (1998) found that grain yield enhancements due to CO$_2$ enrichment among three different Asian cultivars were related to the relative tillering ability of a particular cultivar. In their study, the 'new plant type' cultivar had a relatively low tiller number and was the least responsive to CO$_2$ enrichment. However, in a chamber study with three US rice cultivars, Baker (2004) found that elevated [CO$_2$] did not increase the number of panicles whereas yield increases due to elevated [CO$_2$] were realized mainly by increases in the number of spikelets per panicle. These findings suggest that plasticity with respect to tiller formation and/or spikelet formation may be a factor in optimizing the response of rice to elevated [CO$_2$]. However, Sheehy *et al.* (2001) argued that high tillering ability is not a desirable trait in a high-yield environment because this leads to lodging. Thus, cultivars with moderate tillering capacity but with a higher responsiveness of panicle size to elevated [CO$_2$] (e.g. hybrid cultivars; Table 3.1) should be more favoured in future high-[CO$_2$] environments.

The rice FACE system is better suited to field comparisons among multiple cultivars, but few reports of comparative yield responses from FACE have been published to date. Therefore, we have limited understanding of variation in germplasm response to elevated [CO$_2$] under fully open-air conditions. However, the Chinese FACE rice experiments for the first time investigated the magnitude of yield response of two hybrid varieties to elevated [CO$_2$] (Liu *et al.*, 2008; Yang *et al.*, 2009), and reported an average increase of ca. 32% across three growing seasons (Table 3.1). The yield response of hybrids was not only higher than the FACE results on two *japonica* inbred varieties (Kim *et al.*, 2003a; Yang *et al.*, 2006; Shimono *et al.*, 2008), but was also higher than the average values for *japonica* or *indica* inbred varieties tested in the previous enclosure studies, when the yield data were adjusted to a [CO$_2$] of 550 ppm (see reviews by Horie *et al.*, 2000; Long *et al.*, 2006), suggesting that grain yield of the hybrids would benefit more from an increase in atmospheric [CO_2]. However, to test whether the higher CO_2 responsiveness of hybrid rice is a general phenomenon, we need to evaluate more hybrid rice varieties under higher [CO_2].

Perhaps the most critical results from previous studies are that the relative enhancement of rice yields due to elevated [CO_2] is reduced with increases in air temperature (Kim *et al.*, 1996; Ziska *et al.*, 1996a; Matsui *et al.*, 1997; Moya *et al.*, 1998). This is because elevated [CO_2] reduces transpiration through its effect on stomatal closing, and consequently increases canopy temperatures. Higher canopy temperature can increase pollen sterility, especially under high air temperature. Therefore, without the identification and development of heat-tolerant cultivars, CO_2-induced yield gains will most likely be limited in a possibly warmer, but almost certainly higher-[CO_2], world in the future.

3.6 Control of Insects and Diseases

While we recognize plant–pathogen inter-actions as a factor affecting rice yields, our ability to predict CO_2 impacts on pathogen biology and the effect of subsequent changes on rice yield is limited. The FACE rice experiment is the first and only experiment to show the effect of elevated [CO_2] on rice diseases, and it found that FACE could enhance the occurrence of leaf blast and sheath blight, which are now the most serious and prevalent among rice diseases (Kobayashi *et al.*, 2006b). The decline of silicon (Si) in rice plants due to FACE may enhance susceptibility to blast, and a change in rice canopy structure may accelerate the spread of sheath blight in the field (Kobayashi *et al.*, 2006b). This finding sug-gests that agronomic avenues should include breeding varieties with increased resistance to specific diseases, as well as increasing the rate of Si and/or K fertilizer to reduce the susceptibility of rice to diseases under elevated [CO_2].

3.7 Weed Control

Weeds limit the yield potential of rice crops. Will such limitations increase or decrease in response to future changes in CO_2? To date, only one study has assessed the effect of CO_2 enrichment on weed–rice competition (Alberto *et al.*, 1996), which demonstrated that competitiveness could be enhanced in rice (C_3 species) relative to *Echinochloa glabrescens* (C_4 species) with elevated CO_2 alone, but that simultaneous increases in CO_2 and temperature could still favour *E. glabrescens*. A detailed field assessment of CO_2 effects on weed population biology, rice–weed competition and weed-induced yield losses should be taken into account in developing adaptation technology (Ziska and Bunce, 2007).

3.8 Mitigating Methane Emissions

Methane (CH_4), the second most important greenhouse gas after CO_2, accounts for about 20% of the current increase in global warming. Paddy fields have been regarded as major anthropogenic sources of global CH_4 emissions, with annual estimates ranging from 47 to 60 Tg CH_4 (IPPC, 2001). Although CH_4 emissions from rice paddies will be stimulated by elevated [CO_2] (Inubushi *et al.*, 2003), they could be reduced by changes in agronomic practices for N and organic matter management (Zheng *et al.*, 2006), as well as by the selection of genotypes (Lou *et al.*, 2008). Using FACE techniques, Zheng *et al.* (2006) reported that the stimulatory effects of elevated [CO_2] on seasonal accumulative CH_4 emissions were negatively correlated with the addition rates of decomposable organic carbon, but positively correlated with the rates of N fertilizers applied in either the current rice season or over the whole year. A recent climatron study suggested that screening and breeding rice cultivars with low root exudation and also high allelochemicals will be promising and helpful in mitigating CH_4 emissions from rice paddy fields, while simultaneously keeping grain yield stable under [CO_2] enrichment (Lou *et al.*, 2008).

3.9 Recommendations for Future Research Priorities

1. In adaptation, the main concern is the change in manipulable variables (e.g. cultivar variability, sowing time, plant density) in response to increased $[CO_2]$. If there is no change due to increased $[CO_2]$ in the optimal combination of other variables, no adaptation would occur. However, a satisfactory understanding of how crop management could modify rice yield response to rising $[CO_2]$ has not been forthcoming, as mentioned above, and this research must be included in future experiments.

2. It is becoming increasingly clear that any effect of CO_2 on yield will be modulated by other global change factors, that is, rising day/night temperature, rising tropospheric $[O_3]$, water shortage and decreasing N supply. For example, it has become apparent that higher night temperatures may already be limiting rice yields globally (Peng et al., 2004). Recent chamber studies demonstrated that, as do day temperatures (Kim et al., 1996; Matsui et al., 1997; Moya et al., 1998), high night temperatures also reduce the stimulatory effect of elevated $[CO_2]$ on rice production, owing to increased spikelet sterility (Cheng et al., 2009). However, studies on the interaction of elevated $[CO_2]$ with other global change factors have been noticeably lacking. Therefore, attention must be given to increasing understanding of these interactive effects, especially under the fully open-air conditions of FACE, if we are to be able to forecast with any confidence the future supply of the world's most important food crops in an uncertain and changing environment.

3. Process-based models of plant growth and paddy ecosystems could be a powerful tool in the design of the adaptation of rice production to future environments, but they have to be tested against observations before being reliable in use. FACE would provide a very good opportunity for model testing; however, to date, only limited FACE results have been used to test models of rice growth against observations (Bannayan et al., 2005).

4. Priority here is not limited to grain yield, but also includes grain quality and resistance against biotic and abiotic stresses. However, the problems of grain quality and stress resistance have not been given sufficient importance when assessing the impact of CO_2 on rice, and this requires more research attention.

References

Ainsworth, E.A. (2008) Rice production in a changing climate: a meta-analysis of responses to elevated carbon dioxide and elevated ozone concentration. *Global Change Biology* 14, 1642–1650.

Ainsworth, E.A., Davey, P.A., Hymus, G.J., Osborne, C.P., Rogers, A., Blum, H. et al. (2003) Is stimulation of leaf photosynthesis by elevated carbon dioxide concentration maintained in the long term? A test with *Lolium perenne* grown for 10 years at two nitrogen fertilization levels under free air CO_2 enrichment (FACE). *Plant, Cell and Environment* 26, 705–714.

Alberto, A.M.P., Ziska, L.H., Cervancia, C.R. and Manalo, P.R. (1996) The influence of increasing carbon dioxide and temperature on competitive interactions between a C_3 crop, rice (*Oryza sativa*) and a C_4 weed (*Echinochloa glabrescens*). *Australian Journal of Plant Physiology* 23, 795–802.

Baker, J.T. (2004) Yield responses of southern US rice cultivars to CO_2 and temperature. *Agricultural and Forest Meteorology* 122, 129–137.

Baker, J.T. and Allen, L.H. Jr. (1993) Contrasting crop species responses to CO_2 and temperature: rice, soybean and citrus. *Vegetatio* 104/105, 239–260.

Baker, J.T., Allen, L.H. Jr, Boote K.J. and Pickering, N.B. (1996) Assessment of rice responses to global climate change: CO_2 and temperature. In: Koch, G.W. and Mooney, H.A. (eds) *Carbon Dioxide and Terrestrial Ecosystems*. Academic Press, San Diego, California, pp. 265–282.

Bannayan, M., Kobayashi, K., Kim, H.Y., Lieffering, M., Okada, M. and Miura, S. (2005) Modeling the interactive effects of atmospheric CO_2 and N on rice growth and yield. *Field Crops Research* 93, 237–251.

Bouman, B.A.M., Humphreys, E., Tuong, T.P. and

Barker, R. (2007) Rice and water. *Advances in Agronomy* 92, 187–237.

Cheng, W.G., Sakai, H., Yagi, K. and Hasegawa, T. (2009) Interactions of elevated [CO_2] and night temperature on rice growth and yield. *Agricultural and Forest Meteorology* 149, 51–58.

Coats, B. (2003) Global rice production. In: Smith, C.W. and Dilday, R.H. (eds) *Rice Origin, History, Technology and Production*. Wiley, Hoboken, New Jersey, pp. 247–470.

Du Cloux, H.C., Andre, M., Daguenet, A. and Massimino, J. (1987) Wheat response to CO_2 enrichment, growth and CO_2 exchanges at two plant densities. *Journal of Experimental Botany* 38, 1421–1431.

FAO (2002) *World Agriculture: Towards 2015/2030 Summary Report*. FAO, Rome.

Hedley, M.J., Kirk, G.J.D. and Santos, M.B. (1994) Phosphorus efficiency and the forms of soil phosphorus utilized by upland rice cultivars. *Plant and Soil* 158, 53–62.

Horie, T., Baker, J.T., Nakagawa, H., Matsui, T. and Kim, H.Y. (2000) Crop ecosystem responses to climate change: rice. In: Reddy, K.R. and Hodges, H.F. (eds) *Climate Change and Global Crop Productivity*. CAB International, Wallingford, UK, pp. 81–106.

Imai, K., Coleman, D.F. and Yanagisawa, T. (1985) Increase in atmospheric partial pressure of carbon dioxide and growth and yield of rice (*Oryza sativa* L.). *Japanese Journal of Crop Science* 54, 413–418.

Inubushi, K., Cheng, W., Aonuma, S., Hoque, M.M., Kobayashi, K., Miura, S. *et al.* (2003) Effects of free-air CO_2 enrichment (FACE) on CH_4 emission from a rice paddy field. *Global Change Biology* 9, 1458–1464.

IPCC (2001) *Climate Change: The Scientific Basis*. Cambridge University Press, Cambridge, UK.

IPCC (2007) Climate change 2007: the physical science basis. In: Solomon, S., Qin, D., Manning, M., Chen, Z., Marquis, M., Averyt, K.B. *et al.* (eds) *Contribution of Working Group I to the Fourth Annual Assessment Report of the Intergovernmental Panel on Climate Change*. Cambridge University Press, Cambridge, UK, 996 pp.

Kim, H.Y., Horie, T., Nakagawa, H. and Wada, K. (1996) Effects of elevated CO_2 concentration and high temperature on growth and yield of rice. *Japanese Journal of Crop Science* 65, 644–651 [in Japanese with abstract in English].

Kim, H.Y., Lieffering, M., Miura, S., Kobayashi, K. and Okada, M. (2001) Growth and nitrogen uptake of CO_2-enriched rice under field conditions. *New Phytologist* 150, 223–229.

Kim, H.Y., Lieffering, M., Kobayashi, K., Okada, M., Mitchell, M.W. and Gumpertz, M. (2003a) Effects of free-air CO_2 enrichment and nitrogen supply on the yield of temperate paddy rice crops. *Field Crops Research* 83, 261–270.

Kim, H.Y., Lieffering, M., Kobayashi K., Okada, M. and Miura, S. (2003b) Seasonal changes in the effects of elevated CO_2 on rice at three levels of nitrogen supply: a free air CO_2 enrichment (FACE) experiment. *Global Change Biology* 9, 826–837.

Kimball, B.A., Kobayashi, K. and Bindi, M. (2002) Responses of agricultural crops to free-air CO_2 enrichment. *Advances in Agronomy* 77, 293–368.

Kobayashi, K., Okada, M., Kim, H.Y., Lieffering, M., Miura, S. and Hasegawa, T. (2006a) Paddy rice responses to free-air [CO_2] enrichment. In: Nosberger, J., Long, S.P., Norby, R.J., Stitt, M., Hendrey, G.R. and Blum, H. (eds) *Managed Ecosystems and CO$_2$: Case Studies, Processes and Perspectives*. Ecological Studies Series, vol. 187. Springer-Verlag, Berlin, pp. 87–104.

Kobayashi, T., Ishiguro, K., Nakajima, T., Kim, H.Y., Okada, M. and Kobayashi, K. (2006b) Effects of elevated atmospheric CO_2 concentration on rice blast and sheath blight epidemics. *Phytopathology* 96, 425–431.

Liu, G., Han, Y., Zhu, J.G., Okada, M., Nakamura, H. and Yoshimoto, M. (2002) Rice–wheat rotational FACE platform. I. System construct and control. *Chinese Journal of Applied Ecology* 13, 1253–1258.

Liu, H.J., Yang, L.X., Wang, Y.L., Huang, J.Y., Zhu, J.G., Wang, Y.X. *et al.* (2008) Yield formation of CO_2-enriched hybrid rice cv. Shanyou 63 under fully open-air field conditions. *Field Crops Research* 108, 93–100.

Long, S.P., Ainsworth, E.A., Rogers, A. and Ort, D.R. (2004) Rising atmospheric carbon dioxide: plants FACE the future. *Annual Review of Plant Biology* 55, 591–628.

Long, S.P., Ainsworth, E.A., Leakey, A.D.B. and Morgan, P.B. (2005) Global food insecurity. Treatment of major food crops with elevated carbon dioxide or ozone under large-scale fully open-air conditions suggests recent models may have overestimated future yields. *Philosophical Transactions of the Royal Society B* 360, 2011–2020.

Long, S.P., Ainsworth, E.A., Leakey, A.D.B., Nosberger, J. and Ort, D.R. (2006) Food for thought: lower-than-expected crop yield stimulation with rising CO_2 concentrations. *Science* 312, 1918–1921.

Lou, Y., Inubushi, K., Mizuno, T., Hasegawa, T., Lin,

Y., Sakai, H. *et al.* (2008) CH_4 emission with differences in atmospheric CO_2 enrichment and rice cultivars in a Japanese paddy soil. *Global Change Biology* 14, 2678–2687.

Maclean, J.L., Dawe, D.C., Hardy, B. and Hettel, G.P. (2002) *Rice Almanac: Source Book for the Most Important Economic Activity on Earth*, 3rd edn. International Rice Research Institute, Los Baños, the Philippines.

Matsui, T., Namuco, O.S., Ziska, L.H. and Hone, T. (1997) Effects of high temperature and CO_2 concentration on spikelet sterility in indica rice. *Field Crops Research* 51, 213–219.

McLeod, A.R. and Long, S.P. (1999) Free-air carbon dioxide enrichment (FACE) in global change research: a review. *Advances in Ecological Research* 28, 1–55.

Moya, T.B., Ziska, L.H., Namuco, O.S. and Olszyk, D. (1998) Growth dynamics and genotypic variation in tropical, field-grown paddy rice (*Oryza sativa* L.) in response to increasing carbon dioxide and temperature. *Global Change Biology* 4, 645–656.

Parry, M.L., Rosenzweig, C., Iglesias, A., Livermore, M. and Fischer, G. (2004) Effects of climate change on global food production under SRES emissions and socio-economic scenarios. *Global Environmental Change – Human and Policy Dimensions* 14, 53–67.

Peng, S., Huang, J., Sheehy, J.E., Laza, R.C., Visperas, R.M., Zhong, X. *et al.* (2004) Rice yields decline with higher night temperature from global warming. *Proceedings of the National Academy of Sciences USA* 101, 9971–9975.

Reid, C.D. and Fiscus, E.L. (2008) Ozone and density affect the response of biomass and seed yield to elevated CO_2 in rice. *Global Change Biology* 14, 60–76.

Sakai, H., Hasegawa, T. and Kobayashi, K. (2006) Enhancement of rice canopy carbon gain by elevated CO_2 is sensitive to growth stage and leaf nitrogen concentration. *New Phytologist* 170, 321–332.

Seneweera, S.P. and Conroy, J.P. (1997) Growth, grain yield and quality of rice (*Oryza sative* L) in response to elevated $[CO_2]$ and phosphorus nutrition. *Soil Science and Plant Nutrition* 43, 1131–1136.

Seneweera, S.P., Blakeney, A., Milham, P., Basra, A.S., Barlow, E.W.R. and Conroy, J.P. (1996) Influence of rising atmospheric CO_2 and phosphorus nutrition on the grain yield and quality of rice (*Oryza sativa* cv. Jarrah). *Cereal Chemistry* 73, 239–243.

Sheehy, J.E., Dionora, M.J.A. and Mitchell, P.L. (2001) Spikelet numbers, sink size and potential yield in rice. *Field Crops Research* 71, 77–85.

Shimono, H., Okada, M., Yamakawa, Y., Nakamura, H., Kobayashi, K. and Hasegawa, T. (2007) Lodging in rice can be alleviated by atmospheric CO_2 enrichment. *Agriculture, Ecosystems and Environment* 118, 223–230.

Shimono, H., Okada, M., Yamakawa, Y., Nakamura, H., Kobayash, K. and Hasagawa, T. (2008) Rice yield enhancement by elevated CO_2 is reduced in cool weather. *Global Change Biology* 14, 276–284.

Tubiello, F.N., Amthor, J.S., Boote, K.J., Donatelli, M., Easterling, W., Fischer, G. *et al.* (2007) Crop response to elevated CO_2 and world food supply: a comment on 'Food for thought ...' by Long *et al. Science*, 312: 1918–21, 2006. *European Journal of Agronomy* 26, 215–223.

Weerakoon, W.M.W., Ingram, K.T. and Moss, D.N. (2000) Atmospheric carbon dioxide and fertilizer nitrogen effects on radiation interception by rice. *Plant and Soil* 220, 99–106.

Weerakoon, W.M.W., Ingram, K.T. and Moss, D.M. (2005) Atmospheric CO_2 concentration effects on N partitioning and fertilizer N recovery in field grown rice (*Oryza sativa* L.). *Agriculture, Ecosystems and Environment* 108, 342–349.

Yang, L.X., Huang, J.Y., Yang, H.J., Zhu, J.G., Liu, H.J., Dong, G.C. *et al.* (2006) The impact of free-air CO_2 enrichment (FACE) and N supply on yield formation of rice crops with large panicle. *Field Crops Research* 98, 141–150.

Yang, L.X., Huang, J.Y., Yang, H.J., Dong, G.C., Liu, H.J., Liu, G. *et al.* (2007a) Seasonal changes in the effects of free-air CO_2 enrichment (FACE) on nitrogen uptake and utilization of rice at three levels of nitrogen fertilization. *Field Crops Research* 100, 189–199.

Yang, L.X., Wang, Y.L., Huang, J.Y., Zhu, J.G., Yang, H.J., Liu, G. *et al.* (2007b) Seasonal changes in the effects of free-air CO_2 enrichment (FACE) on phosphorus uptake and utilization of rice at three levels of nitrogen fertilization. *Field Crops Research* 102, 141–150.

Yang, L.X., Wang, Y.L., Kobayashi, K., Zhu, J.G., Huang, J.Y., Yang, H.J. *et al.* (2008) Seasonal changes in the effects of free-air CO_2 enrichment (FACE) on growth, morphology and physiology of rice root at three levels of nitrogen fertilization. *Global Change Biology* 14, 1844–1853.

Yang, L.X., Liu, H.J., Wang, Y.X., Zhu, J.G., Huang, J.Y., Liu, G. *et al.* (2009) Yield formation of CO_2-

enriched inter-subspecific hybrid rice cultivar Liangyoupeijiu under fully open-air field condition in a warm sub-tropical climate. *Agriculture, Ecosystems and Environment* 129, 193–200.

Yoshimoto, M., Oue, H. and Kobayashi, K. (2005) Energy balance and water use efficiency of rice canopies under free-air CO_2 enrichment. *Agricultural and Forest Meteorology* 133, 226–246.

Zheng, X.H., Zhou, Z.X., Wang, Y.S., Zhu, J.G., Wang, Y.L., Yue, J. *et al.* (2006) Nitrogen-regulated effects of free-air CO_2 enrichment (FACE) on methane emissions from paddy rice fields. *Global Change Biology* 12, 1717–1732.

Ziska, L.H. and Bunce, J.A. (2007) Predicting the impact of changing CO_2 on crop yields: some thoughts on food. *New Phytologist* 175, 607–618.

Ziska, L.H., Manalo, P.A. and Ordonez, R.A. (1996a) Intraspecific variation in the response of rice (*Oryza sativa* L.) to increased CO_2 and temperature: growth and yield response of 17 cultivars. *Journal of Experimental Botany* 47, 1353–1359.

Ziska, L.H., Weerakoon, W., Namuco, O.S. and Pamplona, R. (1996b) The influence of nitrogen on the elevated CO_2 response on field-grown rice. *Australian Journal of Plant Physiology* 23, 45–52.

4 Recent Changes in Pampean Agriculture: Possible New Avenues in Coping with Global Change Challenges

E.H. Satorre

4.1 Introduction

Located in the southern cone of South America, the pampas can be considered among the most highly productive areas of the world. Recently, this area has experienced dramatic changes in agricultural production technologies and farm management. In regard to the magnitude of these changes, the main transformations in the pampas include a major increase in overall grain production and sown area: total production has almost tripled, from 35 to nearly 90 million t and arable land has expanded from 14 to 26 million ha since the 1990s. Although this growth has been supported by changing technologies, such as major increases in the use of external subsidies in the form of new varieties, agrochemicals and/or fertilizers, the most important changes have not been linked to crop technologies but to farm management, i.e. farmers began to apply a more systemic approach in their decisions incorporating process and information technologies as effective production tools. All these changes have been dependent on ecological, political and technological factors that have rarely been explored together in the literature. However, new production avenues, capable of coping with global change challenges, must be linked to the ability of farmers to produce adaptive responses. Exploring and understanding changes such as those seen in the pampas can help in understanding new productive avenues being effectively addressed in new ecological and social scenarios.

In this chapter, I will try to address three general objectives: (i) to present the main ecological and extensive agriculture characteristics of the pampean region; (ii) to analyse recent technological and productive transformations; and (iii) to discuss the influence of future possible transformation factors.

4.2 A Brief Overview of Some Ecological Characteristics of the Pampas

The pampas, located in Argentina between parallels 32° and 39°S and meridians 58° and 64°W, comprises an area of approximately 52 million ha. Although it is mostly a flat area, interrupted only by the Tandilia and Ventania hills in its southern portion, internal heterogeneity in climate, soil and vegetation permits the recognition of five main ecozones – rolling pampas, inland pampas, flooding pampas, southern pampas and mesopotamic pampas (Hall *et al.*, 1992).

The climate of the pampas is temperate humid, without a dry season and with a very hot summer (Hall *et al.*, 1992). Annual rainfall diminishes from east to west in the range 1000–500 mm. The general pattern of rainfall distribution describes a minimum in winter and maximum in summer, with intermediate levels in spring and autumn. The summer rainfall pattern is very similar among the various ecozones in the pampas, but winter rainfall varies considerably. Winter rainfall is close to monsoonal in the north-west of the pampas but tends to be isohigrous in the south-east. The pampas is under the influence of El Niño southern oscillation (ENSO), which partially explains this inter-annual variability in rainfall (Spescha et al., 1997; Messina, 1999; Vargas et al., 1999; Grimm et al., 2000). The rainfall pattern is frequently altered by ENSO signals, particularly in the spring/summer. There is a significant increase in rainfall amount during November, December and January due to hot 'El Niño' events, and below-average rainfall during October, November and December due to cold 'La Niña' events (Magrín et al., 1998; Podestá et al., 1999). Moreover, a low-frequency pattern of rainfall variability was also described in the last century – from approximately 1967 up to the present time a humid cycle has become established in the region, leading to an increase of 180 mm in the long-term average rainfall (Messina, 1999) and the displacement of the 600 mm isohyet almost 100 km to the west from its long-term average position (Sierra et al., 1994).

Average temperature decreases along a north–south direction; however, thermal amplitude increases from east to west as one moves away from oceanic influences, into the continent. The temperature regime for the region shows that June and July are the coldest months and January the hottest. The frost-free period varies between 180 and 260 days, from north-east to south-west in the region. The inter-annual variation in the date of last frost is large, and this greatly influences winter crop yields and summer crop sowing dates (Damario and Pascale, 1988). Frost damage is among the most important factors determining wheat yield and maize establishment risks in the region.

Soils are mollisols, formed over loessic sediments originating in the Andes. Typic argiudolls are more frequent in the rolling, southern pampas (Hall et al., 1992), these having a Black A surface horizon of 18–35 cm and loam or clay–loam texture; topsoil organic matter content ranges from 2 to 6%. The B horizon usually extends from 35 to 80 cm, having a high clay content (30–57%) with illite as the predominant clay mineral (Senigagliesi et al., 1996). Typic and entic

hapludolls and haplustolls are most frequent in inland pampas; these soils have low topsoil organic matter content (1–3%) and a less developed B horizon, with low clay content. They are usually deep, with a predominantly sandy and loamy texture. In the southern pampas a petrocalcic subsoil layer may be present at 40–90 cm on typic argiudolls, hapludolls and haplustolls; while a buried horizon (thapto soil) may be recognized in the flooding and inland pampas. In the mesopotamic pampas vertic argiudolls and vertisols are predominant soils. In this region, soil clay contents are over 50% with expansive montmorillonite as the most frequent clay mineral. Natracualf soils and hapludolls dominate the flooding pampas. No soils in the pampas freeze in winter and, with the exception of the mesopotamic pampas, together with the high silt content in the soils, this determines a low internal ability to regenerate aggregates and porosity structure.

The rolling pampas soils show marked nitrogen deficiencies that increase from east to west, following the organic matter content pattern; for this reason, cereals are frequently fertilized with nitrogen. With the exception of sandy soils (i.e. > 55% sand), all pampean soils tend to show marked available phosphorus deficiency following a west-to-east pattern (Darwich, 1983, 1994). Potassium levels are high throughout, with the exception of northern mesopotamic pampas soils, while sulfur and magnesium deficiencies have recently been identified in crops in some areas (Melgar, 1997).

4.3 Recent Transformations in Pampas Dryland Agricultural Production Systems

The pampas is a market-oriented producing area with soybean, maize, wheat and sunflower as the main grain crops; the landscape has been designed under the influence of socio-economic and political factors. Economic and political factors such as the price of grains, internal export taxes on agricultural products or trade barriers, have played and are playing a central role in determining the spatial and temporal dynamics of the pampean landscape. Variations in soils and climate – ecological factors – and new technologies have also played a central role in determining various patterns of agricultural production within each ecozone in the region.

Traditionally, the pampas are identified with extensive mixed crop–cattle farms; however, pampean agricultural production systems have changed markedly since the 1990s. On the one hand, simple and specialized cropping systems were developed, with soybean as the main crop; cereal species were replaced by oilseed crops and the total cultivated area has increased. Although beef cattle and crops can still be found almost everywhere within the pampas (Solbrig and Viglizzo, 1999), the spatial distribution of the traditional mixed crop–cattle production systems in the region has been modified. Beef cattle production was moved to the less productive soils while arable expanded on those of medium and high productivity. On the other hand, production technologies have greatly changed, contributing to a more intensified land use and productive agriculture. Taken together, the above-mentioned changes have contributed to the modification of farm management, which is now more professional, and land tenancy. New actors developed in the agricultural sector, such as the contractors and pooled producers on rented land. These changes have extended throughout the pampas at a rapid rate.

4.3.1 A brief overview of production systems

In the pampas, arable, calf, beef and dairy are still the most frequent production systems. Lucerne-based perennial pastures and multi-species pastures are still sown, although animal production is also carried out under improved natural grassland or old pastures. When animal and grain production systems share the same land, the positive effect of pasture on soil aggregation, nitrogen fixation and yield stability has been recognized (Puricelli, 1996). However, crop and animal production systems have been intensified since the 1990s and the area under pasture reduced.

There are well-documented examples in the rolling pampas and some parts of the flat and southern pampas of a steady reduction in land area under beef cattle production, followed by an increase in the cultivated area (Satorre, 2001). The main driving force of this process was the need for greater farm income attainable with crop harvests, followed by better ecological conditions for grain production (see climate description above). Therefore, in most regions of the pampas, the usual rotation of crops with pasture was replaced by that in a permanent cropping sequence. However, capital-intensive production systems are now more frequently seen, and confined cattle production – namely in the form of feedlots – has appeared. The replacement and transformation of the above-mentioned activities modified the landscape of whole areas within the pampas, conforming to a new scenario where annual species play a central role.

Grain crops are rotated in this cropping sequence, including at least three to five different species in each ecozone. Wheat, barley and oilseed rape are the main winter crops, while soybean, maize and sunflower are the main summer crops. At present, the predominant crop rotation in the pampas includes full-season soybean–wheat double-cropped with late-sown soybean–maize, i.e. four crops every 3 years. Usually, two full-season soybean crops or a full-season soybean and a sunflower crop are sown before the maize crop, mainly in the western pampas, i.e. five crops in 4 years. Variations of these predominant rotations depend on the ecozone; for example, in the southern pampas the most common rotation tends to be maize–sunflower–wheat. In some areas, such as in the centre of Santa Fe province (northern rolling pampas), monocultures of soybean sporadically rotated with maize or wheat–double-cropped soybean are common, particularly on small farms.

These changes in production systems were accompanied by technological transformations which contributed to consolidation of production growth in grain crops.

4.3.2 A brief overview of technological changes in agriculture

The pampas has experienced profound technological changes since the 1990s. Conventional tillage systems, with either chisel or mouldboard plough, have gradually been replaced by no-tillage cropping, in order to maintain soils covered with plant residues. At present, it is estimated that 16 million ha (almost 65% of the sown land in Argentina) are cultivated under no-tillage production systems. Rotations with cereals and soil cover maintenance with crop residues are crucial for the efficient use of water, while conserving soil resources and contributing to management of pests, weeds and diseases. Water is the main environmental limiting factor for crop yield under the dryland systems of the pampas, and plant residues covering the soil increase water infiltration and reduce water losses. Moreover, these conserve soil structure in the top few centimetres, reducing wind and water erosion and regulating soil temperature, thus increasing faunal activity.

Fertilizer use has increased 11-fold in the last decade, to 2.3 million t; however, the average amount of fertilizer used in the pampas is still very low. Nitrogen and phosphorus are the nutrients most frequently applied; granular urea (N:P:K relative composition – 46:00:00) and urea–ammomnium–nitrate solutions (UAN, 32:00:00) are the main sources of nitrogen, while mono-ammonium (11:52:00) and diammonium (18:46:00) phosphate are the main sources of phosphorus fertilizer. A large proportion of the nitrogen and phosphorus fertilizers are applied to wheat and maize; inoculated full-season soybean and sunflower crops may be fertilized only with phosphorous sources. Fertilizer use decision criteria have helped in the adoption of fertilization techniques, and are based on various approaches experimentally developed by research groups in the different ecozones. Regarding nitrogen and phosphorus, most approaches are based on soil estimates of the amount of available nitrogen and extractable phosphorus (Kurtz and Bray method). Since fertilizer costs represent an important proportion of total variable production costs of any extensive crop in the pampas, fertilizer use efficiency and economic return are carefully considered when any recommendation for application is considered.

Grain yield increases due to sulfur application have been observed in various areas, particularly with wheat and soybean. However, the use of this nutrient is still very rare (Melgar, 1997). Research on sulfur, as on micronutrients (zinc, boron), in the pampas is still limited and results are conflicting (Ratto de Miguez and Diggs, 1990; Ratto de Miguez and Fatta, 1990).

Genetically modified crops are now widely adopted in the pampas; in particular, the use of glyphosate-resistant varieties of soybean has been surprisingly high. At present nearly 98% of the area sown with soybeans includes one of the various genetically modified varieties available within maturity groups II and VIII. The spread of this technology has allowed an average soybean grain yield increase of almost 0.5 t/ha, huge reductions in cost and enhanced weed control efficacy (Satorre, 2006). Moreover, the adoption of glyphosate-resistant cultivars has contributed to easing the management and increasing the efficiency of no-tillage, direct-drill systems. These combined technologies (direct-drill plus glyphosate-resistant cultivars) are contributing greatly to both soil conservation and the expansion of soybean cultivation in the traditional producing areas in the rolling and inland pampas in the north of Buenos Aires and the south of Santa Fe and Córdoba provinces. Successful soybean crops are now found in the western, mesopotamic and even in the flooding pampas of Argentina, previously considered marginal areas with highly variable yields.

Biotechnology has also contributed, through new germplasm in other crops sown in the pampas. Genetically modified maize hybrids have been registered to improve crop protection against insects: Bt, MG, TD and HX genotypes are sown to protect the crop against Lepidoptera, mainly *Diatraea saccharalis*, *Helicoverpa zea* and *Spodoptera frugiperda*. Also, GM maize hybrids were registered and are now in use to improve weed control: ammonium gluphosinate and glyphosate herbicide-resistant genotypes have been released. Recently, stack hybrids with combined insect- and weed-transgenic genes have been released on the Argentine market. At present, the contribution of genetically modified crops to the success of agriculture has been recognized among producers and agronomists by their widespread and rapid adoption. Some concern has been voiced, however, based on the negative reaction of some markets to biotechnologically derived products.

Precision agriculture technology has been evaluated and used commercially in the pampas since the 1990s. The recognition of environmental heterogeneity in the field is helping to improve crop planning and management. It is widely accepted that precision agriculture may contribute to increased sustainability, mainly by improving input use efficiency (Bragachini *et al.*, 1997), but the gap between such sophisticated technologies and agronomic knowledge in decision making has been made clear.

The expansion of the area under soybean crop may be considered a technological transformation for a pampas cereal-producing region: from the 28 million ha previously sown with grain crops almost 20 million are now sown with soybean. Although soybean has greatly transformed the pampas, huge transformations were also derived from the adoption of this crop in Argentine regions beyond the pampas – in fact, the greatest changes were experienced in non-pampas regions. For example, in the north-east and north-west of the country, grain production increases recorded were 800 and 400%, respectively, in the period 1990–2003 (Satorre, 2005). Moreover, in these regions 73% (north-east) and 67% (north-west) of the land is sown with soybean.

4.4 Driving Forces of the New Pampas Agriculture

In the local context of the pampas, technical decisions on crop management have emphasized yield increases. This was viewed as the simplest and most direct way of increasing farm income, since fixed farm costs are diluted in a greater level of production per unit area. The production of specialities instead of commodities has only recently been considered by some farms as an alternative to overcoming the low and highly variable price of grains.

From an ecological point of view the transformations in land use commented upon in the previous section may have negatively affected the natural resource background of the region. The reduction of land under pasture in the most productive areas, the reduction of carbon and soil organic matter productivity due to the increase in soybean area, the increase in agrochemical use, and soil deterioration have been included among the greatest ecological risks. However, within the context in which these transformations have occurred, the risk of negative ecological impacts has been reduced (Satorre, 1998, 2001). This occurred mainly due to both the availability of new production technologies and scientifically based production knowledge. Technologies such as no-tillage and crop rotation planning, the new varieties and hybrids released and the increase in fertilizer use are today considered as relevant contributions to the successful transformations of pampas agriculture to improved economic results for farmers and the greater stability and sustainability of extensive production systems.

Climate variations have influenced the agricultural systems of the pampas, as previously mentioned. Farmers responded with adaptive measures to increased rainfall and to the changes in distribution patterns: (i) they modified the allocation of productive factors, mainly land and capital, increasing grain production; (ii) they modified crop priorities, moving from cereals to oil and protein crops; and (iii) they introduced new technologies in order to mitigate the effects of climatic changes and bring stability to the farm in a changing scenario. Global climate change has the potential to cause huge variations in the climate of the various pampean regions; according to some scenarios, rainfall could be diminishing in the pampas, although there is great uncertainty in this respect. Therefore, climate is one of the crucial risk factors considered by farmers when planning dryland crops.

Genetically modified varieties (GMO) may also be considered among the important driving forces of change. Soybean GMO varieties were extensively adopted since their release during 1996 in Argentina. At present their impact on farm improvement has been recognized by farmers and agronomists, since this technology has helped to consolidate the expansion and growth of other technologies; for example, GMO soybean varieties contributed to expansion of no-tillage cropping systems by providing an effective and economical way to control weeds. Direct production costs were reduced by means of this technology and most annual or perennial weeds were effectively controlled, increasing yields and the value of the land. Biotechnology and GMO have greatly influenced soybean, maize, sunflower and, recently, wheat varieties in the pampas. The development of varieties with special attributes such as better nutritional value has been reduced. However, they may promote an important change in the productive and business scenario of the region.

The adoption of no-tillage crops may be attributed to ecological reasons (better water use efficiency, see below), but management, farm labour organization and land tenure were also important reasons for its wide and rapid adoption in the pampas. First, no-tillage cropping was adopted following the expansion of double-cropped soybean after wheat. In the early 1990s the sequence of wheat–double-cropped soybean became consolidated as a stable and economic activity among farmers. The performance of soybean directly drilled on wheat stubble was ideal due to the higher yields found when it was sown and established early in the season. Moreover, wheat residues contributed to keeping soils covered and to improving crop water economy by reducing water losses through evaporation and runoff, and by increasing water infiltration and water use efficiency. It

is also known that plant residues left on topsoil reduce water and wind erosion when covering at least 30% of the soil surface. Although in some environments, such as in the mesopotamic pampas, terraces may be needed to prevent water erosion, in areas with moderate risks no-tillage cropping proved sufficient to stop the problem.

These aspects were crucial for dryland cropping systems, mainly in sub-humid and semi-arid areas of the pampas. By reducing the negative effects of erosion and the dependence of yields and yield variability on rainfall, no-tillage crops began to drive the use of higher rates of fertilizer and better, usually more expensive, seeds that together contributed to increased yields. At the same time, more effective herbicides became available at a relatively low cost (namely, glyphosate- and imidazolinone-derived herbicides, among others), and drillers were developed and continuously improved to sow seeds on top of previous crop residues. These allowed better weed control and improved physical conditions for crop germination and establishment. A positive feedback process was then gradually developed and expanded all over the pampas, yield and erosion risks decreased, soil physical conditions improved and results became less variable and more predictable. Despite the fact that the positive feedback productive process was ecologically and technologically driven, there is ample consensus that, perhaps more importantly, other determinants also operated on the agriculture transformation of the pampas.

On-farm managerial decisions were key, mainly because farm organization and functioning were modified by no-tillage cropping and yield improvement. No-tillage cropping frees up more time for crop production, i.e. a greater area may be prepared and sown with less time input from human labour, thus helping to increase farmers' operative capacity. Reduction in tillage and crop establishment time allowed many farmers to increase the land they were sowing by renting it and to open new business and services opportunities. In addition, crop management became more independent of weather and soil conditions, allowing the introduction of lands previously considered marginal into production.

A systemic approach was also a characteristic of pampas agriculture. The evaluation of crops as individual activities has slowly been replaced by the view that the rotation and production system may become an analytical unit. Chronologically, crop sequence was therefore valued in several regions such that it was considered by farmers as an important decision factor (Bert *et al.*, 2006). Crop rotations when properly organized produce a crucial contribution to soil conservation and organic matter balance, water use efficiency of crops and weed, insect and plant disease management. Undoubtedly crop rotations including winter and summer species, cereals, legumes and oilseeds and various functional characteristics in both time and space have increased the potential to capture positive agronomic interactions. The rotation of soybean with maize is considered an effective tool for increasing the yield of both species; soybean following maize tends to yield up to 0.5 t/ha more than vice versa (Satorre, 2006). On the other hand, there is evidence of yield reductions due to soybean monoculture when compared with rotated fields. The unexpected negative impact of diseases (e.g. cancrosis in soybean) and pests (e.g. nematodes in soybean) in some areas has been associated with monoculture and a long agricultural history in many areas within the pampas. Although maize contributes to improved crop diversity and economic returns, the area under maize has steadily been reduced due to low international prices, its high direct costs, the loss of financial capital, the low and variable yield of this crop in degraded soils and the short duration of rent contracts that motivate rapid economic returns.

In the past 10 years precision agriculture has amalgamated various decision and management tools with highly sophisticated satellite technology, with the aim of recognizing and managing within-farm and field environmental heterogeneity. Agronomic problems are geo-referenced (GPS), giving a spatial dimension to decisions that were previously made at the field level. GPS technology is used in controlling the application of fertilizers, herbicides, pesticides and to manage decisions regarding sowing rates or crop varieties at the field level. By means of this technology crop management opens opportunities effectively to incorporate more knowledge and research findings into practice. Moreover, this technology has been useful in creating a new image and developing a farm manager rather than a farmer approach to a modern, more complex and intellectually demanding agriculture.

In the pampas, the development and transference of technology are the result of close collaboration between public institutions such as INTA (Instituto Nacional de Tecnología Agropecuaria [National Institute of Agricultural Technology, Argentina]), universities (e.g. the Faculty of Agronomy at the University of Buenos Aires) and non-governmental organizations (NGOs) of private farmer groups and private industry. Among the NGOs, AACREA (Argentine Association of Agricultural Experimentation Consortiums: www.aacrea.org.ar) has been developed over the last 50 years, a complete and efficient network of experience interchange and adaptive research, aimed at promoting the development of a more sustainable agriculture. Farmers who are members of AACREA represent the entire farming community, though in some regions they belong to the larger producers sector, according to the size of their farms. The CREA farmers' organization was first proposed by a group of producers from the west of Buenos Aires province (western pampas). They decided to combine their efforts and exchange experiences, seeking new ways of finding solutions to their environmental (mainly soil erosion) and farm managerial problems. In each consortium 8–10 farmers cooperate, mutually analysing and evaluating their technologies and management decisions from the agronomic, economic and social viewpoints. At present 205 CREA groups from various regions make up AACREA. AAPRESID (Asociación Argentina de Productores en Siembra Directa; www.aapresid.org.ar)

should be mentioned as another relevant NGO, having been operating in the agricultural sector of Argentina since the late 1980s. Producers from these organizations have experienced changes in their production systems since the 1990s, as seen from the foregoing and, at the same time, were agents of those changes – i.e. they greatly contributed to increased yields and to creating more intensified, productive and efficient land use.

In the pampas, the search for productivity and crop management efficiency has seen much attention from professional farmers, agronomists and organizations devoted to research and development with transference to the production sector. In this scenario, intensified cropping systems appeared as a technological and economical option. Intensified cropping systems were evaluated as options to sustain the transformations experienced by the extensive agricultural model, to increase its efficiency in resource management and to increase farm income. In this process of system intensification (Satorre, 2005), agriculture was slowly adopting more complex criteria and productive patterns, while incorporating a larger number of crop species, relationships and interactions in decision making. To this end, knowledge and information technologies are playing an increasingly important role, taking into account a wide, systemic and more integral view of the agribusiness.

4.5 Final Comments

As will be seen from the foregoing, the extensive agriculture of the pampas has experienced huge transformations since the 1990s. The support of scientific and technological knowledge has been relevant in this process; moreover, the concept of sustainability has been central as a condition for change and growth. Overall, in the pampas there exists a possibly well-founded feeling that the ecological impact of agriculture to date has been low. Modern technologies have contributed to simultaneously increased productivity and

sustainability in the cropping systems of some ecological regions. However, the adoption of better management practices is still needed in some areas to fully sustain the functioning and structure of the pampas agroecosystems. Technologies such as better fertilizer management, new genotypes and better crop protection techniques are, however, an essential part of more productive crop systems and are, at the same time, more environmentally friendly.

Industrialized or intensified animal production systems and intensified cropping systems have been developed since the 1990s in the pampas, which may raise environmental concerns for producers. For example, partly as a consequence of the changes that have occurred (see above), feedlot production systems are now an alternative to the extensive, traditional grass-fed systems (Basso, 1999). In these newer systems faecal contamination represents one cause of potential environmental damage: the excretion of nitrogen and phosphorus and the production of methane gas from cattle contribute to both locally and globally negative effects (Voorburg, 1993 cited in Basso, 1999). Moreover, the change from a rural lifestyle to one of commercialized and industrialized farms has introduced new managerial concepts. Grain production systems are very competitive, and the amount of grain produced per unit cost is relatively high for the various crops sown. As a consequence, scale and efficiency will eliminate a large number of medium and small farmers, concentrating most of the agricultural production in fewer hands. Several technologies and variables operating in the transformation of the pampas can be seen as risk or opportunity factors. A general perception is that environment problems are now within the jurisdiction of the ordinary farmer, who has recognized that their decisions can contribute to soil damage in particular and the environment in general. The interest in environmentally related subjects has not only filled the agendas of government, industry and NGOs, but also of individual producers. For example, within organizations such as AACREA or AAPRESID, farmers have discussed the topics of sustainable production systems and production and environment at various well-attended conferences (virtually annually since 1995), and also within smaller groups. It is expected that regulations and a greater awareness of the ecological impacts of agriculture will help formulate better decision procedures in the near future. However, we recognize that farmers' decisions are still largely under the influence of market and macroeconomic factors, which are out of their control.

New technologies and intensification may contribute to sustainable production systems with low environmental impact on some farms. However, it must be stressed that in agricultural production and natural resource management the attitude of producers is crucial; producers themselves are involved in the sustainability of their systems. Certainly, producers' attitudes have been considered among the most important sustainability indicators (Soriano, 1996). However, low-productive, marginal areas need to be carefully managed and, particularly in these regions of the pampas, the need to integrate many technologies rather than adopting a single one should be emphasized. Areas previously considered marginal are the most vulnerable to environmental change such as global climate change, since land use has been greatly affected in these areas.

In an unfavourable climatic scenario (i.e. lower rainfall), producers may react according to educational, economic and social factors. In marginal areas with a less favourable climate, lower crop yields and farm income may certainly increase the economic risks associated with various crops, affecting land value and reducing investment and productive capital in these areas. Future scenarios such as this must be met with adaptive on-farm responses to reduce the negative effects, while promoting new production measures. Technological changes (e.g. new varieties more tolerant to drought), social measures (e.g. continuing education)

and the construction of within-farm social capital (e.g. good professional teams) are essential components of successful adaptive responses. If we look at the margins of response available to pampas farmers under various plausible scenarios based on global climate change, with the present technologies, individual productive changes (reduced density, modified crop sowing dates, etc.) may offer quite effective adaptive responses in various areas. However, the negative impacts of an extended dry period can seldom be mitigated by present technologies in some pampas marginal areas, as seen during the 2008/2009 harvest in Argentina. In these areas (i.e. western, mesopotamic and southern pampas), only the changing of production systems could offer an effective response to mitigating the negative effects and reducing on-farm income risks arising from a low rainfall pattern. It is now recognized that there are major doubts as to what a positive adaptive response might be, and to how new technologies might help to mitigate negative scenarios arising from global climate change. Nonetheless, it is also recognized that under complex scenarios science and education can provide strategic resources to solve new problems. Factors such as global climate change impose new working strategies: these must become much more interdisciplinary, building knowledge networks and strengthening professional teams in various disciplines. In this context, new on-farm technologies require complementary skills seldom found in single individuals. The organization of groups and teams among farms becomes even more relevant in uncertain scenarios. The agriculture growth in the pampas and the role played by various factors discussed above clearly show that farmers may support outstanding adaptive responses if clear signals, either ecological, technological, economic or political, are given. Global climate change signals must be properly communicated if new avenues are to be either found or explored to sustain agricultural systems' ability to produce food for the world.

References

Basso, L.R. (1999) Alimentación Animal e Impacto Ambiental. Sustentabilidad de los sistemas mixtos agrícolas-ganaderos. In: *Proceedings 22 Congreso Argentino de Producción Animal*, Río Cuarto, Córdoba, Argentina.

Bert, F.E., Satorre, E.H., Ruiz Toranzo, F. and Podestá, G.P. (2006) Climatic information and decision making in maize crop production systems of the Argentinean pampas. *Agricultural Systems* 88, 180–204.

Bragachini, M., Bongiovanni, R. and Martellotto, E. (1997) *Agricultura de precisión. Manejo localizado de los nutrientes del cultivo*. Proyecto IPG, EEA INTA Manfredi, Argentina.

Damario, E.A. and Pascale, A.J. (1988) Características agroclimáticas de la región pampeana. *Revista de la Facultad de Agronomía* 9, 41–69.

Darwich, N. (1983) *Niveles de Fósforo Asimilable en los Suelos Pampeanos*. IDIA 409–412, pp. 1–5.

Darwich, N. (1994) Siembra directa y ambiente edáfico. In: *Cuaderno de Actualización Técnica No 54 – Siembra Directa (AACREA)*, pp. 25–28.

Grimm, A.M., Barros, V.R. and Doyle, M.E. (2000) Climate variability in southern South America associated with El Niño and la Niña events. *Journal of Climate* 13, 53–58.

Hall, A.J., Rebella, C.M., Ghersa, C.M. and Culot, J. Ph. (1992) Field-crop systems of the pampas. In: Pearson, C.J. (ed.) *Field Crop Systems. Ecosystems of the World*, Vol. 18. Elsevier, Amsterdam, pp. 413–450.

Magrín, G.O., Grondona, M.O., Travasso, M.I., Boullón, D.R., Rodriguez, G.R. and Messina, C.D. (1998) *Impacto del Fenómeno "El Niño" sobre la Producción de Cultivos en la Región pampeana*. Instituto de Clima y Agua, INTA Reporte. Dirección de Comunicaciones INTA, Buenos Aires, Argentina.

Melgar, R.J. (1997) Potasio, azufre y otros nutrientes necesarios para considerar en una fertilización. *Revista Fertilizar – Suplemento Trigo* 17–24.

Messina, C.D. (1999) El Niño southern oscillation y la productividad de cultivos en la zona pampeana: Evaluación de estrategias para mitigar el riesgo climático. MSc thesis. Faculty of Agronomy, University of Buenos Aires.

Podestá, G.P., Messina, C.D., Grondona, M.O. and Magrín. G.O. (1999) Associations between grain crops yield in central-eastern Argentina and El Niño southern oscillation. *Journal of Applied Meteorology* 38, 1488–1498.

Puricelli, C.A. (1996). La sustentabilidad de los sistemas de producción mixtos. In: *Proceedings Congreso CREA Zona Oeste*, Mar del Plata, Argentina, pp. 20–28.

Ratto de Miguez, S. and Diggs, C.A. (1990) Niveles de Boro en suelos de la Pradera pampeana. Aplicación al cultivo de girasol. *Ciencia del Suelo* 8, 93–100.

Ratto de Miguez, S. and Fatta, N. (1990) Disponibilidad de micronutrimentos en suelos del área maicera núcleo. *Ciencia del Suelo* 8, 9–15.

Satorre, E.H. (1998) Aumentando los rendimientos de manera sustentable en la pampa. Aspectos generales. In: Solbrig, O.T. and Vainesman, L. (eds) *Hacia una Agricultura mas Productiva y Sostenible en la Pampa Argentina*. Harvard University–CPIA–Banco de la Nación Argentina. Orientación Gráfica Editora, Buenos Aires, pp. 72–98.

Satorre, E.H. (2001). Production systems in the Argentine pampas and their ecological impact. In: Solbrig, O.T., Paalberg, R. and Di Castri, F. (eds) *Globalization and the Rural Environment*. Harvard University Press, Cambridge, Massachusetts, pp. 79–102.

Satorre, E.H. (2005) Cambios tecnológicos en la agricultura argentina actual. *Ciencia Hoy* 15, 24–31.

Satorre, E.H. (2006) Intensificación productiva y los nuevos horizontes tecnológicos para la agricultura argentina. In: *Actas del Congreso Mundo Agro 2006*, Buenos Aires, pp. 3–6.

Senigagliesi, C.A., Ferrari, M. and Ostojic, J. (1996) La degradación de los suelos en el partido de Pergamino. In: Morello, J. and Solbrig, O.T. (eds) *¿Argentina Granero del Mundo: Hasta Cuando?* Orientación Gráfica Editora SRL, Buenos Aires, pp. 137–155.

Sierra, E.M., Hurtado, R.H. and Spescha, L. (1994) Corrimiento de las isoyetas anuales medias decenales en la región pampeana 1941–1990. *Revista de la Facultad de Agronomía* 14, 139–144.

Solbrig, O.T. and Viglizzo, E. (1999) Sustainable farming in the Argentine pampas: history, society, economy and ecology. Paper No. 99/00-1, DRCLAS (Working papers on Latin America), Harvard University, Cambridge, Massachusetts, 40 pp.

Soriano, A. (1996) Agricultura sustentable: Estado actual y perspectivas de la cuestión. In: *Proceedings Congreso CREA Zona Oeste*, Mar del Plata, Argentina, pp. 72–77.

Spescha, L., Beltrán, A., Messina, R. and Hurtado, C. (1997) Efectos del ENSO (El Niño – Southern

Oscillation) sobre la producción agrícola argentina. In: *Proceedings Workshop on "Efectos de El Niño sobre la variabilidad climática, agricultura y recursos hídricos en el sudeste de Sudamérica"*, Montevideo, Uruguay, pp. 19–22.

Vargas, W.M., Penalba, O.C. and Minetti, J.L. (1999) Las precipitaciones mensuales en zonas de la Argentina y el ENOS. Un enfoque hacia problemas de decisión. *Meteorológica* 24, 2–22.

5 Global Change Challenges for Horticultural Systems

C. Ramos, D.S. Intrigliolo and R.B. Thompson

5.1 Introduction

The horticultural industry is important in many parts of the world, not so much in terms of agricultural surface occupied by horticultural crops but because of its contribution to the agricultural gross product. In the USA, the market value of horticultural crops was about 43% of the total value of crops in 2002 and 35% in 2007 (USDA, 2009), and in the European Union (EU-27) the equivalent figure for 2007–2008 was 39% (European Commission, 2009). Therefore, assessing the effects of climate change on horticultural crops is important. However, this task is difficult due to the many variables involved, not only agronomic or physiological (effects of increasing atmospheric CO_2 and changes in the mean and extreme temperatures and water availability, amongst others), but also the socio-economic and technological factors that affect land use and which are continuosly changing and that interact in a complex way (Tubiello and Rosenzweig, 2008).

There is limited information on the relationships between climate change and horticultural crops (vegetables, fruits, herbs and flowers and greenhouse production). These crops are usually cultivated using intensive agricultural practices: irrigation, and high labour, fertilizer and agrochemical inputs. The need for irrigation makes them more dependent on water resource availability, which is expected to decrease in many parts of the world (IPCC, 2007). We need also to consider how agriculture affects climate change: it was estimated that 13.5% of the total effect of anthropogenic greenhouse gas (GHG) emissions in 2004 could be attributed to agriculture (IPCC, 2007).

To our knowledge there are no data on the specific contribution of horticulture to climate change. Many of the aspects considered in our review, such as the effects of increased atmospheric CO_2 concentration or temperature on crop yield in C_3 species (most horticultural crops are C_3 plants), or the possible agronomic alternatives to deal with reduced water availability for crop yields and quality, are not specific to horticultural crops but affect all crops. However, the marketable yield of many horticultural crops such as vegetables and fruits is probably more sensitive to climate change than that of grain and oilseed crops (Hatfield et al., 2008).

To make the horticultural industry more sustainable under expected climate change, crop management will have to take into account the environmental footprint of the available management options and the other services that horticulture can provide besides food supply, including carbon sequestration (Robertson and Swinton, 2005).

Some of the major likely impacts of climate change on horticultural crops include the following:

- Increased temperature will produce changes in the growing periods of crops, which will in turn require changes in varieties and/or locations.
- Increased temperature will result in greater evaporative demand, which will require improved irrigation practices.
- More problems in pollination will be experienced due to increased heat stress during flowering.

- Fruit trees that have a given winter chilling requirement for optimum flowering and fruit set may have problems in a warmer climate.
- Decreases in yield and quality of some horticultural crops may occur due to temperatures outside the optimal range.
- Yields of well-managed crops, not negatively affected by temperature increase, may be higher due to the increase in atmospheric $[CO_2]$.
- Heavier rainfall, in some locations, will increase field flooding, problems with field operations, soil compaction and crop losses due to anoxic conditions for roots, and disease problems associated with wet conditions.
- Pathogens, pests and weed problems will change due to the temperature changes.
- Adjustments in N fertilizer management will be needed for adapting to the changes in crop yields and soil N dynamics.

There are numerous publications on the effects of climate change on agriculture (Reddy and Hodges, 2000; Singh, 2009), but their main emphasis is on extensive agriculture (arable crops) and there is much less information available on horticultural crops. We found especially useful the reviews by Cavagnaro et al. (2006) and Hatfield et al. (2008). In most of these publications, the effects of agriculture on climate change are also considered. This chapter deals more specifically with horticultural crops that may be more sensitive than arable crops to climate and global change due to specific temperature requirements and their need for irrigation and high nitrogen inputs.

Next, we will review the scientific evidence on the effects of the expected changes of temperature, atmospheric $[CO_2]$ and water availability on the yield and quality of horticultural crops. In addition, we will consider the likely effects of these changes on the N dynamics in the plant–soil system and the available options to mitigate and adapt to these impacts through water and N fertilization management. We will not deal with the effects of climate change on plant diseases, which have been thoroughly reviewed by Gregory et al. (2009).

5.2 Temperature Effects on Horticultural Crops

Temperature is the major environmental factor modulating many plant growth and developmental traits. At the cellular level, temperature directly affects many enzyme- and membrane-controlled processes. In addition, whole-plant physiology is considerably affected by temperature.

Anticipated changes in global air temperature, due to climate change, assuming moderate economic growth are an average increase of 1.2°C over the next 30 years (IPCC, 2007), with night-time minima increasing more than daytime maxima, and winter temperatures generally increasing more than summer temperatures. It is anticipated that there will be an increase in the frequency of extreme events, such as individual or series of days with unseasonably high temperatures.

This temperature increase will positively affect atmospheric evaporative demand by increasing vapour pressure deficit (Allen et al., 1998). Kimball (2007) calculated a 3.4% increase in annual reference ET (evapotranspiration) for a 1°C increase in air temperature while maintaining absolute humidity and all other relevant parameters equal for a hypothetical lucerne crop grown in Arizona, USA.

5.2.1 Vegetables

Vegetable crops can be grouped into several broad categories of climate classification based on crop response to temperature (Tables 5.1 and 5.2). The suitability of a given species to a climatic zone is largely determined by its minimum growing temperature, effective growing temperature range and optimal temperature range for yield (Krug, 1997). Other temperature-dependent considerations such as vernalization requirements, winter dormancy, germination responses and frost sensitivity are major determinants of where specific vegetable crops can be grown (Krug, 1997; Peet and Wolfe, 2000). The effects of rising temperatures on vegetable crops have been reviewed by Peet and Wolfe (2000), and this is a primary source of information for this section.

Considering vegetable crops in their current locations, increasing temperature will affect these through a general acceleration of growth and phenological development, and also through various effects on reproductive development. Temperature effects will occur as a consequence of the ongoing increase in air temperature and through the increased incidence of extreme temperature events (Porter and Semenov, 2005). In the general context of agricultural crops, Porter (2005) and Hatfield *et al.* (2008) suggested that the acceleration of crop life cycles induced by rising temperature will generally result in smaller plants, shorter periods of reproductive development and smaller yields. Krug (1997) commented that both growth rate and the rate of phenological development of vegetable crops will be accelerated and suggested that: (i) relatively larger effects on reproductive development than growth will result in smaller yields; and (ii) equivalent relative effects on both growth and reproductive development could result in a net effect of maintaining similar yields.

As subsequently discussed, phenological development of vegetable crops is commonly sensitive to temperature and consequently is likely to be a major determinant of responses to increasing temperature. Effects of temperature on phenological development of vegetable crops will be subsequently examined in terms of: (i) effects on flowering and dormancy; and (ii) reproductive development after flowering. In some species, flowering will be affected by effects on vernalization and, in others, by direct inducement of flowering.

For cool season vegetable crops with a vernalization requirement, warmer winter temperatures may affect flowering. For broccoli and cauliflower, the timing of flowering is important to ensure that large heads can be supported (Rubatzky and Yamaguchi, 1999; Peet and Wolfe, 2000). Increasing winter temperature effects on vernalization requirements of broccoli and cauliflower may require new cultivars and/or changing planting dates and locations. For cabbage, onion and celery, suppression of flowering is a requirement (Rubatzky and Yamaguchi, 1999; Peet and Wolfe, 2000); changing planting dates and/or locations

Table 5.1. Climate classification categories and associated temperature ranges for vegetable crops (adapted from Krug, 1997).

Climate classification	Minimum temp. for growth (°C)	Acceptable temp. growth range (°C)	Optimal temp. for yield (°C)
Hot	15	18–35	25–27
Warm	10	12–35	20–25
Cool–hot	5	7–30	20–25
Cool–warm	5	7–25	18–25

Table 5.2. Climate classification for major vegetable species (adapted from Krug, 1997).

Climate classification	Examples
Hot	Watermelon, melon, okra, sweet potato, capsicum spp.
Warm	Tomato, sweet pepper, cucumber, aubergine, phaseoulus spp., pumpkin, squash
Cool–hot	Globe artichoke, onion, shallot, garlic
Cool–warm	Pea, cauliflower, broccoli, cabbage, lettuce, broad bean, spinach, turnip, carrot, celery, asparagus, potato

could affect the interactions between day length and winter temperatures. Warmer winter temperatures may also affect perennial vegetable crops such as asparagus and rhubarb that require low winter temperatures to overcome dormancy for the resumption of vegetative growth in the spring (Krug, 1997; Rubatzky and Yamaguchi, 1999; Peet and Wolfe, 2000). New cultivars or movement to colder growing locations may be required for these species.

Higher temperatures during the vegetative growth phase can induce flowering in lettuce and spinach (Krug, 1997; Rubatzky and Yamaguchi, 1999). The development of seed stalks, an effect known as 'bolting' (Krug, 1997; Rubatzky and Yamaguchi, 1999), considerably affects product quality in leaf lettuce and spinach, and makes head lettuce unsaleable (Krug, 1997; Peet and Wolfe, 2000). Higher temperatures can delay flowering in bean and increase the ratio of male to female flowers in cucumber, causing reduced yield (Rubatzky and Yamaguchi, 1999; Peet and Wolfe, 2000). Development of adapted cultivars or movement of production to cooler zones may be required for these species.

For vegetable crops, effects of increasing temperature on reproductive development, after flowering, are fundamental considerations (Peet and Wolfe, 2000; Hatfield et al., 2008; Hedhly et al., 2009). Increased temperatures below maximum critical values increase the rate of reproductive development, and this shortens both the fruit filling and the fruit maturation periods. Shorter fruit filling and maturation periods are likely to result in smaller fruit and consequently smaller yield. Temperatures above maximum critical temperatures can severely affect reproductive events (pollen release, fruit set) and their efficacy (pollen viability) (Hedhly et al., 2009). In many vegetable crops, fruit set is particularly sensitive to temperatures above maximum threshold values, with substantial effects being noted at only several degrees above optimal temperatures for reproductive development (Peet and Wolfe, 2000;

Hatfield et al., 2008). Studies in tomato with higher than critical temperature have demonstrated relatively larger pre-anthesis effects (reduced pollen release and impaired pollen viability) than post-anthesis effects (fruit set, seed set) on yield (Peet et al., 1998). Where conditions were optimal during pre-anthesis, subsequent effects on fruit set and seed set affected yield (Peet et al., 1998).

While the projected temperature increase of 1.2°C over the next 30 years (IPCC, 2007) suggests that previously described effects on reproductive development of greater than maximum critical temperatures may be somewhat infrequent in the short term, increased frequencies of extreme heat events during reproductive development could have important impacts on yields of vegetable crops (Porter and Semenov, 2005; Hedhly et al., 2009). To reduce the likelihood of such occurrences, it may be necessary to alter dates of growing seasons and/or to transfer production to relatively cooler regions.

The relative impact of increased temperature on marketable yield of different vegetable crops will also depend on: (i) whether crops are indeterminate or determinate; (ii) number of harvests; and (iii) harvesting method (Peet and Wolfe, 2000). Indeterminate vegetable crops have longer flowering periods than determinate plants, and where harvested at seed maturity will have had opportunities to compensate for heat stress events during flowering and early reproductive development; examples include pumpkin, dry bean and winter squash. Many indeterminate vegetable crops with lengthy flowering periods are harvested before seed maturity, e.g. tomato, pepper and cucumber. For crops that are mechanically harvested once (e.g. processing tomato), heat stress events may cause large reductions in harvestable yield because there will not be opportunities for compensatory fruit growth. Similarly, for indeterminate crops that have multiple hand-harvests (e.g. fresh tomato), heat stress events are likely to extend the harvesting period, affecting economic viability through increased harvesting costs and possible loss of market windows (Peet and Wolfe, 2000).

For vegetable crops, product quality is a fundamental consideration, as blemished produce can be unsaleable. Increasing temperature is associated with a number of physiological disorders of vegetable crops (Table 5.3). Increased night-time temperature, through increased respiration, is thought to reduce sugar content of fruit such as strawberry, melon and pea (Peet and Wolfe, 2000). A major challenge for vegetable growers and horticultural researchers will be to limit the incidence of high temperature-induced physiological disorders as air temperatures rise and days of extreme high temperature become more frequent. The combination of increased temperature-induced fruit blemishes and the anticipated effects on growth and development suggest that there will be large incentives to shift production of various vegetable species to cooler climates.

In general terms, as temperatures continue to increase, there is likely to be increasing relocation of vegetable production to cooler regions. This will favour currently cooler regions, but will be detrimental to warmer regions that currently have important vegetable production systems.

5.2.2 Fruit trees and vines

In woody perennial crops, the appearance and duration of each phenological phase is in general well predicted by the temperature regime alone (Kozlowski and Pallardy, 1997). However, it has been shown that the strongest alteration in plant development in response to increasing air temperature occurs in the early spring phases, which are more susceptible to changes in the temperature regime than the later phases of seasonal tree growth (Chmielewski *et al.*, 2004; Sadras and Soar, 2009).

In deciduous fruit tree crop species, winter dormancy and timing of the following season bud-break are developmental traits related to winter and early spring temperature regimes (Balandier *et al.*, 1993). Insufficient chilling causes abnormal patterns in bud-break and development in temperate zone fruit trees cultivated in warm climates (Bonhomme *et al.*, 2005). This is particularly important because most of the high-yielding and premium-quality fruit

Table 5.3. Physiological disorders of vegetable crops caused or exacerbated by high temperatures (adapted from Peet and Wolfe, 2000).

Vegetable crop	Disorder(s)	Aggravating factor(s)
Asparagus	High fibre content in stalks	High temperatures
Asparagus	Feathering and lateral branch growth	Temperatures >32°C, particularly if picking frequency not increased
Bean	High fibre in pods	High temperatures
Carrot	Low carotene content	Temperatures >20°C
Cauliflower, broccoli	Hollow stem, leafy heads, no heads, bracting	High temperatures
Cole crops, lettuce	Tip burn, bolting, loose puffy heads	Drought, especially if combined with high transpiration
Onion	Bulb splitting	High temperatures
Pepper	Sun scald	High temperatures
Tomato	Fruit cracking, sun scald	High temperatures
Tomato, pepper, watermelon	Blossom end rot	High temperatures, especially combined with drought; high transpiration

tree-cultivating areas of the world (California, Mediterranean countries and Chile) often have temperature regimes that are marginal for winter chill requirements. In this sense, Luedeling et al. (2009a, b) showed that many areas of California and the Arabian Peninsula may no longer support some of the main tree crops currently being grown, because of insufficient winter chill under some of the predicted temperature scenarios. In those areas, the tree crop industry will probably need to develop agricultural adaptation measures (e.g. low-chill varieties) to cope with these expected changes. In this respect, it has been shown that fruit tree varieties with similar winter chill requirements might have different responses to low-chill regimes (Gariglio et al., 2006). Breeding programmes are already targeting low winter chill requirements as an important physiological trait (Byrne, 2005).

Some research has shown that in the last 50 years flowering time in many fruit tree crops has advanced by 5–10 days (Wolfe et al., 2005). Since climate predictions include not only an increase in mean temperatures in temperate latitudes but also greater temperature variation (Rigby and Porporato, 2008), earlier blooming, particularly for stone fruit trees, may increase the probability of spring frost damage. However, studies conducted to test this hypothesis have produced contrasting results (Cannell and Smith, 1986; Kramer, 1994).

Another effect of increasing temperature is on fruit development. Temperature regime determines the length of time of the early phase of fruit growth (i.e. cell division), which is crucial for determining fruit growth potential expressed during the rest of the season (Austin et al., 1999). For instance, Lopez and DeJong (2007) found that in peach fruit weight at harvest was lower in those seasons with high early spring temperatures. Since large individual fruit weight is a critical determinant of the economic value of fruit production, in the future fruit tree growers may have to make heavier fruit thinnings to maintain large individual fruit weights.

Citrus trees are extremely sensitive to freezing, and the major production areas are located where there is low risk of freezing. A warmer temperature scenario would shift citrus production towards currently cooler regions (Rosenzweig et al., 1996). In current production areas where fruit quality is greatly improved by a dormancy period induced by cold winter temperatures, excessively warm temperatures during fruit set may increase fruit abscission (Moss, 1970). Rosenzweig et al. (1996) predicted a future decline in yield in some regions of southern Florida and Texas due to warmer temperatures. In addition, high temperatures during ripening can reduce fruit quality, decrease the time the mature fruit can be on the tree and increase rind re-greening (Ben Mechlia and Carrol, 1989).

In grapes for wine production, research has shown that vintages have advanced by 5–10 days in the last 15 years (Jones and Davies, 2000). This has implications not only for crop physiology and fruit composition but also for cellar logistic operations; in many cases, such as the Australian viticulture industry, vines are reaching veraison earlier (Soar et al., 2008). This implies that the berry ripening period is occurring under a warmer temperature regime, which is particularly important in warm climate viticulture that is already suffering low wine acidity and where high pH problems in wine production will probably be exacerbated. Interestingly, it has been shown that the sensitivity of grapevine phenology to increased air temperature might be highly cultivar dependent, with some varieties (e.g. Cabernet Sauvignon) showing more tolerance than others (e.g. Merlot). In some regions of southern Spain, the future environment might be too warm for premium red wine production, and northern regions may become more suitable (Kenny and Harrison, 1992). Overall, the expected climate change scenarios in various regions may influence the varieties planted and the wines produced (see Jones et al., 2005 for a review).

However, a recent study on the effects of warm temperatures on final grape yield reported no effect of the temperature regime (Sadras and Soar, 2009). This probably occurred because, although air temperature increased by up to 4°C, canopy temperature was only 1.0°C warmer. Vines may be able to adapt physiologically by increasing transpiration cooling (Soar et al., 2009). This might not occur in the case of rain-fed conditions or when water is a limiting factor. It has also been shown that a heat spell occurring just after flowering reduced berry number per cluster, but that the larger berry weights obtained at harvest compensated for the reduction in fruit set (Ezzahouani, 2003).

5.3 Effects of Elevated [CO_2]

Atmospheric CO_2 concentration is a crucial component for plant performance, because it directly affects photosynthesis rates and may indirectly determine water gas exchange through changes in stomatal conductance and canopy transpiration surface area. As of 2010, atmospheric [CO_2] is approximately 385 µmol/mol; by the middle of this century it is projected to surpass 500 µmol/mol, and is expected to reach 700 µmol/mol by the end of the century (Prentice et al., 2001).

Similar to any other plant, in horticultural species leaf net CO_2 fixation rates will increase at the predicted atmospheric [CO_2] for the middle of this century (Leakey et al., 2009). This will occur despite the decrease in the carboxylation rate of rubisco activity that occurs at higher atmospheric [CO_2]. In addition, it is important to note that, to maximize crop production benefits from increased atmospheric [CO_2], it is necessary to have conditions of unrestricted root growth, optimum fertility and excellent control of weeds, insects and disease (Hatfield et al., 2008). These conditions commonly occur in intensive vegetable and woody crop production. On the other hand, because in intensive horticultural systems plants are not grown in isolation, there are spatial restrictions that will limit the potentially higher growth rate from elevated [CO_2] (Körner, 2006). In order to translate into higher yields with the beneficial effect of higher atmospheric [CO_2] on growth,, it will be necessary that the expected negative temperature effects do not predominate.

There is a strong consensus among researchers that transpiration efficiency (dry matter produced per unit of water transpired) increases with elevated [CO_2] (see review by Idso and Idso in http://www.co2science.org/articles/V4/N50/EDIT.php). This will benefit horticultural crops grown in arid regions.

Biomass production and yields of vegetable crops increase with increased CO_2 (Peet and Wolfe, 2000). Artificial enrichment with CO_2 is an established and profitable practice in commercial vegetable production in intensively managed greenhouses such as in the Netherlands (Sánchez-Guerrero et al., 2009). In commercial practice, CO_2 concentrations of 700–800 µmol/mol are used with vegetable crops such as tomato and cucumber (Sánchez-Guerrero et al., 2009). Sánchez-Guerrero et al. (2009) reported a 19% yield increase with cucumber from enriching greenhouse CO_2 concentration to 700 µmol/mol.

In the longest-term study carried out to date in a horticultural crop, Kimball et al. (2007) reported a spectacular 70% increase in biomass production in response to a 300 ppm increase in [CO_2] in orange trees. In pear trees, long-term CO_2 enrichment had a much weaker effect of about 20% increase in dry matter (Ito et al., 1999). In grapevine, Bindi et al. (2001) reported a 40–45% increase in dry matter production with [CO_2] up to 550 µmol/mol.

It is generally accepted that increasing [CO_2] will reduce stomatal conductance, at leaf level, and that this may result in increased crop water productivity (see review by Leakey et al., 2009). However, the overall effect on the whole-plant water use efficiency and water use is difficult to predict since vegetative growth is also stimulated by increased [CO_2], and also because of temperature interactions. In this sense, some of the results obtained in field crops reviewed by Idso (2001), Hatfield et al. (2008) and Leakey et al. (2009) indicate that an increase in [CO_2] results in greater water use efficiency. On the other hand, Allen et al. (2003) reported that increased [CO_2] resulted in reduced water use in soybean at medium temperatures (28/18°C) but that there was no effect at higher temperatures (40/30°C), and Centritto *et al.* (2002) showed that peach trees growing under elevated [CO_2] did not show any decrease in water use.

The effects of elevated [CO_2] on dry matter partitioning have been less studied in horticultural woody perennial crops. This is crucial for fruit tree growers, where an optimum balance between vegetative and reproductive growth is needed to ensure high yield by avoiding excessive tree growth, which in turn reduces the labour costs associated with pruning and harvest operations. In this sense, Schaffer *et al.* (1997) and Kimball *et al.* (2007) reported no effect on dry matter partitioning in mango and citrus trees, respectively. On the other hand Kriedemann *et al.* (1976) observed that in vines with enriched [CO_2] shoot dry matter distribution was modified in favour of root growth.

In horticultural crops it is also important to consider the effect of elevated [CO_2] on fruit quality. However, there have been very few relevant studies; the available studies on grapevines and pear trees did not find any clear effects of increased [CO_2] on fruit composition (Ito *et al.* 1999; Bindi *et al.*, 2001). A recent investigation also concluded that wine composition was not affected by [CO_2] (Gonçalves *et al.*, 2009). In intensive greenhouse-based vegetable production such as in the Netherlands, CO_2 addition is routinely used in commercial production; obviously negative fruit quality effects have not been a major issue there.

5.4 Combined Effects of Increased Temperature and CO2

Generally, it is expected that increasing temperatures and increasing CO_2 will have contrasting effects on crop yield of horticultural crops (Hatfield *et al.*, 2008). Increasing temperatures will generally tend to reduce yields of vegetable crops by shortening crop cycles and increasing respiration losses, and through the effects of higher temperatures and high-temperature events during reproductive development. In contrast, increasing CO_2 will tend to enhance biomass production and yields.

The available data for field crops suggest that there are no combined beneficial effects of increasing temperature and CO_2 where reproductive processes influence crop yield and with crops having closed canopies (Hatfield *et al.*, 2008), which includes many vegetable species. There are no reports of combined beneficial effects of increasing CO_2 and increasing temperature for a range of major field crops (Hatfield *et al.*, 2008). There are, however, reports of negative interactions where increased CO_2 apparently lowered maximum threshold temperatures (Hatfield *et al.*, 2008).

It appears that, generally, the potential benefits of increased CO_2 on production of vegetable crops may not be realized where negative temperature effects occur. Where negative temperature effects are minimized through different planting dates, movement to different zones, cultivar selection, etc., beneficial effects of increased CO_2 on growth may occur.

5.5 Adaptation to Reduced Water Availability for Irrigation

In most global change scenarios, water scarcity will be a major determinant on agricultural land (Bates *et al.*, 2008). Irrigation in agriculture already accounts for about 70% of the total water use worldwide. Additionally, ongoing degradation of water resources and competition from other sectors such as industry, domestic use and tourism will also contribute to a reduced supply of water for irrigation of horticultural crops.

Increasing crop water productivity, i.e. crop marketable yield per unit of water used, may be achieved by genetic modification (Peterhansel *et al.*, 2008; Murchie *et al.*, 2009). However, improvements in irrigation management seem more feasible in the shorter term. There are three components to improving water productivity in agricultural production: (i) reducing losses prior to field application; (ii) improving application efficiency (i.e. increasing the proportion of applied water that reaches and remains in the root zone); and (iii) optimizing the use of applied water by crops. Any serious effort to improve productivity of water use for a given irrigated agricultural system needs to address these three components. The following sections will review some of the options available for optimizing irrigation management in horticultural crops.

5.5.1 Improving water application

Irrigation application efficiency refers to the proportion of applied water that is retained in the root zone on a field scale. Because of the generally intensive management, high product value and generally smaller field size, horticultural systems are well suited to the use of improved irrigation methods. The technical application efficiencies of different irrigation methods (assuming good management) are well established. Typically, well-managed traditional furrow irrigation systems have application efficiency values in the order of 60% (Postel, 1999). The use of surge valves with furrow irrigation systems, to apply pulses of water, can increase application efficiency to 80% (Postel, 1999). Low-pressure sprinkler systems can have application efficiency values in the order of 80%, which can be increased to values of 90–95% with low-pressure precision application sprinklers positioned just above the soil surface. Drip irrigation, with good management, has an application efficiency of 95% (Postel, 1999).

The technical efficiency of water application can be improved by land preparation procedures such as laser levelling to reduce runoff and enhance infiltration, and to ensure adequate water flow and distribution along furrows in furrow irrigation (Fereres *et al.*, 2003). Additionally, in furrow irrigation, determination of adequate furrow length is a factor in optimizing efficiency. The improved irrigation methods require an investment in irrigation equipment (drip tape, sprinkler systems), the associated equipment such as pumps for pressurized irrigation systems (drip, sprinkler) and, where relevant, for land preparation. The

economics of such investments are influenced by the price of water, which will presumably increase with more competition for reduced supplies. There has been a rapid and extensive adoption of improved irrigation methods in horticulture since the 1960s (Fereres *et al.*, 2003). The adoption of improved irrigation application systems has resulted in substantial savings in water use (Postel, 1999; Fereres *et al.*, 2003). Under the probable future scenario of reduced and more expensive water supply, there will be further adoption of improved irrigation systems in horticulture.

5.5.2 Improving irrigation scheduling

For the efficient use of water in agriculture, proper irrigation scheduling protocols are crucial for matching water application to plant water needs. Nowadays, irrigation scheduling is often based on the FAO method where crop evapotranspiration (ETc) is estimated using the reference evapotranspiration (ETo) × the crop coefficient (Kc), according to the procedure suggested by FAO (Allen *et al.*, 1998). Developments in information technology enable provision of both current and forecast ETo estimations (Postel, 1999).

The FAO approach for estimating ETc, despite its widespread use, has some uncertainties, particularly for fruit trees. An important example that affects the calculation of ETo, particularly in tall tree orchards, is the high degree of coupling between tree evapotranspiration and rapidly changing air humidity. This is in contrast to the reference grass used for ETo measurement, where ETo is relatively uncoupled from the characteristics of the bulk air and is primarily dependent on net radiation (Annandale and Stöckle, 1994). In addition, in woody crops, water use might change as a function of orchard and tree characteristics that affect the amount of light intercepted by a tree (Consoli *et al.*, 2006), soil management (Allen *et al.*, 1998) or tree yield level (Naor, 2006).

The estimation of ETc with the FAO approach only provides information on the amount of water to apply during a certain period. Determining frequency on the basis of the theoretical soil water balance requires information on rooting depth, soil water characteristics, species response to soil water deficit, etc. that is not commonly available. At a practical level, irrigation frequency will be influenced by crop, management and soil characteristics; commonly, irrigation scheduling is based on grower experience. Data of plant and/or soil water status can provide a very useful complement to ETc estimation, enabling irrigation frequency to be based on specific crop requirements.

The measurement of soil water status can assist with on-farm irrigation management. The sensors and data interpretation can be managed by farmers. At a scientific level, it has been shown for a given soil water status that the evaporative demand can be the major determinant of crop response (Sadras et al., 1993). However, there is no doubt that soil water status can generally provide a good indication of the likely crop performance with respect to irrigation.

Soil moisture sensors enable irrigation management to be adjusted to the particular characteristics of individual crops and fields. Soil moisture sensors can be used as a 'stand-alone' method or can be combined with the FAO method for estimating crop water requirements or as a supplement to irrigation management based on experience. They are useful for the implementation of deficit irrigation strategies. The use of soil moisture sensors for irrigation scheduling is particularly well suited to horticultural systems because of the widespread use of advanced irrigation systems (e.g. drip irrigation), small field sizes and the high value of the crops. The combination of advanced irrigation methods, estimation of crop water requirements using the FAO approach and the use of soil moisture sensors provides a technology and management package for the optimization of on-farm irrigation management. In future scenarios of reduced supply of irrigation water, this package could contribute to

meeting the challenge of increasing horticultural production with less water.

The current soil sensor technologies are described by Charlesworth (2005) and IAEA (2008). The use of these sensors for irrigation scheduling is discussed by Hanson et al. (2000), Thompson and Gallardo (2005) and Evett (2007). Sensors measure soil matric potential (SMP) or volumetric soil water content (VSWC). A useful and simple sensor determines the arrival of the wetting front at a given depth (Charlesworth, 2005).

For SMP and VSWC sensors, lower-limit values are used to identify when to irrigate and upper-limit values to identify when irrigation is sufficient (Thompson and Gallardo, 2005). Most commonly, fixed irrigation volumes are applied; automatic cessation requires very frequent measurement with sensors that have rapid response times. Tendencies over time can be used to adjust irrigation volumes. Irrigation scheduling with SMP is relatively straightforward; lower-limit values for given crops are generally applicable for a range of soil types and conditions. Lower-limit SMP values are commonly available in scientific and extension literature (e.g. Shock et al., 2007); adjustments to standard lower-limit SMP values may be required for evaporative demand and soil texture (Hanson et al., 2000). Limits for VSWC need to be determined in situ or at least locally (Thompson et al., 2007a); relevant procedures are described by Charlesworth (2005) and Thompson et al. (2007b).

On the other hand, measurements of plant water status integrate both soil water available to plants and the climatic conditions, and might therefore provide a better prediction of tree responses to water supply (Intrigliolo and Castel, 2006). At the same time, the coupling of the plant to atmospheric evaporative demand can result in crop water status being responsive to several fluctuating environmental variables (Hincley and Bruckerhoff, 1975). This means that a single measurement of plant water status may be meaningless if not compared with a reference value from plants without soil water limitations.

Leaf water potential measured with the pressure chamber, either predawn or at midday, has long been used as a plant water stress indicator. More recently, the use of water potential of bag-covered leaves, termed stem water potential (Ψstem) (Shackel *et al.*, 1997), has been adopted because of its high sensitivity to plant water deficit (McCutchan and Shackel, 1992) and its good prediction of the yield response to deficit irrigation (Naor, 2000). In fruit trees, Ψstem has been used to modify the irrigation regime, thereby preventing mild plant water stress from becoming severe (Lampinen *et al.*, 2001). Ψstem is considered the reference measurement of plant water status but it is quite time and labour consuming, which often precludes its use, particularly for crop management applications.

The major drawback of the above-mentioned techniques for determining plant water status is that they provide only an indication of the water status of a single tree. However, it is well known that, within an orchard, there can be appreciable variation between individual trees in water status (Naor *et al.*, 2006). Therefore, it would be more useful to provide growers with tools for determining the water status of the whole orchard. In this sense, new advances in the field of remote sensing offer some promising possibilities (Moran *et al.*, 2005). Canopy temperature seems to be the most useful indicator for remote sensing of plant water stress (Sepulcre-Cantó *et al.*, 2006). However, because satellite information does not yet have sufficient resolution to be used in commercial applications for horticultural crops, the use of an unmanned aerial vehicle could be an alternative to quantifying the water status of up to 500 ha of irrigated fields in a single flight (Berni *et al.*, 2009).

5.5.3 Regulated or deficit irrigation

When the available water is insufficient for crop requirements, irrigation can be reduced during the whole crop period (deficit irrigation) or only in those phenological periods in which yield is relatively less sensitive to soil water deficits (regulated deficit irrigation (RDI). This last option was developed in the 1980s as a strategy to reduce vegetative growth of vigorous trees and to save water (Behboudian and Mills, 1997). Many experiments, recently reviewed by Naor (2006) and Fereres and Soriano (2007), have shown that, by applying the most appropriate RDI strategy, it is possible to reduce water applications and plant transpiration by approximately 10–20% without any yield loss in the short term. However, the long-term effects of irrigation using RDI strategies are still relatively unknown.

The usefulness of RDI practices is strongly dependent upon avoiding water stress during those periods in which marketable crop yield is particularly sensitive to water stress. A clear separation between vegetative and fruit growth phases is often needed (Hueso and Cuevas, 2008) to guarantee that fruit growth will not be reduced by water restrictions. For instance, in citrus trees under a Mediterranean climate the first phase of fruit growth after the June fruit drop is probably the most appropriate period for applying water restrictions (González-Altozano and Castel, 1999). Citrus trees appear to have compensatory growth following alleviation of water deficit (Cohen and Goell, 1988). To take advantage of this mechanism, it is necessary that recovery of optimum water status occurs well before harvest. The intensity of autumn rainfalls and the irrigation regime have a crucial importance for the rapid recovery of tree water status.

In stone fruit trees, phase II of fruit growth (i.e. pit hardening) has been identified as an appropriate period for RDI. During this period, fruit growth is minimal and therefore generally not affected by drought, while shoot growth can be reduced (Chalmers *et al.*, 1981). However, in early-maturing cultivars this phase is very short, enabling only small water savings. In these cultivars there is a long postharvest period, which is more suitable for reduced irrigation (Johnson *et al.*, 1994). However, caution is required to prevent negative effects of after harvest drought on flower bud development (Johnson and Handley, 2000). In mid-season-maturing cultivars, the best approach is a combined deficit irrigation strategy, with water restrictions applied during postharvest and moderate reductions during stage II of fruit growth, in order to reduce shoot growth and eventually increase fruit quality (Girona *et al.*, 2005).

In pear and apple trees, experiments conducted in Australia (Chalmers *et al.*, 1986) showed that it was possible to apply RDI during the early phase of fruit growth. Pear fruit showed some compensatory growth later in the growing season. However, different results were obtained by Marsal *et al.* (2000) with container-grown trees; in this case, RDI pear fruit were not able to catch up with the fully irrigated fruit after the imposition of water stress. Overall, this apparent contradiction highlights some of the limitations of RDI. The final effects of water stress on crop performance seem to be very dependent on several orchard, environmental and tree characteristics. These aspects are currently limiting the widespread adoption of this promising technique for dealing with water scarcity in the expected scenario of global change.

Regulated deficit irrigation is possible in fruit trees and vines because these species have periods of growth when yield is not sensitive to an ET deficit. These insensitive growth stages are of sufficient duration for reduced irrigation during these periods to result in a noticeable reduction in the volume of applied irrigation water. There are limited possibilities for RDI in vegetable crops because there are no clearly defined growth stages of sufficient duration during which deficit irrigation can be applied without affecting yield. In general terms, in many vegetable crops there is a linear relationship between yield and ET, and any reduction in ET will linearly affect yield and farmer income (Fereres *et al.*, 2003). Additionally, vegetable crops are generally shallow rooted and rapidly experience effects of irrigation at less than potential ETc (Stanley and Maynard, 1990). Irrigation of high-value vegetable crops is normally conducted to avoid water stress to ensure maximum yield (Stanley and Maynard, 1990; Fereres *et al.*, 2003).

There is some interest in the use of deficit irrigation with vegetable crops to increase fruit quality. Deficit irrigation during fruit production – particularly during ripening – can enhance fruit quality by increasing the soluble solids content in the fruit of species such as tomato and melon (Fabeiro *et al.*, 2002; Shock *et al.*, 2007).

5.5.4 Additional agricultural practices to mitigate drought effects

Some additional irrigation strategies to mitigate reductions in water availability are summarized in Table 5.4.

Water consumed by horticultural crop systems is due to soil evaporation and plant transpiration. While this last component is needed to ensure optimum plant productivity, soil water evaporation should be reduced as much as possible in order to achieve high irrigation efficiency. Woody perennial crops often have incomplete ground cover, leaving part of the soil directly receiving a high radiation regime that increases the evaporative component of orchard evapotranspiration (Fereres *et al.*, 2003). Subsurface drip irrigation systems eliminate most of the soil evaporation component from the system water balance, and these could be used in cases of water scarcity. In addition, when the volume of applied water needs to be restricted in relation to potential ETc, decreasing the volume of soil wetted by the drip system by reducing the number of emitters per tree can be a useful way of increasing the irrigation efficiency (Girona *et al.*, 2010).

Weeds – and particularly cover crops – compete with the main crop of an agricultural ecosystem for water and nitrogen. It is important to reduce this competition. Where cover crops are used for reducing soil erosion, less competitive cover crop species can be adopted, such as legume species with growth rates and/or cycles that minimize competition with the main crop.

In fruit trees, there is evidence that high fruit load may enhance the sensitivity of

Table 5.4. Selected drought mitigation strategies involving irrigation design, orchard soil and crop management and orchard environment.

Irrigation management	Regulated deficit irrigation	Subsurface drip irrigation	Under deficit irrigation, reducing the wetted soil surface by eliminating drippers	
Orchard soil and crop management	Eliminate cover crops and weed growth	Heavier fruit thinning	Reduce tree light interception by summer pruning, de-suckering and shoot top trimming	Use of training system and row orchard planting direction to decrease light interception during the warmest hours of the day
Orchard environment management	Shading nets to reduce radiation load	Use of artificial windbreaks		

fruit growth to water stress (Berman and DeJong, 1996), and hence reducing fruit load has been used to mitigate the negative effects of plant water stress, though with important yield penalties (López *et al.*, 2006; Marsal *et al.*, 2008). Under low crop demand conditions, a reduced plant photosynthesis rate due to water stress is less detrimental since fruit represent the major sink for carbohydrates, particularly during stage III of fruit growth. In addition, lowering fruit load has been shown to reduce plant water use because of a reduction in stomatal conductance via feedback mechanisms (Hansen, 1971).

Eliminating part of the actively transpiring canopy surface area (such as whole branches) can also be used in encouraging tree survival under extreme drought conditions (Marsal *et al.*, 2006). It is obvious that this practice has major consequences for current year tree performance, but at least it can guarantee plant survival. In addition, innovative canopy forms can be designed in order to optimize light interception, reducing tree transpiration under soil water-limiting conditions. In this sense, Intrigliolo and Lakso (2011) have shown that, in north–south oriented

vineyard rows, leaning vines towards the west can be used to increase slightly (+8%) overall water use efficiency, because vine light interception decreased during the warmer hours of the day when evaporative demand is higher. Under a future scenario of reduced water availability, new orchard and vineyard designs might alleviate the impact of drought and heat spells.

The use of shading nets has proved useful in increasing water use efficiency (Alarcón *et al.*, 2006) and even crop performance (Cohen *et al.*, 1997). The installation of a shading net is expensive but, for high-value crops, its use may be profitable.

5.5.5 Substrate growing systems

Substrate growing systems with recirculation of drainage water have high water productivity (Savvas, 2002; Pardossi *et al.*, 2004). In these systems, vegetable crops are grown in substrate (e.g. rockwool, perlite, coconut fibre) in greenhouses and are fertigated with complete nutrient solutions. Drainage water (generally 20–30% of irrigation) is collected and reapplied as nutrient solution. Nutrient composition is maintained by regular addition of fresh water and nutrients and periodic partial replacement of the recirculating solution because of salt accumulation, particularly sodium and chloride.

Recirculation is widely used in the highly productive greenhouses of the Netherlands. These are intensive, high-input, high-output cropping systems characterized by high capital costs of the glass greenhouses and associated infrastructure. They are energy-intensive systems that require heating and lighting, apart from the energy costs associated with construction. An alternative model of greenhouse-based substrate growing system is that being used in semi-arid climates in south-eastern Spain, which is currently expanding rapidly in central America (Pardossi *et al.*, 2004). This is a simpler and cheaper system, with plastic greenhouses that generally have no heating or additional lighting. In this greenhouse model, substrate-growing systems are 'open' in that the 20–30% of irrigation that is drained is generally not collected. Comparative studies of recirculating and open systems have shown large savings in water and nutrient use with recirculation (Savvas, 2002). However, unlike north-western Europe where high-quality irrigation water derived from rainfall is used, in these semi-arid regions irrigation water is obtained mostly from aquifers. The higher electrical conductivity (EC) of aquifer water, compared with rainwater, results in more frequent replacement of recirculating water. This may be a major impediment to the adoption of recirculation in drier regions (Magán *et al.*, 2008). Any consideration of substrate-growing systems in the context of climate change should take into account the carbon costs of heating, lighting, etc. Increasing temperatures will reduce heating costs in intensive greenhouse systems such as those in the Netherlands; however, they will further increase the requirement for cooling during the warmer months in greenhouses located in warmer climates.

In summary, substrate-growing systems with recirculation have very high water productivity, but the quality of the irrigation water is a major consideration influencing the economic viability of recirculation. Also, the associated carbon costs of the greenhouse system may be an important consideration in the context of climate change.

5.6 Effects of Climate Change on Plant and Soil N Dynamics in Horticultural Systems

Nitrogen is a key element in crop production, but also plays a relevant role in the global CO_2 balance because of carbon–nitrogen interactions in agricultural ecosystems (Socolow, 1999). Recognizing this, the IPCC Fifth Assessment Report (scheduled to be finalized in 2014) will include the interaction between the carbon and nitrogen cycles (IPCC, 2009).

There are two aspects to consider in the relationship between agricultural N use and climate change: (i) how climate change may affect N cycling in agricultural soils; and (ii) how N management in horticultural crops can decrease or mitigate the production of GHGs. Figure 5.1 shows the main plant and soil interactions with climate change in horticultural production.

5.6.1 Effects on plant N

At the field level, free-air CO_2 enrichment (FACE) experiments have shown that an increase in atmospheric CO_2 produces an increment in crop yield, although not as high as previously predicted (Long *et al.*, 2006). This increase in growth will modify crop N uptake due to the linkage of C and N metabolism (Stitt and Krapp, 1999).

There is evidence that increases in atmospheric $[CO_2]$ produce a reduction in leaf N due to a decrease in leaf rubisco (Taub and Wang, 2008), since this enzyme may account for up to 50% of leaf N (Spreitzer and Salvucci, 2002). It was previously thought that this reduction in leaf protein N might be diverted to other uses and therefore that the metabolic N use efficiency could be

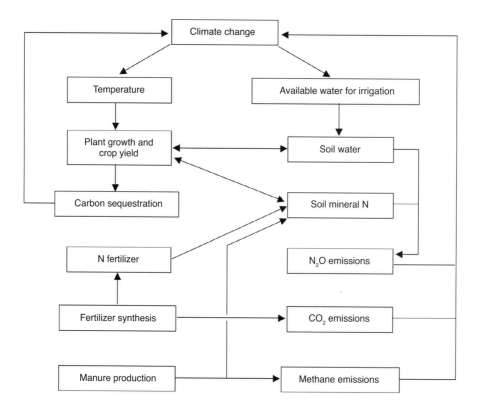

Fig. 5.1. Main plant–soil–climate change relationships in horticultural production.

increased. However, results from FACE experiments showed that the estimated N savings in leaf N are small: about 3–4% for C3 herbaceous crops and 0.6% for trees (Leakey *et al.*, 2009).

5.6.2 Effects on soil N

The most important expected changes in agricultural soils as a result of climate change are those related to soil organic matter content (Johnston *et al.*, 2009). Organic matter is important for soil fertility and is also strongly linked to the soil organic N content, since the C:N ratio in soils varies over a relatively narrow range, and mineralization of organic matter is usually an important source of mineral N for crop uptake in horticultural systems (Jarvis *et al.*, 1996). Increased atmospheric $[CO_2]$ is

expected to affect soil organic matter content due to an increase in crop residues derived from higher crop growth, and to a lesser degree through a greater amount of root exudates and root residues that supply C and N to soil organisms (Drigo *et al.*, 2008).

Soil N in intensive horticultural systems depends on the fertilizer and manure inputs and the N cycle processes: mineralization, immobilization, denitrification, crop uptake, nitrate leaching and gaseous losses. All these processes are temperature and soil moisture dependent and, therefore, climate change will affect them. However, since in horticultural systems soil moisture and fertilizer N input are relatively well controlled by the farmer, it is difficult to ascertain how much the expected climate change will affect irrigation and N fertilization management.

5.7 Nitrogen and GHG Emissions in Horticultural Systems

Agricultural nitrogen management affects the emission of nitrous oxide (N_2O) and CO_2 from soils (Mosier, 1998). In the USA, agriculture was responsible in 2007 for about 5.8% of total global GHG emissions in terms of CO_2 equivalents (EPA, 2009). N_2O and CH_4 were the main GHGs emitted by agricultural activities, and soil management activities such as fertilizer application and other cropping practices accounted for 67% of total N_2O sources (all values as CO_2 equivalents). Xiong and Khalil (2009) reviewed greenhouse gas emissions from crop fields. To our knowledge there is no information on the contribution of horticultural crops to total GHG emissions from agriculture.

Horticulture contributes directly to N-related GHG emissions through:

- fertilizer and manure application;
- irrigation;
- crop residues; and
- soil management.

Smith et al. (2008), after reviewing the GHG mitigation potential of agricultural management practices, concluded that about 90% of the total GHG mitigation potential in agriculture corresponds to soil C sequestration. Carbon sequestration research is an active field and there are many reviews on the subject (Jarecki and Lal, 2003; Ghani et al., 2009), but it is outside the scope of this chapter. The effects of N fertilizer application on C sequestration is a controversial subject, and has recently been reviewed by Schils et al. (2008).

5.7.1 Soil nitrous oxide emissions

Nitrous oxide (N_2O) is a greenhouse gas that is produced in soil as part of the denitrification or nitrification processes (Bremner, 1997). The concentration of N_2O in the troposphere has increased over the last two centuries from approximately 270 ppb to about 320 ppb, and contributes around 6% of the total GHG warming effect

(Forster et al., 2007). Smith et al. (2007) reported that around 60% of the anthropogenic emissions of N_2O can be attributed to agriculture. Vegetable crops have a high potential contribution to N_2O emissions since they use relatively large amounts of fertilizer N and manures, and both inputs have a direct effect on nitrous oxide production. Nitrous oxide emission rates from soil vary considerably due to oxygen and organic carbon availability, soil pH, temperature and soil mineral N (Coyne, 2008). Two recent reviews on the influence of different factors on N_2O emissions from crop fields are those by Majumdar (2009) and Millar et al. (2010).

Many studies show a roughly proportional relationship between N input (i.e. fertilizers, manure) and the amount of N lost as N_2O (Bouwman, 1996). The Intergovernmental Panel on Climate Change (IPCC, 2006) proposed the estimation of regional N_2O emissions as 1% of the N input in fertilizers and manures when no experimental data are available. Fertilizer type can also influence N_2O emissions. For example, Clayton et al. (1997) found that N_2O emissions from nitrate-containing fertilizers were much higher than those from ammonium-containing fertilizers. Mulvaney et al. (1997) found that under waterlogging conditions alkaline-hydrolysing fertilizers (liquid anhydrous NH_3 and urea) induced higher emission of N_2O than the more acidic ones (such as $(NH_4)_2SO_4$, NH_4NO_3 and $NH_4H_2PO_4$), and that a large denitrification loss can occur when urea is applied in combination with ammonium nitrate due to the high concentration of NO_3^- and the rise in soil pH from urea hydrolysis. These findings are important for horticultural production, where it is quite common to use urea and urea–ammonium nitrate solutions in drip irrigation, and where soil moisture content in the wet bulb is relatively high, but to our knowledge no studies under field conditions to test this hypothesis have been conducted.

Maximum denitrification rates are commonly observed when water content is greater than about 90% of soil porosity, and maximum N_2O losses occur at water content equivalent to 60–70% of soil porosity

(Bateman and Baggs, 2005); irrigation is clearly a factor influencing nitrous oxide emissions (Rolston *et al.*, 1982; Sanchez-Martín *et al.*, 2008). Nitrous oxide emission can be predicted using different simulation models (Del Grosso *et al.*, 2000). A simple empirical model, developed by Conen *et al.* (2000) to predict N_2O emission, used only three soil parameters: soil mineral N in the topsoil, soil water-filled pore space and soil temperature; this model gave satisfactory predictions of seasonal N_2O fluxes in several grassland sites, and in cereal and oilseed rape crops, but grossly underestimated these emissions from two vegetable crops (potato and broccoli).

Nitrous oxide emission rates measured in several horticultural crops are presented in Table 5.5.

5.7.2 Management options to reduce nitrous oxide emissions in horticultural systems

Measures to reduce general N_2O emissions include:

- increasing N use efficiency;
- managing crop residues;
- good irrigation management; and
- improving soil management.

Next we will review these options for horticultural systems.

Increasing N use efficiency

In horticultural crops, improving fertilizer use requires an appropriate N fertilizer and irrigation management. Good fertilizer practice requires applying the correct N rate, at the right time and in the most adequate chemical form. Good irrigation management implies reducing drainage losses to a minimum compatible with an adequate salt balance in the root zone. Reviews of the different approaches to increased N use efficiency in vegetable and field crops can be found in Tremblay *et al.* (2001) and Robertson and Vitousek (2009), respectively.

In vegetable and other horticultural crops, fertilizer requirements are mostly based on plant or soil measurements. For plant measurements there are several options: leaf analysis, nitrate sap analysis,

Table 5.5. N_2O emissions measured in selected horticultural crops.

Crop	N fertilizer input (kg/ha)	N_2O emission (kg/ha)	Period	Reference
Vegetables				
Lettuce, celery, broccoli, cauliflower, artichoke	290–665	20–42	1 year	1
Radish	323	1.3	139 days	2
Celery	796	5.8	81 days	2
Lettuce	129	2.9	47 days	2
Potato	50–150	8–16	1 year	3
Onion	0–52	0.9–3.8	260 days	4
Tomato	230–600	0.3–0.7	6–8 months	5
Chinese cabbage	0–250	0.12–0.85	82 days	6
Fruit trees				
Apple	310	3.2	1 year	7
Ornamental plants				
Pelargonium zonale (potted)	–	2.4	1 year	8

1, Ryden and Lund (1980); 2, Xiong *et al.* (2006); 3, Ruser *et al.* (1998); 4, van der Weerden *et al.* (2000); 5, Hosono *et al.* (2006); 6, Cheng *et al.* (2006); 7, Pang *et al.* (2009); 8, Agner (2003).

chlorophyll meters and remote-sensing techniques. Samborski *et al.* (2009) reviewed the available sensing tools for N status assessment and pointed out that one of the major reasons for the limited adoption of these techniques is the requirement for procedures to convert sensor readings to N fertilizer recommendations under a variety of soil and weather conditions. Nevertheless, examples of successful use of chlorophyll meters on N management include Rodrigo and Ramos (2007) in artichoke, Gianquinto *et al.* (2006) in tomato and Porro *et al.* (2001) in apple and grapevine.

Nitrate sap analysis has been recommended for N testing in vegetable crops (Hochmuth, 1994), but Hartz (2006) found a lack of response of sap nitrate to soil nitrate content and discouraged its use as an N management tool.

In relation to the methods based on soil measurements, one of the most commonly used is the N_{min} method (Wehrmann *et al.*, 1988). This method determines the fertilizer requirement based on the crop requirement minus the mineral N content of soil down to a certain depth just before planting, and has been used in many European countries for vegetable crops (Feller and Fink, 2002). In the USA a similar method has been introduced, which uses the soil nitrate content just prior to side-dressing time (Hartz *et al.*, 2000). These methods are especially useful in vegetable crops where the N mineral content at planting time can be very high (Neeteson *et al.*, 2003).

There is also an ongoing effort to develop simulation models that integrate the various biological processes and, in some cases, also economic aspects, and that can guide N fertilizer management (Fink and Scharpf, 1993; Rahn *et al.*, 2010). In fruit orchards there are few simulation model approaches, because of the difficulties involved in modelling N remobilization within the plant from one year to the next, but some models are available (e.g. Nesme *et al.*, 2006).

Crop residues management

Vegetable crops can leave behind in the soil high levels of crop residues contain-ing considerable amounts of nitrogen (Wehrmann and Scharpf, 1989), which can be mineralized within a few weeks if soil temperature and moisture are adequate (De Neve and Hofman, 1996). This mineralized N can be leached or denitrified if there is no active plant N uptake and/or there is high rainfall or water application. To reduce the rapid mineralization of vegetable crop residues with low C:N ratios, Rahn *et al.* (2003) used several amendments (molasses, paper waste, green compost, etc.) and found that the amounts of N mineralized decreased when the concentration of cellulose and lignin in the amendment materials increased. These amendments had a variable effect on nitrous oxide emissions. In a similar study, Chaves *et al.* (2005) found that the addition of several organic biological waste materials to the soil significantly reduced N_2O emissions from celery crop residues in most cases.

Other mitigation techniques

Irrigation management can reduce nitrous oxide emissions. For example, Sanchez-Martín *et al.* (2008) found that, in a melon crop under drip irrigation, N_2O losses were reduced by 70% in comparison with those under furrow-irrigated treatment. This was probably due to the different wetting and drying cycles in the two irrigation systems, which affected the determining microbial processes (Burger *et al.*, 2005).

In relation to the effect of soil management on nitrous oxide emissions, there are some studies on the effect of tillage in field crop rotations (Six *et al.*, 2004; Halvorson *et al.*, 2008), with variable results. To our knowledge, there are no similar studies recorded on vegetable crops or in fruit orchards.

Some authors have proposed using nitrification inhibitors to reduce N_2O emissions. For instance, Clough *et al.* (2007) found that adding the nitrification inhibitor dicyandiamide (DCD) to the soil reduced N_2O emissions by 72%, and proposed its use since it does not leave persistent residues in the soil; Meijide *et al.* (2007) found that adding DCD to pig slurry decreased N_2O

emissions by 64% in a maize crop. Hyatt *et al.* (2010) reported that applying polymer-coated urea to a potato crop reduced N_2O emissions significantly without reducing yields, and Halvorson *et al.* (2008) found similar results in maize. De Klein *et al.* (2001) reviewed the effect of nitrification inhibitors and slow-release fertilizers on N_2O emissions and showed that both substantially reduce these emissions.

Conflicts among mitigating strategies

There is evidence that some measures that decrease GHGs by sequestering carbon in soils can increase N_2O emissions to the point of offsetting the initial benefits. While conducting simulations, Li *et al.* (2005) found that some agricultural practices that increase C sequestration by soil – such as reduced tillage, crop residue incorporation and farmyard manure application – also increased N_2O emissions, resulting in some cases in a net production of CO_2 equivalents. Similarly, Rochette *et al.* (2008) found that no-tillage in a heavy clay soil increased nitrous oxide emissions that exceeded (in carbon equivalents) the increase in C sequestration. Six *et al.* (2004), in a review of the effects of no-tillage management on the net global warming potential in humid and dry temperate regions, concluded that no-tillage increased N_2O emissions but that this effect was offset by an increase in C sequestration. However, these authors concluded that the net effect on global warming potential varied with time and differed for the humid and dry regions, This possibility of conflict between mitigating GHG strategies has led some authors to propose the need for an integrated systems approach (Robertson, 2004).

5.8 Priority Areas for Future Research

Most of the studies reviewed in this chapter have analysed short-term plant responses to both increased temperature and $[CO_2]$, probably precluding the appearance of physiological mechanisms that plants might be able to activate to cope with a changing environment. There is a need to carry out longer-term experiments under more gradual changes of temperature and $[CO_2]$ regimes. In addition, both factors should be evaluated together, because plant response to elevated $[CO_2]$ can vary according to the temperature regime. These experiments are required to better understand and quantify plant responses to the global change scenario, in order to provide growers with appropriate management and cultural tools to better respond to global change.

In addition to the acquisition of physiological information to better understand the effects of increasing temperature and $[CO_2]$ on horticultural crops, studies are required to identify the most heat-tolerant varieties and genetic material. There is clearly a requirement to initiate and expand breeding programmes to prepare for horticultural production under the conditions of climate change.

Under a global change scenario of reduced water availability for horticultural crops, deficit irrigation strategies will have to be tested during several growing seasons. However, most of the experiments carried out with woody crops, to date, have lasted only 2–4 years. Long-term responses to reduced irrigation are little known, and should be now evaluated under warmer temperatures and elevated $[CO_2]$.

In relation to N management and climate change interactions in horticulture, there are several priority areas for research. One of the most important is the need to develop better and cheaper methods to improve N use efficiency. These improvements may come from better N management based on the use of more sensitive methods for assessing crop N status, and by the application of robust simulation models of the N dynamics in horticultural systems capable of simulating the amount and distribution of soil mineral N, and also N_2O losses. Crop genetic engineering to obtain plants with a higher N use efficiency is a promising field (Masclaux-Daubresse *et al.*, 2010).

Another area where more research is needed is the study of the net warming potential of the different irrigation and N management options in horticultural production, integrating environmental and socio-economic impacts (Brentrup *et al.*, 2004; Mosier *et al.*, 2006; Lillywhite *et al.*, 2007).

References

Agner, H. (2003) Denitrification in cultures of potted ornamental plants. Thesis, University of Hanover, Germany.

Alarcón, J.J., Ortuño, M.F., Nicolás, E., Navarro, A. and Torrecillas, A. (2006) Improving water-use efficiency of young lemon trees by shading with aluminised-plastic nets. *Agricultural Water Management* 82, 387–398.

Allen, L.H. Jr, Deyun Panb, Booteb, K.J., Pickering, N.B. and Jones, J.W. (2003) Carbon dioxide and temperature effects on evapotranspiration and water use efficiency of soybean. *Agronomy Journal* 95, 1071–1081.

Allen, R.G., Pereira, L.S., Raes, D. and Smith, M. (1998) *Crop Evapotranspiration. Guidelines for Computing Crop Water Requirements*. FAO Irrigation and Drainage paper No. 56, Rome, Italy.

Annandale, J.G. and Stockle, C.O. (1994) Fluctuation of crop evapotranspiration coefficients with weather. A sensitivity analysis. *Irrigation Science* 15, 1–7.

Austin, P.T., Hall, J., Gandar, P.W., Warrington, I.J., Fulton, T.A. and Halligan, E.A. (1999) A compartment model of the effect of early-season temperatures on potential size and growth of 'Delicious' apple fruits. *Annals of Botany* 83, 129–143.

Balandier, P., Bonhomme, M., Rageau, R., Capitan, F. and Parisot, E. (1993) Leaf bud endodormancy release in peach trees: evaluation of temperature models in temperate and tropical climate. *Agriculture and Forest Meteorology* 67, 95–113.

Bateman, E.J. and Baggs, E.M. (2005) Contributions of nitrification and denitrification to N_2O emissions from soils at different water-filled pore space. *Biology and Fertility of Soils* 41, 379–388.

Bates, B.C., Kundzewicz, Z.W., Wu, S. and Palutikof, J.P. (eds) (2008) *Climate Change and Water*. Technical Paper of the Intergovernmental Panel on Climate Change, IPCC Secretariat, Geneva, Switzerland, 210 pp.

Behboudian, M.H. and Mills T.M. (1997) Deficit irrigation in deciduous orchards. *Horticultural Reviews* 21, 125–131.

Ben Mechlia, N. and Carrol, J.J. (1989) Agroclimatic modelling for the simulation of phenology, yield and quality of crop production. 2. Citrus model implementation and verification. *International Journal of Biometeorology* 33, 52–65.

Berman, M.E. and DeJong, T.M. (1996) Water stress and crop load effects on fruit fresh and dry weights in peach (*Prunus persica*). *Tree Physiology* 16, 859–864.

Berni, J.A.J., Zarco-Tejada, P.J., Suarez, L. and Fereres, E. (2009) Thermal and narrow-band multispectral remote sensing for vegetation monitoring from an unmanned aerial vehicle. *IEEE Transactions on Geoscience and Remote Sensing* 47, 722–738.

Bindi, M., Fibbi, L. and Miglietta, F. (2001) Free air CO_2 enrichment (FACE) of grapevine (*Vitis vinifera* L.): II. Growth and quality of grape and wine in response to elevated CO_2 concentrations. *European Journal of Agronomy* 14, 145–155.

Bonhomme, M., Regeau, R., Lacointe, A. and Gendraud, M. (2005) Influences of cold deprivation during dormancy on carbohydrate contents of vegetative and floral primordia and nearby structures of peach buds (*Prunus persica* L. Batch). *Scientia Horticulturae* 105, 223–240.

Bouwman, A.F. (1996) Direct emission of nitrous oxide from agricultural soils. *Nutrient Cycling in Agroecosystems* 46, 53–70.

Bremner, J.M. (1997) Sources of nitrous oxide in soils. *Nutrient Cycling in Agroecosystems* 49, 7–16.

Brentrup, F., Küsters, J., Lammel, J., Barraclough, P. and Kuhlmann, H. (2004) Environmental impact assessment of agricultural production systems using the life cycle assessment (LCA) methodology. II. The application to N fertilizer use in winter wheat production systems. *European Journal of Agronomy* 20, 265–279.

Burger, M, Jackson, L., Lundquist, E., Louie, D., Miller, R., Rolston, D. *et al.* (2005) Microbial responses and nitrous oxide emissions during wetting and drying of organically and conventionally managed soil under tomatoes. *Biology and Fertility of Soils* 42, 109–118.

Byrne, D.H. (2005) Trends in stone fruit cultivar development. *HortTechnology* 15, 494–500.

Cannell, M.G.R. and Smith, R.I. (1986) Climatic warming, spring budburst, and frost damage on trees. *Journal of Applied Ecology* 23, 177–191.

Cavagnaro, T.R., Jackson, L.E. and Scow, K.M. (2006) Climate change: challenges and solutions for California agricultural landscapes.

Public Interest Energy Research Program Report. California Energy Commission, California. Available at www.energy.ca. gov/2005publications/CEC-500-2005-189/ CEC-500-2005-189-SF.PDF (accessed 18 June 2010).

Centritto, M., Lucas, M.E. and Jarvis, P.G. (2002) Gas exchange, biomass, whole-plant water use efficiency and water uptake of peach (*Prunus persica*) seedlings in response to elevated carbon dioxide concentration and water availability. *Tree Physiology* 22, 699–706.

Chalmers, D.J, Mitchell, P.D. and van Heek, L. (1981) Control of peach tree growth and productivity by regulated water supply, tree density and summer pruning. *Journal of the American Society for Horticultural Science* 106, 307–312.

Chalmers, D.J., Burge, G., Jerie, P.H. and Mitchell, P.D. (1986) The mechanisms of regulation of Bartlett pear fruit growth and vegetative growth by irrigation withholding and regulated deficit irrigation. *Journal of the American Society for Horticultural Science* 114, 15–19.

Charlesworth, P. (2005) *Soil Water Monitoring: an Information Package*, 2nd edn. Land & Water Australia, Canberra, ACT, Australia. Available at http://www.precirieg.net/documentacion/ soilwater.pdf (accessed 14 June 2010).

Chaves, B., De Neve, S., Cabrera, M., Boeckx, P., Van Cleemput, O. and Hofman, G. (2005) The effect of mixing organic biological waste materials and high-N crop residues on the short-time N_2O emission from horticultural soil in model experiments. *Biology and Fertility of Soils* 41, 411–419.

Cheng, W., Sudo, S., Tsuruta, H., Yagi, K. and Hartley, A. (2006) Temporal and spatial variations in N_2O emissions from a Chinese cabbage field as a function of type of fertilizer and application. *Nutrient Cycling in Agroecosystems* 74, 147–155.

Chmielewski, F.M., Muller, A. and Bruns, E. (2004) Climate change and trends in phenology of fruit trees and field crops in Germany. *Agricultural and Forest Meteorology* 121, 69–78.

Clayton, H., McTaggart, I.P., Parker, J., Swan, L. and Smith, K.A. (1997) Nitrous oxide emission from fertilised grassland: A 2-year study of the effects of N fertiliser form and environmental conditions. *Biology and Fertility of Soils* 25, 252–260.

Clough, T., Di, H., Cameron, K., Sherlock, R., Metherell, A., Clark, H. *et al.* (2007) Accounting for the utilization of a N_2O mitigation tool in the IPCC inventory methodology for agricultural soils. *Nutrient Cycling in Agroecosystems* 78, 1–14.

Cohen, A. and Goell, A. (1988) Fruit growth and dry matter accumulation in grapefruit during periods of water withholding and after reirrigation. *Australian Journal of Plant Physiology* 15, 633–639.

Cohen, S., Moreshet, S., LeGuillou, L., Simon, J.C. and Cohen, M. (1997) Response of citrus trees to modified radiation regime in semi-arid conditions. *Journal of Experimental Botany* 48, 35–44.

Conen, F., Dobbie, K.E. and Smith, K.A. (2000) Predicting N_2O emissions from agricultural land through related soil parameters. *Global Change Biology* 6, 417–426.

Consoli, S., O´Connell, N. and Snyder, R. (2006) Measurement of light interception by navel orange orchard canopies: Case study of Lindsay, California. *Journal of Irrigation and Drainage Engineering* 132, 9–20.

Coyne, M.S. (2008) Biological denitrification. In: Schepers, J.S. and Raun, W. (eds) *Nitrogen in Agricultural Systems*. ASA-CSSSA-SSSA Agronomy Monograph 49, Madison, Wisconsin, pp. 197–249.

De Klein, C.A.M., Sherlock, R.R., Cameron, K.C. and van der Weerden, T.J. (2001) Nitrous oxide emissions from agricultural soils in New Zealand – a review of current knowledge and directions for future research. *Journal of the Royal Society of New Zealand* 31, 543–574.

Del Grosso, S.J., Parton, W.J., Mosier, A.R., Ojima, D.S., Kulmala, A.E. and Phongpan, S. (2000) General model for nitrous oxide and N_2 gas emissions from soils due to denitrification. *Global Biogeochemical Cycle* 14, 1045–1060.

De Neve, S. and Hofman, G. (1996) Modelling N mineralization of vegetable crop residues during laboratory incubation. *Soil Biology and Biochemistry* 28, 1451–1457.

Drigo, B., Kowalchuk, G.A. and van Veen, J.A. (2008) Climate change goes underground: effects of elevated atmospheric CO_2 on microbial community structure and activities in the rhizosphere. *Biology and Fertility of Soils* 44, 667–679.

EPA (2009) Inventory of U.S. Greenhouse Gas Emissions and Sinks: 1990–2007. Available at http://www.epa.gov/climatechange/emissions/ usinventoryreport.html (accessed 1 March 2010).

European Commission (2009) *Agricultural Statistics. Main Results 2007–2008*. Eurostat, Luxembourg.

Evett, S.R. (2007) Soil water and monitoring technology. In: Lascano, R.J. and Skoja, R.E. (eds) *Irrigation of Agricultural Crops*. American Society of Agronomy, Madison, Wisconsin, pp. 25–84.

Ezzahouani, A. (2003) Behaviour study of 'Danlas' grapevines grown under plastic cover. *Journal International des Sciences de la Vigne et du Vin* 37, 117–122.

Fabeiro, C., Martín de Santa Olalla, F. and de Juan, J.A. (2002) Production of muskmelon (*Cucumis melo* L.) under controlled deficit irrigation in a semi-arid climate. *Agricultural Water Management* 54, 93–105.

Feller, C. and Fink, M. (2002) N_{min} target values for field vegetables. *Acta Horticulturae* 571, 195–201.

Fereres, E. and Soriano, M.A. (2007) Deficit irrigation for reducing agricultural water use. *Journal of Experimental Botany* 58, 147–159.

Fereres, E., Goldhamer, D.A. and Parsons L. (2003) Irrigation water management of horticultural crops. *HortScience* 38, 1036–1042.

Fink, M. and Scharpf, H.C. (1993) N-expert – a decision support system for vegetable fertilization in the field. *Acta Horticulturae* 339, 67–74.

Forster, P., Ramaswamy, V., Artaxo, P., Berntsen, T., Betts, R., Fahey, D.W. *et al.* (2007) Changes in atmospheric constituents and in radiative forcing. In: Solomon, S., Qin, D., Manning, M., Chen, Z., Marquis, M., Averyt, K.B. *et al.* (eds) *Climate Change 2007: The Physical Science Basis.* Contribution of Working Group I to the Fourth Assessment Report of the Intergovernmental Panel on Climate Change. Cambridge University Press, Cambridge, UK and New York, pp. 129–234.

Gariglio, N., González Rossia, D., Mendowa, M., Reig, C. and Agustí, M. (2006) Effect of artificial chilling on the depth of endodormancy and vegetative and flower budbreak of peach and nectarine cultivars using excised shoots. *Scientia Horticulturae* 108, 371–377.

Ghani, A., Mackay, A., Clothier, B., Curtin, D., and Sparling, G. (2009) A literature review of soil carbon under pasture, horticulture and arable land uses. Report prepared for AGMARDT, 119 pp. Available at www.agresearch.co.nz/science/docs/AGMARDT-Soil-Carbon-Review.pdf (accessed 2 June 2010).

Gianquinto, G., Sambo, P. and Borsato, D. (2006) Determination of SPAD threshold values for the optimisation of nitrogen supply for processing tomato. *Acta Horticulturae* 700, 159–166.

Girona, J., Gelly, M., Mata, M., Arbonés, A., Rufat, J. and Marsal, J. (2005) Peach tree response to single and combined deficit irrigation regimes in deep soil. *Agricultural Water Management* 72, 97–108.

Girona, J., Behboudian, M.H., Mata, M., Del Campo, J. and Mansal, J. (2010) Exploring six

reduced irrigation options under water shortage for 'Golden Smoothee' apple: Responses of yield components over three years. *Agricultural Water Management* 98, 370–375.

Gonçalves, B., Falco, F., Moutinho-Pereira, H., Bacelar, E., Peixoto, F. and Correia, C. (2009) Effects of elevated CO_2 on grapevine (*Vitis vinifera* L.): volatile composition, phenolic content, and *in vitro* antioxidant activity of red wine. *Journal of Agricultural and Food Chemistry* 57, 265–273.

González-Altozano, P. and Castel, J.R. (1999) Regulated deficit irrigation in "Clementina de Nules" citrus trees. I: Yield and fruit quality effects. *Journal of Horticultural Science and Biotechnology* 74, 706–713.

Gregory, P.J., Johnson, S.N., Newton, A.C. and Ingram, J.S.I. (2009) Integrating pests and pathogens into the climate change/food security debate. *Journal of Experimental Botany* 60, 2827–2838.

Halvorson, A.D., Del Grosso, S.J. and Reule, C.A. (2008) Nitrogen, tillage, and crop rotation effects on nitrous oxide emissions from irrigated cropping systems. *Journal of Environmental Quality* 37, 1337–1344.

Hansen, P. (1971) The effect of fruiting upon transpiration rate and stomatal opening in apple leaves. *Physiologia Plantarum* 25, 181–183.

Hanson, B.R., Orloff, S. and Peters, D.W. (2000) Monitoring soil moisture helps refine irrigation management. *California Agriculture* 54, 38–42.

Hartz, T.K. (2006) Vegetable production best management practices to minimize nutrient loss. *HortTechnology* 16, 398–403.

Hartz, T.K., Bendixen, W.E. and Wierdsma, L. (2000) The value of pre-sidedress soil nitrate testing as a nitrogen management tool in irrigated vegetable production. *HortScience* 35, 651–656.

Hatfield, J., Boote, K., Fay, P., Hahn, L., Izaurralde, C., Kimball, B.A. *et al.* (2008) Agriculture. In: *The Effects of Climate Change on Agriculture, Land Resources, Water Resources, and Biodiversity in the United States.* A Report by the U.S. Climate Change Science Program and the Subcommittee on Global Change Research. Washington, DC, USA. Available at http://www.usda.gov/oce/global_change/files/CCSP FinalReport.pdf (accessed 18 June 2010).

Hedhly, A., Hormaza, J.I. and Herrero, M. (2009) Global warming and sexual plant reproduction. *Trends in Plant Science* 14, 30–36.

Hincley, T.M. and Bruckerhoff, D.M. (1975) The effect of drought on water relations and stem shrinkage of *Quercus alba*. *Canadian Journal of Botany* 53, 62–72.

Hochmuth, G.J. (1994) Efficiency ranges for nitrate nitrogen and potassium for vegetable petiole sap quick test. *HortTechnology* 4, 218–222.

Hosono, T., Hosoi, N., Akiyama, H. and Tsuruta, H. (2006) Measurements of N_2O and NO emissions during tomato cultivation using a flow-through chamber system in a glasshouse. *Nutrient Cycling in Agroecosystems* 75, 115–134.

Hueso, J.J. and Cuevas, J. (2008) Loquat as a crop model for successful deficit irrigation. *Irrigation Science* 26, 269–276.

Hyatt, C.R., Venterea, R.T., Rosen, C.J., McNearney, M., Wilson, M.L. and Dolan, M.S. (2010) Polymer-coated urea maintains potato yields and reduces nitrous oxide emissions in a Minnesota loamy sand. *Soil Science Society of America Journal* 74, 419–428.

IAEA (2008) Field estimation of soil water content. A practical guide to methods, instrumentation, and sensor technology, Training Course Series No. 30, International Atomic Energy Agency, Vienna. Available at www-pub.iaea.org/MTCD/publications/PDF/TCS-30_web.pdf (accessed 14 June 2010).

Idso, C.D. (2001) Earth's rising atmospheric CO_2 concentration: Impacts on the biosphere. *Energy and Environment* 12, 287–310.

Intrigliolo, D.S. and Castel, J.R. (2006) Performance of various water stress indicators for prediction of fruit size response to deficit irrigation. *Agricultural Water Management* 83, 173–180.

Intrigliolo, D.S. and Lakso, A.N. (2011) Effects of amount of light interception and canopy orientation to the sun on grapevine water status and canopy gas exchange. *Acta Horticulturae* 889, 99–104.

IPCC (2006) N_2O emissions from managed soils, and CO_2 emissions from lime and urea application. In: Eggleston, S., Buendía, L., Miwa, K., Ngara, T. and Tanabe, K. (eds) *Guidelines for National Greenhouse Gas Inventories*, Vol. 4. IPCC, Geneva, Switzerland, pp. 11.1–11.54.

IPCC (2007) *Climate Change 2007: The Physical Science Basis*. Contribution of Working Group I to the Fourth Assessment Report of the Intergovernmental Panel on Climate Change. Solomon, S., Qin, D., Manning, M., Chen, Z., Marquis, M., Averyt, K.B. (eds) Cambridge University Press, Cambridge, UK and New York, 996 pp. Available at http://www.ipcc.ch/ipccreports/ar4-wg1.htm (accessed 24 June 2010).

IPCC (2009) *Chapter Outline of the Working Group I*. Contribution to the IPCC Fifth Assessment Report (AR5). Available at http://www.ipcc-wg1.unibe.ch/docs/Doc.19-WGI-Outline.pdf (accessed 25 June 2010).

Ito, J., Hasegawa, S., Fujita, K., Ogasawara, S. and Fujiwara, T. (1999) Effects of CO_2 enrichment on fruit growth and quality in Japanese pear (*Pyrus serotina* Rehder cv. Kosui). *Soil Science and Plant Nutrition* 45, 385–393.

Jarecki, M.K. and Lal, R. (2003) Crop management for soil carbon sequestration. *Critical Reviews in Plant Science* 22, 471–502.

Jarvis, S.C., Stockdale, E.A., Shepherd, M.A. and Powlson, D.S. (1996) Nitrogen mineralization in temperate agricultural soils: Processes and measurement. *Advances in Agronomy* 57, 187–235.

Johnson, R.S. and Handley, D.F. (2000) Using water stress to control vegetative growth and productivity of temperate fruit trees. *HortScience* 35, 1048–1050.

Johnson, R.S., Handley, D.F. and Day, K.R. (1994) Postharvest water stress of an early maturing plum. *Journal of Horticultural Science and Biotechnology* 69, 1035–1041.

Johnston, A.E., Poulton, P.R. and Coleman, K. (2009) Soil organic matter: Its importance in sustainable agriculture and carbon dioxide fluxes. *Advances in Agronomy* 101, 1–57.

Jones, G.V. and Davies, R.E. (2000) Climate influences on grapevine phenology, grape composition and wine production and quality for Bordeaux, France. *American Journal of Enology and Viticulture* 51, 249–261.

Jones, G.V., White, M.A., Cooper, O.R. and Storchmann, K. (2005) Climate change and global wine quality. *Climate Change* 73, 319–343.

Kenny, G.J. and Harrison, P.A. (1992) The effects of climate variability and change on grape suitability in Europe. *Journal of Wine Research* 3, 163–183.

Kimball, B.A. (2007) Global change and water resources. In: Lascano, R.J. and Skoja, R.E. (eds) *Irrigation of Agricultural Crops*. American Society of Agronomy, Madison, Wisconsin, pp. 627–653.

Kimball, B.A., Idso, S.B., Johnson, S. and Rillig, M.C. (2007) Seventeen years of carbon dioxide enrichment of sour orange trees: Final results. *Global Change Biology* 13, 2171–2183.

Körner, C. (2006) Plant CO_2 responses: an issue of definition, time and resource supply. *New Phytologist* 172, 393–411.

Kozlowski, T.T. and Pallardy, S.G. (1997) *Physiology of Woody Plants*. Academic Press, San Diego, California.

Kramer, K. (1994) A modeling analysis of the

effects of climatic warming on the probability of spring frost damage to tree species in The Netherlands and Germany. *Plant, Cell, and Environment* 17, 367–377.

Kriedemann, P.E., Sward, R.J. and Downton, W.J.S. (1976) Vine response to carbon dioxide enrichment during heat therapy. *Australian Journal of Plant Physiology* 3, 605–618.

Krug, H. (1997) Environmental influences on development, growth and yield. In: Wien, H.C. (ed.) *The Physiology of Vegetable Crops*. CAB International, Wallingford, UK, pp. 101–108.

Lampinen, B.D., Shackel, K.A., Southwick, S.M. and Olson, B. (2001) Deficit irrigation strategies using midday stem water potential in prune. *Irrigation Science* 20, 47–54.

Leakey, A.D.B., Ainsworth, E.A., Bernacchi, C.J., Rogers, A., Long, S.P. and Ort, D.R. (2009) Elevated CO_2 effects on plant carbon, nitrogen and water relations: six important lessons from FACE. *Journal of Experimental Botany* 60, 2859–2876.

Li, C., Frolking, S. and Butterbach-Bahl, K. (2005) Carbon sequestration in arable soils is likely to increase nitrous oxide emissions, offsetting reductions in climate radiative forcing. *Climate Change* 72, 321–338.

Lillywhite, R., Chandler, D., Grant, W., Lewis, K., Firth, C., Schmutz, U. *et al.* (2007) Environmental footprint and sustainability of horticulture (including potatoes) – A comparison with other agricultural sectors. Warwick HRI, University of Warwick, report for DEFRA, UK. Available at www2.warwick.ac.uk/fac/sci/whri/research/nitrogenandenvironment/footprint/environmental_footprint.pdf (accessed 4 June 2010).

Long, S.P., Ainsworth, E.A., Leakey, A.D.B., Nösberger, J. and Ort, D.R. (2006) Food for thought: lower-than-expected crop yield stimulation with rising CO_2 concentration. *Science* 312, 1918–1921.

Lopez, G. and DeJong, T.M. (2007) Spring temperatures have a major effect on early stage of peach fruit growth. *Journal of Horticultural Sciences and Biotechnology* 82, 507–512.

López, G., Mata, M., Arbonés, A., Solans, J.R., Girona, J. and Marsal, J. (2006) Mitigation of effects of extreme drought during stage III of peach fruit development by summer pruning and fruit thinning. *Tree Physiology* 26, 469–477.

Luedeling, E., Gebauer, J. and Buerkert, A. (2009a) Climatic changes effects on winter chill for tree crops with chilling requirements on the Arabian peninsula. *Climatic Change* 96, 219–237.

Luedeling, E., Zhang, M. and Girvetz, E.H. (2009b) Climatic changes lead to declining winter chill

for fruit and nut trees in California during 1950–2099. *PloS ONE* 4(7), e6166.

Magán, J.J., Gallardo, M., Thompson, R.B. and Lorenzo, P. (2008) Effects of salinity on fruit yield and quality of tomato grown in soil-less culture in greenhouses in Mediterranean climatic conditions. *Agricultural Water Management* 95, 1041–1055.

Majumdar, D. (2009) Nitrous oxide emission from crop fields and its role in atmospheric radiative forcing. In: Singh, S.N. (ed.) *Climate Change and Crops*, Springer, Berlin, pp. 147–190.

Marsal, J., Rapoport, H.F., Manrique, T. and Girona, J. (2000) Pear fruit growth under regulated deficit irrigation in container-grown trees. *Scientia Horticulturae* 85, 242–259.

Marsal, J., López, G., Mata, M. and Girona, J. (2006) Branch removal and defruiting for the amelioration of water stress effects on fruit growth during Stage III of peach fruit development. *Scientia Horticulturae* 108, 55–60.

Marsal, J., Mata, M., Rabones, A., Del Campo, J., Girona, J. and López, G. (2008) Factors involved in alleviating water stress by partial crop removal in pear trees. *Tree Physiology* 28, 1375–1382.

Masclaux-Daubresse, C., Daniel-Vedele, F., Dechorgnat, J., Chardon, F., Gaufichon, L. and Suzuki, A. (2010) Nitrogen uptake, assimilation and remobilization in plants: challenges for sustainable and productive agriculture. *Annals of Botany* 105, 1141–1157.

McCutchan, H. and Shackel, K.A. (1992) Stem water potential as a sensitive indicator of water stress in prune trees (*Prunus domestica* L. cv. French). *Journal of the American Society of Horticultural Science* 117, 607–611.

Meijide, M., Díez, J.A., Sánchez-Martín, L., López-Fernández, S. and Vallejo, A. (2007) Nitrogen oxide emissions from an irrigated maize crop amended with treated pig slurries and composts in a Mediterranean climate. *Agriculture, Ecosystems and Environment* 121, 383–394.

Millar, N., Robertson, G.P., Grace, P.R., Gehl, R.J. and Hoben, J.P. (2010) Nitrogen fertilizer management for nitrous oxide (N_2O) mitigation in intensive corn (maize) production: an emissions reduction protocol for US Midwest agriculture. *Mitigation and Adaptation Strategies for Global Change* 15, 185–204.

Moran, S., Zarco-Tejada, P.J. and Clarke, T. (2005) Crop water stress detection using remote sensing. In: Lehr, X. and Jay, H. (eds) *Water Encyclopedia, Vol. 4: Surface and Agricultural Water*. John Wiley & Sons, London, pp. 719–724.

Mosier, A.R. (1998) Soil processes and global

change. *Biology and Fertility of Soils* 27, 221–229.

Mosier, A.R., Halvorson, A.D., Reule, C.A. and Liu, X.J. (2006) Net global warming potential and greenhouse gas intensity in irrigated cropping systems in northeastern Colorado. *Journal of Environmental Quality* 35, 1584–1598.

Moss, G.I. (1970) The influence of temperature on fruit set in cuttings of sweet orange (*Citrus sinensis* L. Osbeck). *Horticultural Research* 10, 97–107.

Mulvaney, R.L., Khan, S.A. and Mulvaney, C.S. (1997) Nitrogen fertilizers promote denitrification. *Biology and Fertility of Soils* 24, 211–220.

Murchie, E.H., Pinto, M. and Horton, P. (2009) Agriculture and the new challenges for photosynthesis research. *New Phytologist* 181, 532–552.

Naor, A. (2000) Midday stem water potential as a plant water stress indicator for irrigation scheduling in fruit trees. *Acta Horticulturae* 537, 447–454.

Naor, A., (2006) Irrigation scheduling and evaluation of tree water status in deciduous orchards. *Horticultural Reviews* 32, 111–166.

Naor, A., Gal, Y. and Peres, M. (2006) The inherent variability of water stress indicators in apple, nectarine and pear orchards, and the validity of a leaf-selection procedure for water potential measurements. *Irrigation Science* 24, 129–135.

Neeteson, J.J., Langeveld, J.W.A., Smit, A.L. and de Haan, J.J. (2003) Nutrient balances in field vegetable production systems. *Acta Horticulturae* 627, 13–23.

Nesme, T., Brisson, N., Lescourret, F., Bellon, S., Crété, X., Plénet, D. *et al.* (2006) Epistics: a dynamic model to generate nitrogen fertilisation and irrigation schedules in apple orchards, with special attention to qualitative evaluation of the model. *Agricultural Systems* 90, 202–225.

Pang, J., Wang, X., Mu, Y., Ouyang, Z. and Liu, W. (2009) Nitrous oxide emissions from an apple orchard soil in the semiarid Loess Plateau of China. *Biology and Fertility of Soils* 46, 37–44.

Pardossi, A., Tognoni, F. and Incrocci, L. (2004) Mediterranean greenhouse technology. *Chronica Horticulturae* 44, 28–34.

Peet, M.M. and Wolfe, D. (2000) Crop ecosystem responses to climate change: vegetable crops. In: Reddy, K.R. and Hodges, H.F. (eds) *Climate Change and Global Crop Productivity*. CAB International, Wallingford, UK, pp. 213–243.

Peet, M.M., Sato, S. and Gardner, R.G. (1998) Comparing heat stress effects on male-fertile and male-sterile tomatoes. *Plant Cell and Environment* 21, 225–231.

Peterhansel, C., Niessen, M. and Kebeish, R.M. (2008) Metabolic engineering towards the enhancement of photosynthesis. *Photochemistry and Photobiology* 84, 1317–1323.

Porro, D., Dorigatti, C., Stefanini, M. and Ceschini, A. (2001) Use of SPAD meter in diagnosis of nutritional status in apple and grapevine. *Acta Horticulturae* 564, 243–252.

Porter, J.R. (2005) Rising temperatures are likely to reduce crop yields. *Nature* 436, 174.

Porter, J.R. and Semenov, M.A. (2005) Crop responses to climatic variation. *Philosophical Transactions of the Royal Society B – Biological Sciences* 360, 2021–2035.

Postel, S. (1999) *Pillar of Sand: Can the Irrigation Miracle Last?* W.W. Norton and Co., New York.

Prentice, I.C., Farquhar, G.D. and Fasham, M.J.R. (2001) The carbon cycle and atmospheric carbon dioxide. In: Houghton, J.T., Ding, Y., Griggs, D.J., Noguer, M., van der Linden, P.J. and Xiaosu, D. (eds) *Climate Change 2001: the Scientific Basis.* Contribution of Working Group I to the Third Assessment Report of the Intergovernmental Panel on Climate Change. Cambridge University Press, UK and New York, pp. 183–239.

Rahn, C., Bending, G.D., Turner, M.K. and Lillywhite, R. (2003) Management of N mineralization from crop residues of high N content using amendment materials of varying quality. *Soil Use Management* 19, 193–200.

Rahn, C.R., Zhang, K., Lillywhite, R., Ramos, C., Doltra, J., de Paz, J.M. *et al.* (2010) EU-Rotate_N – a decision support system to predict environmental and economic consequences of the management of nitrogen fertiliser in crop rotations. *European Journal of Horticultural Science* 75, 20–32.

Reddy, K.R. and Hodges, H.F. (2000) *Climate Change and Global Crop Productivity.* CABI Publishing, Wallingford, UK.

Rigby, J.R. and Porporato, A. (2008) Spring frost risk in a changing climate. *Geophysical Research Letters* 35, L12703.

Robertson, G.P. (2004) Abatement of nitrous oxide, methane, and the other non-CO_2 greenhouse gases: The need for a systems approach. In Field, C.B. and Raupach, M.R. (eds) *The Global Carbon Cycle.* Island Press, Washington, DC, pp. 493–506.

Robertson, G.P. and Swinton, S.M. (2005) Reconciling agricultural productivity and environmental integrity: a grand challenge for agriculture. *Frontiers in Ecology and the Environment* 3, 38–46.

Robertson, G.P. and Vitousek, P.M. (2009) Nitrogen in agriculture: Balancing the cost of an essential resource. *Annual Review of Environment and Resources* 34, 97–126.

Rochette, P., Angers, D.A., Chantigny, M.H. and Bertrand, N. (2008) Nitrous oxide emissions respond differently to no-till in a loam and a heavy clay soil. *Soil Science Society of America Journal* 72, 1363–1369.

Rodrigo, M.C. and Ramos, C. (2007) Nitrate sap analysis as a tool to assess nitrogen nutrition in artichoke. *Acta Horticulturae* 730, 251–256.

Rolston, D.E., Sharpley, A.N., Toy, D.W. and Broadbent, F.E. (1982) Field measurement of denitrification. III: Rates during irrigation cycles. *Soil Science Society of America Journal* 46, 289–296.

Rosenzweig, C., Phillips, J., Goldberg, R., Carrol, J. and Hodges, T. (1996) Potential impacts of climate change on citrus and potato production in the US. *Agricultural Systems* 52, 455–479.

Rubatzky, V.E. and Yamaguchi, M. (1999) *World Vegetables, Principles, Production, and Nutritive Values*. 2nd edn, Aspen Publishers. Inc., Gaithersberg, Mississippi.

Ruser, R., Flessa, H., Schilling, R., Stemdl, H. and Beese, F. (1998) Soil compaction and fertilization effects on nitrous oxide and methane fluxes in potato fields. *Soil Science Society of America Journal* 62, 1587–1595.

Ryden, J.C. and Lund, L.J. (1980) Nitrous-oxide evolution from irrigated land. *Journal of Environmental Quality* 9, 387–393.

Sadras, V.O. and Soar, C.J. (2009) Shiraz vines maintain yield in response to a 2-4°C increase in maximum temperature using an open-top heating system at key phenostages. *European Journal of Agronomy* 31, 250–258.

Sadras, V.O., Villalobos, F. and Fereres, E. (1993) Leaf expansion in field grown sunflower in response to soil and leaf water status. *Agronomy Journal* 85, 564–670.

Samborski, S.M., Tremblay, N. and Fallon, E. (2009) Strategies to make use of plant sensors-based diagnostic information for nitrogen recom-mendations. *Agronomy Journal* 101, 800–816.

Sánchez-Guerrero, M.C., Lorenzo, P., Medrano, E., Baille, A. and Castilla, N. (2009) Effects of EC-based irrigation scheduling and CO_2 enrichment on water use efficiency of a greenhouse cucumber crop. *Agricultural Water Management* 96, 429–436.

Sanchez-Martín, L., Arce, A., Benito, A., Garcia-Torres, L. and Vallejo. A. (2008) Influence of drip and furrow irrigation systems on nitrogen oxide emissions from a horticultural crop. *Soil Biology and Biochemistry* 40, 1696–1706.

Savvas, D. (2002) Nutrient solution recycling. In: Savvas, D. and Passam, H. (eds) *Hydroponic Production of Vegetables and Ornamentals*. Embryo Publications, Athens, pp. 299–343.

Schaffer, B., Whiley, A.W., Searle, C. and Nissen, R.J. (1997) Leaf gas exchange, dry matter partitioning and mineral element concentrations in mango as influenced by elevated CO_2 and root restriction. *Journal of the American Society for Horticultural Science* 122, 849–855.

Schils, R., Kuikman, P., Liski, J., van Oijen, M., Smith, P., Webb, J. *et al.* (2008) Review of existing information on the interrelations between soil and climate change. ClimSoil. European Commission. Final Report. Available at http://ec.europa.eu/environment/soil/review_en.htm (accessed 30 January 2010).

Sepulcre-Cantó, G., Zarco-Tejada, P.J., Jiménez-Muñoz, J.C., Sobrino, J.A., de Miguel, E. and Villalobos, F.J. (2006) Detection of water stress in an olive orchard with thermal remote sensing imagery. *Agricultural and Forest Meteorology* 136, 31–44.

Shackel, K.A., Ahmadi, H., Biasi, W., Buchner, R., Goldhamer, D., Gurusinghe, S. *et al.* (1997) Plant water status as an index of irrigation need in deciduous fruit trees. *HortTechnology* 7, 23–29.

Shock, C.C., Pereira, A.B., Hanson, B.R. and Cahn, M.D. (2007) Vegetable irrigation. In: Lascano, R.J. and Skoja, R.E. (eds) *Irrigation of Agricultural Crops*. 2nd edn, American Society of Agronomy, Madison, Wisconsin, pp. 535–606.

Singh, S.N. (2009) *Climate Change and Crops.* Springer, Berlin.

Six, J., Ogle, S.M., Breidt, F.J., Conant, R.T., Mosier, A.R. and Paustian, K. (2004) The potential to mitigate global warming with no-tillage management is only realized when practiced in the long term. *Global Change Biology* 10, 155–160.

Smith, P., Martino, D., Cai, Z., Gwary, D., Janzen, H., Kumar, P. *et al.* (2007) Agriculture. In: Metz, B., Davidson, O.R., Bosch, P.R., Dave, R. and Meyer, L.A. (eds) *Climate Change 2007: Mitigation.* Contribution of Working Group III to the Fourth Assessment Report of the Intergovernmental Panel on Climate Change Cambridge University Press, Cambridge, UK and New York, pp. 497–540.

Smith, P., Martino, D., Cai, Z., Gwary, D., Janzen, H.H., Kumar, P. *et al.* (2008) Greenhouse gas mitigation in agriculture. *Philosophical Trans-actions of the Royal Society* B 363, 789–813.

Soar, C.J., Sadras, V.O. and Petrie, P.R. (2008) Climate drivers of red wine quality in four contrasting Australian wine regions. *Australian Journal of Grape and Wine Research* 14, 78–90.

Soar, C.J., Collins, M.J. and Sadras, V.O. (2009) Irrigated Shiraz vines up-regulate gas exchange and maintain berry growth under short spells of high maximum temperature in the field. *Functional Plant Biology* 36, 801–814.

Socolow, R.H. (1999) Nitrogen management and the future of food: lessons from the management of energy and carbon. *Proceedings of the National. Academy of Sciences USA* 96, 6001–6008.

Spreitzer, R.J. and Salvucci, M.E. (2002) Rubisco: structure, regulatory interactions, and possibilities for a better enzyme. *Annual Review of Plant Biology* 53, 449–475.

Stanley, C.D. and Maynard, D.N. (1990) Vegetables. In: Stewart, B.A. and Nielsen, D.R. (eds) *Irrigation of Agricultural Crops*. American Society of Agronomy, Madison, Wisconsin, pp. 922–950.

Stitt, M. and Krapp, A. (1999) The interaction between elevated carbon dioxide and nitrogen nutrition: the physiological and molecular background. *Plant Cell and Environment* 22, 583–621.

Taub, D.R. and Wang, X. (2008) Why are nitrogen concentrations in plant tissues lower under elevated CO_2? A critical examination of the hypothesis. *Journal of Integrative Plant Biology* 50, 1365–1374.

Thompson, R.B. and Gallardo, M. (2005) Use of soil sensors for irrigation scheduling. In: Fernández-Fernández, M., Lorenzo-Minguez, P. and Cuadrado Gómez, M.I. (eds) *Improvement of Water Use Efficiency in Protected Crops*. Dirección General de Investigación y Formación Agraria de la Junta de Andalucía, Seville, Spain, pp. 351–376.

Thompson, R.B., Gallardo, M., Valdez, L.C. and Fernández, M.D. (2007a) Using plant water status to define soil water thresholds for irrigation management of vegetable crops using soil moisture sensors. *Agricultural Water Management* 88, 147–158.

Thompson, R.B., Gallardo, M., Valdez, L.C. and Fernández, M.D. (2007b) Establishing lower limits for on-farm irrigation management with volumetric soil water content sensors by using in-situ assessments of apparent crop water uptake. *Agricultural Water Management* 92, 13–28.

Tremblay, N., Scharpf, H.C., Weier, U., Laurence, H. and Owen, J. (2001) *Nitrogen Management in Field Vegetables. A Guide to Efficient Fertilization.* Agriculture and Agri-Food, Canada. Available at http://dsp-psd.pwgsc.gc.ca/Collection/A42-92-2001E.pdf (accessed 24 June 2010).

Tubiello, F.N. and Rosenzweig, C. (2008) Developing climate change impact metrics for agriculture. *The Integrated Assessment Journal* 8, 165–184.

USDA (2009) *Census of Agriculture 2007.* United States. Summary and State Data. Vol. 1, Part 51. AC-07-A-51. National Agricultural Statistics Service, Washington, DC.

van der Weerden, T.J., Sherlock, R.R., Williams, P.H. and Cameron, K.C. (2000) Effect of the three contrasting onion (*Allium cepa* L.) production systems on nitrous oxide emissions from soil. *Biology and Fertility of Soils* 31, 334–342.

Wehrmann, J. and Scharpf, H.C. (1989) Reduction of nitrate leaching in a vegetable farm – Fertilization, crop rotation, plant residues. In: Germon, J.C. (ed.) *Management Systems to Reduce Impact of Nitrates*. Elsevier Applied Sciences, London, pp. 99–109.

Wehrmann, J., Scharpf, H.C. and Kuhlmann, H. (1988) The Nmin method – an aid to improve nitrogen efficiency in plant production. In: Jenkinson, D.S. and Smith, K.A. (eds) *Nitrogen Efficiency in Agricultural Soils*. Elsevier Applied Science, London, pp. 256–268.

Wolfe, D.W., Schwartz, M.D., Lakso, A.N., Otsuki, Y., Pool, R.M. and Shaulis, N. (2005) Climate change and shifts in spring phenology of three horticultural woody perennials in northeastern USA. *International Journal of Biometeorology* 49, 303–309.

Xiong, Z. and Khalil, M.A.K. (2009) Greenhouse gases from crop fields. In: Singh, S.N. (ed.) *Climate Change and Crops*. Springer, Berlin, pp. 113–132.

Xiong, Z.Q., Xie, Y.X., Xing, G.X., Zhu, Z.L. and Butenhoff, C. (2006) Measurements of nitrous oxide emissions from vegetable production in China. *Atmospheric Environment* 40, 2225–2234.

6 The Impact of High CO_2 on Plant Abiotic Stress Tolerance

M.S. Lopes and C.H. Foyer

6.1 Introduction

Atmospheric concentrations of CO_2 have risen from about 270 µl/l in pre-industrial times to over 380 µl/l at present, and are predicted to reach between 530 and 970 µl/l by the end of this century (IPCC, 2007). Over the same period, atmospheric O_2 concentrations have declined to an even greater extent (an average of 1.4 µl/l O_2 decrease per 1.0 µl/l CO_2 increase), because (i) the combustion of fossil fuels consumes more O_2 than it releases CO_2; and (ii) a portion of the CO_2 released during combustion or respiration dissolves in large bodies of water. Recent evidence suggests, however, that the capacity of the earth's oceans to absorb CO_2 may be reaching saturation and that this may cause an even greater problem than the predicted rise in temperature that will accompany climate change (Khatiwala et al., 2009).

Average global temperatures have increased by about 0.76°C over the last 150 years and they are likely to increase by another 1.7–3.9°C during this century. These changes in atmospheric composition and the temperature of the earth will directly influence the balance between C_3 carbon fixation and photorespiration. Currently, the atmosphere has a CO_2:O_2 ratio of 0.0018. Gaseous CO_2, however, is much more soluble in water than O_2, and thus the CO_2:O_2 ratio in the aqueous environment if the chloroplast stroma is about 0.026 at 25°C. The enzyme ribulose-1, 5-bisphosphate carboxylase/oxygenase (rubisco) has between a 50-fold (in cyanobacteria) to 100-fold (in higher plants) greater affinity for CO_2 than O_2. These relative concentrations and affinities dictate that rubisco catalyses about two to three cycles of ribulose-1, 5-bisphosphate carboxylation (i.e. the carboxylation step of the Benson–Calvin cycle) for every cycle of ribulose-1, 5-bisphosphate oxidation (i.e. the reaction initiating the photorespiratory pathway). Thus, during photosynthesis carbon is partitioned accordingly between the Benson–Calvin cycle and the photorespiratory pathways. The anticipated changes in the earth's environment will therefore influence the partitioning between these pathways, and as a consequence influence the distribution of plants with different carbon fixation pathways. Some authors consider that the increasing CO_2 and declining O_2 levels will favour C_3 plants relative to C_4 plants, particularly in environments where NH_4^+ is most available as a nitrogen source. Conversely, higher temperatures will tend to favour C_4 plants compared with C_3 plants. By the end of this century, the earth's atmosphere may have a CO_2:O_2 ratio as high as 0.0047. The predicted changes in the balance between atmospheric composition and temperature will directly influence the balance between C_3 carbon fixation and photorespiration.

It has long been recognized that short-term exposures to such atmospheres initially stimulate photosynthesis, increasing quantum yields and net CO_2 fixation rates in C_3 plants (Makino and Mae, 1999). However, this enhancement is often not sustained and acclimatization occurs, such that, after days or weeks of exposure to elevated CO_2 or low O_2 concentrations, photosynthetic rates decline and growth rates slow (Sage et al., 1988). In such cases, leaf nitrogen and protein concentrations often decrease by more than 10% under CO_2 enrichment. Together, these trends define the process of 'CO$_2$ acclimatization'.

In spite of the acclimatization of photosynthesis with time following exposure to CO_2, experiments with elevated CO_2 – both in controlled environments (Kimball, 1983; Wu et al., 2004) and in the field using Free Air Carbon Dioxide Enrichment (FACE) facilities – have shown that C_3 and C_4 crop plants can benefit in particular in terms of improved water use efficiencies, increased growth rates and yields (Mauney et al., 1994; Makino and Mae, 1999). Such effects are most marked in C_3 plants under optimal conditions of nitrogen availability and irrigation (Patterson and Flint, 1980; Carter and Peterson, 1983; Morison and Gifford, 1984; Cure and Acock, 1986; Leakey et al., 2006a). Far fewer experiments have been performed with C_4 species, which can show either greatly enhanced growth rates (Soares et al., 2007) or little effect on productivity or yield (Leakey et al., 2006b). Water use efficiency is increased in both C_3 and C_4 species at elevated CO_2, a feature that has the potential to lower water requirements and thus irrigation of crops (Conley et al., 2001; Wullschleger et al., 2002; Wu et al., 2004).

The general consensus of opinion based on modelling climate change is that global food supply may show little change as a result of elevated atmospheric CO_2 levels. This positive view of future food security is possible because the direct effects of rising CO_2 levels in enhancing photosynthesis while decreasing photorespiration and transpiration will stimulate yield in all C_3 crops (Long et al., 2005). Furthermore, no negative effects of elevated atmospheric CO_2 levels have been observed in C_4 crops, which benefit from only higher growth temperatures and rising CO_2. However, a major accompanying reaction that will tend to oppose the beneficial effects of high-CO_2 species is an increase in global temperatures. The positive effects of elevated CO_2 on photosynthesis and plant productivity may be offset by the predicted increases in temperature, particularly in C_3 plants (Kandel, 1981).

While some predictions of plant responses to increasing atmospheric CO_2 levels are less dramatic than others, present models that predict the impacts of global change on plant productivity remain imperfect and lack accuracy. In this chapter, we will review the state-of-the-art and current concepts regarding plant responses to elevated CO_2 particularly in terms of whole-plant morphology, molecular physiology and development.

6.1.1 Effects of elevated CO_2 on plant growth

Effects of elevated CO_2 on shoot growth

From a historical perspective, plant scientists have long been interested in the effects of elevated CO_2 on plant growth, and bibliographic reports on this topic have appeared for well over 100 years, with the earliest studies referenced in 1902. A meta-analysis of 460 observations on 37 different species conducted by Kimball (1983) revealed that a doubling of the atmospheric CO_2 concentration of the earth would probably increase crop yields by 33%. Many of the early studies reporting beneficial effects of elevated CO_2 on plant growth were undertaken largely in the glasshouse or under controlled environmental conditions (Witter and Robb, 1964). Such observations resulted in the widespread introduction and use of elevated CO_2 to stimulate the growth and production of glasshouse crops such as tomatoes.

Wheat yields were enhanced by as much as 31% when ambient air CO_2 concentrations were increased from 350 to 700 ppm under optimal growth conditions (Amthor, 2001). While the effects of elevated CO_2 on wheat yields were somewhat smaller when mineral nutrient availability was limited, they were stimulated by elevated CO_2 compared with those of air-grown controls under conditions of drought stress (Amthor, 2001). However, a modest increase in temperature of between 1 and 4°C tended to counteract the positive effects of high CO$_2$ on wheat yields (Amthor, 2001). Such comprehensive studies of the effects of elevated CO$_2$ in relation to other environmental variables illustrate the inherent problems associated with the accurate prediction of the effects of high CO$_2$ on the yield of crops such as wheat (Amthor, 2001). Similar complexities in the prediction of effects on yield are also apparent for other C$_3$ species (Poorter *et al.*, 1997). Far fewer data are available on the effects of high CO$_2$ on the yield of C$_4$ plants, and they are largely derived from studies on maize, where few or no increases in plant growth in response to elevated CO$_2$ have been observed under field conditions. Early studies suggested that maize plants grew taller under elevated CO$_2$ (Ford and Thorne, 1967). We initially reported a similar effect of high CO$_2$ on the height of maize plants (Driscoll *et al.*, 2006), but later studies have revealed that this is not a consistent trend (Prins *et al.*, 2011).

An analysis of data collected from 156 species revealed that the high CO$_2$-dependent stimulation of growth was larger in C$_3$ compared with C$_4$ plants, even though a stimulation of growth was also observed in some C$_4$ species (Poorter, 1993). As maize is a major crop species, projections regarding future global food security also have to consider available data from this C$_4$ species, where photosynthesis, productivity and yield have been shown to be largely unaffected by high CO$_2$ atmospheric concentrations in FACE experiments conducted under optimum conditions of irrigation (Leakey *et al.*, 2006b). However, it must be noted that maize leaves show acclimatization to high CO$_2$, for example in terms of parameters such as stomatal densities and leaf chlorophyll and protein accumulation (Driscoll *et al.*, 2006). Moreover, photosynthetic rates were transiently stimulated under high CO$_2$ during periods of drought (Leakey *et al.*, 2004), an effect that appears to be related to the improved water relations under elevated CO$_2$. Hence, maize plants can achieve improved growth and photosynthesis compared with plants grown at air levels of CO$_2$ under conditions of water deficit (Leakey *et al.*, 2006b).

The responses of C$_4$ photosynthesis to drought stress are complex, with recent evidence supporting roles for both stomatal and metabolic limitations (Ghannoum, 2009). It is proposed that a decline in leaf water status has three successive effects on C$_4$ photosynthesis (Ghannoum, 2009). The initial stage is characterized by a decline in stomatal conductance, which may or may not be accompanied by reduced rates of CO$_2$ assimilation. This is possible because the CO$_2$-concentrating mechanism of C$_4$ plants provides saturating levels of CO$_2$ for photosynthetic CO$_2$ assimilation even at relatively low C$_i$. As leaf water status drops further, photosynthesis is regulated by a combination of stomatal and non-stomatal factors. In the final stage, a decline in photosynthesis occurs at very low leaf water status, and this is mainly caused by non-stomatal factors (Ghannoum, 2009).

Effects of elevated CO$_2$ on root growth

Studies on the effects of elevated CO$_2$ on root growth are not as abundant as those on shoot growth and yield, although a few examples can be found in the literature. Almost without exception the growth of the roots of crop plants is enhanced in CO$_2$-enriched atmospheres (Rogers *et al.*, 1994). For example, Stulen and den Hertog (1993) observed that root length was increased in plants under elevated CO$_2$ and it was suggested that carbon allocation to the roots was increased under those conditions (Hungate *et al.*, 1997). In another example, the most pronounced effect of elevated CO$_2$ in soybean plants was the increase in root length and root biomass measured as a dry weight increase (Rogers *et al.*, 1992). Moreover, under glasshouse and controlled environment conditions, roots generally exhibit a far greater growth enhancement than shoots (Witter, 1978; Norby *et al.*, 1992; Rogers *et al.*, 1992; Kimball *et al.*, 2002), a trend that has also been confirmed in field studies (Prior *et al.*, 1993; Wechsung *et al.*, 1999). These observations are explained by the need to balance whole-plant carbon/nitrogen ratios in situations where shoot carbon gain is considerably increased by growth at high CO_2. Enhanced root growth may be required to increase nutrient uptake and balance the whole carbon/nitrogen budgets. The metabolic and morphological adjustments would tend to favour lower shoot:root ratios in plants grown with CO_2 enrichment. However, a comprehensive analysis of literature data has not provided support to a general effect of elevated CO_2 on root:shoot ratio (Norby, 1994; Ainsworth et al., 2002). When differences in root:shoot ratio have been noted between elevated CO_2 and air treatments (Rogers et al., 1996), they are often associated with other effects of elevated CO_2 on plant ontogeny. Plants grow faster under elevated CO_2 and root:shoot ratios change during plant development, and thus some comparisons may be subject to error because plants can be at different developmental stages when grown under air and high CO_2.

There are relatively few studies of the effects of elevated CO_2 on dry matter allocation (Norby, 1994). In C_4 species such as sorghum (Chaudhuri et al., 1986; Pritchard et al., 2006) and maize (Whipps, 1985) root growth and length were increased under elevated CO_2. Taken together, this evidence suggests that, even if shoot growth, shoot:root ratios and crop yields are not greatly changed by growth under elevated CO_2, the plants benefit in terms of stronger root systems, which may provide better anchorage and enhanced capacity to explore deeper soil profiles for water and nutrients.

6.2 Elevated CO₂-dependent Effects on Photosynthesis

There is a general consensus of opinion that growth with CO_2 has stimulating effects on photosynthesis because of decreased photorespiration, particularly in C_3 plants. A recent meta-analysis of data on the responses of photosynthesis, canopy properties and plant production to rising atmospheric CO_2 in FACE experiments revealed that light-saturated carbon uptake rates, diurnal carbon assimilation, growth and above-ground production were all increased at higher CO_2 levels, while specific leaf area and stomatal conductance were decreased in this condition (Ainsworth and Long, 2005). This analysis largely confirmed results from previous chamber experiments and a meta-analysis of literature data by Kimball (1983). Studies documenting the long-term effects of high CO_2 on photosynthesis (Idso and Kimball, 1992; Gunderson et al., 1993; Davey et al., 1999; Ainsworth et al., 2003) have revealed a steady stimulation of net photosynthesis and dark respiration rates.

Photorespiration may be defined as the light-dependent release of CO_2 that is sensitive to atmospheric $CO_2:O_2$ ratios and that drives the synthesis and metabolism of compounds through the glycolate pathway (Wingler et al., 2000). The photorespiratory pathway is initiated with the oxygenation of ribulose-1, 5-bisphosphate by rubisco, a reaction that produces phophoglycolate, which is rapidly hydrolysed by a phosphatase in the chloroplasts to glycolate, which is then exported to the peroxisomes for further metabolism (Foyer et al., 2009). Carbon dioxide and oxygen are competitive substrates for rubisco, and the relative partial pressures of these gases in the atmosphere affects the relative rates of ribulose-1, 5-bisphosphate carboxylation and oxygenation (Farquhar et al., 1980). It is generally considered that the most significant regulator of flux-through between photosynthesis and photorespiration is the relative concentrations of CO_2 and O_2 at the rubisco-active site (Jordan and Ogren, 1984). Photorespiration is thus negatively affected by an increase in atmospheric $CO_2:O_2$ ratios and so photorespiratory CO_2 loss is diminished when plants are grown at high CO_2 (Bowes, 1991). Increases in atmospheric CO_2 concentrations will thus favour reduced carbon loss and will improve photosynthetic carbon assimilation, which essentially becomes more efficient at high CO_2 in terms of carbon gain and carbohydrate production, particularly in C_3 plants. While the photorespiratory pathway operates in C_4 plants (Foyer et al., 2009), the overall capacity for photorespiration to act as an alternative electron sink for the protection of the photochemical apparatus during stress is limited compared with C$_3$ photosynthesis (Ghannoum, 2009). Limitations on the capacity for photorespiration may explain why C$_4$ photosynthesis can show similar sensitivities to stress to C$_3$ photosynthesis (Ghannoum, 2009).

Short-term exposures to elevated CO$_2$ cause an immediate increase in photosynthesis and plant biomass production in C$_3$ species (Kramer, 1981; Cure and Acock, 1986). However, this is not always the case and overall photosynthetic capacity can be decreased in plants after long-term exposures to elevated CO$_2$, a process that is known as 'acclimatization' (Paul and Foyer, 2001). Photosynthetic acclimatization is caused by the responses of metabolism and gene expression to the increases in photosynthetic rates that occur as a result of the initial exposure to elevated CO$_2$. This phenomenon is generally observed in the days and weeks following the initial point of exposure to elevated CO$_2$ or lowered O$_2$ concentrations (Paul and Foyer, 2001). As discussed below in detail, the stimulatory effects of elevated CO$_2$ are considered to be offset by adjustments in metabolic regulation and gene expression that attenuate CO$_2$-dependent increases in carbon gain in line with the acquisition of essential nutrients, particularly nitrogen and phosphate, and the overall requirements for plant growth and development (Raper and Peedin, 1978; Kramer, 1981; DeLucia et al., 1985).

Several hypotheses have been put forward to explain the process of acclimatization to high CO$_2$. The first hypothesis concerns 'sink limitation' of photosynthesis (Paul and Foyer, 2001). It has been proposed that plants grown under CO$_2$ enrichment become enriched in carbohydrate and that they do not have sufficient capacity to make and store fixed carbon as sucrose, starch and other carbohydrates. They thus develop a 'carbohydrate sink limitation' where the enhanced rates of photosynthesis under CO$_2$ enrichment exceed their capacity for carbohydrate production and storage. In such circumstances, photosynthesis is down-regulated, mainly through decreased

carbon fixation capacity attained by decreasing the levels of rubisco protein or its activation state, in order to restore the balance between carbon supply and demand/storage. However, this type of photosynthetic acclimatization may not necessarily be a selective response to CO$_2$ enrichment, because enhanced photosynthesis and carbon acquisition rates place extra demands on the uptake of other essential nutrients, particularly nitrogen and phosphorus. Rubisco is the major leaf protein, comprising about 50% of the total protein of photosynthetic cells. Hence, the decreased protein rubisco levels that are often observed after long periods of CO$_2$ enrichment may reflect adjustments in the whole-plant carbon–nitrogen balance. Total protein and the accumulation of nitrogen compounds can be much lower in plants grown under CO$_2$ enrichment.

The second hypothesis is related to the first but it is defined by 'progressive nitrogen limitation'. In this hypothesis, it is considered that, under CO$_2$ enrichment, nitrogen uptake from soils fails to keep pace with photosynthesis and carbon acquisition. Thus, carbon:nitrogen ratios increase and there is a progressive nitrogen deficit, particularly in the leaves. These changes in plant metabolism impact on the soil and on litter quality, which tends to decline under CO$_2$ enrichment. A decline in litter quality has several effects, including (i) increased microbial immobilization because of high carbon:nitrogen ratios; (ii) decreased soil nitrogen availability; (iii) nitrogen limitation to the plant; and (iv) lower photosynthesis rates. Yet another hypothesis concerning CO$_2$ acclimatization suggests that elevated CO$_2$ inhibits shoot nitrate assimilation in C$_3$ plants but not in C$_4$ plants, because nitrate assimilation in C$_3$ plants depends on photorespiration. Photorespiration rates are decreased under elevated CO$_2$ and thus C$_3$ plants that rely on nitrate as the sole nitrogen source will tend to suffer from nitrogen deprivation under such concentrations (Foyer et al., 2009). The relative availability of nitrate and ammonium in soils varies considerably between locations and with time (seasons, years). Such variations may account for observed differences in the responses of C_3 plants to CO_2 enrichment and nitrogen fertilization regimes.

The acclimatization of CO_2 assimilation caused by prolonged periods of exposure to elevated CO_2 is frequently associated with the increased abundance of starch and soluble carbohydrate concentrations in source leaves and a decreased abundance of transcripts encoding several photosynthetic enzymes (Delucia et al., 1985; Sasek et al., 1985; Peet et al., 1986; Sage et al., 1988; Yelle et al., 1989a, b; Stitt, 1991). The alterations in gene expression patterns in response to elevated CO_2 is often attributed to altered sugar signalling arising from the carbohydrate-rich state of the leaves which, in turn, alters hormone signalling pathways that interact with sugar signalling to control plant growth and development (Prins et al., 2009). Acclimatization of photosynthesis is rapidly observed when plants are supplied with exogenous sugars, for example by direct supply to leaves through the transpiration stream (Krapp et al., 1991).

The initial CO_2-dependent increases in photosynthetic observed in C_3 species are followed by a decline in leaf rubisco protein and activity (Sage et al., 1988; Chen et al., 2005; Rowland-Bamford et al., 2006). An increase in photosynthesis was observed in barley and wheat plants grown in the field under high (700 ppm) CO_2 (Sicher and Bunce, 1997). In this study a concomitant decrease in the initial activity of rubisco was observed (Sicher and Bunce, 1997). Similarly, decreases in apparent rubisco activity were observed in Agrostis capillaris, Lolium perenne and Trifolium repens plants grown for 2 years in open-top chambers with elevated CO_2 (Davey et al., 1999). Moreover, these CO_2-dependent decreases in rubisco activity were associated with reduction in the total leaf nitrogen content on a unit leaf area basis (Davey et al., 1999). This study highlights the effect of elevated CO_2 on leaf photosynthetic nitrogen use efficiency, i.e. the rate of carbon assimilation per unit leaf nitrogen (Davey et al., 1999). This parameter is important in considerations of acclimatization of photo-

synthesis to elevated CO_2, because high rates of photosynthesis are attained at lower leaf nitrogen contents. The stimulatory effects of elevated CO_2 on photosynthesis are much less marked in C_4 plants such as Amaranthus, maize, Sorghum bicolor and Paspalum dilatatum (Maroco et al., 1999; Ziska and Bunce, 1999; Driscoll et al., 2006; Soares et al., 2007) than in C_3 plants. Frequently, little or no stimulation of CO_2 assimilation rates is observed (Wand et al., 2001; Leakey et al., 2006b).

Similarly, the leaves of C_4 plants do not show a CO_2-dependent decline in photosynthesis rates after prolonged growth at elevated CO_2. No significant effects of elevated CO_2 on the photosynthesis or activities of key photosynthetic enzymes such as phosphoenolpyruvate (PEP) carboxylase (PEPc), pyruvate orthophosphate dikinase (PPDK) and rubisco were found in maize plants over the growing season (Leakey et al., 2006b). However, a decrease in photosynthesis was observed in sorghum leaves, together with a reduction in phosphoenolpyruvate carboxylase but not rubisco content, when the plants were grown under elevated CO_2 in controlled environment chambers (Watling et al., 2000). Later studies on sorghum conducted under FACE conditions revealed no effect of elevated CO_2 on the cell-specific localization of rubisco or PEPc at any stage of leaf development (Cousins et al., 2003). While the relative ratios of rubisco to PEPc remained constant during leaf development under conditions of high CO_2, the oldest tissues in the leaf blades showed reduced total activities of rubisco and PEPc (Cousins et al., 2003). Although these data are inconclusive, they indicate that the leaves of C_4 plants can show acclimatization of photosynthesis to high CO_2 (Cousins et al., 2003). Moreover, acclimatization to elevated CO_2 is observed in terms of stomatal patterning and protein content and composition (Driscoll et al., 2006; Soares et al., 2007). In addition, growth at high CO_2 results in specific changes to the leaf transcriptome and metabolite profiles that reveal altered hormone signalling and decreased oxidative stress (Prins et al., 2011).

In summary, growth under elevated

atmospheric CO$_2$ levels favours higher photosynthetic rates, particularly in C$_3$ species. In some cases, this CO$_2$-dependent stimulation of photosynthesis can be attenuated over the longer term because of 'acclimatization processes', particularly in conditions of sink limitations or low nutrient availability. Long-term exposure to elevated CO$_2$ can have the following consequences: (i) 'carbohydrate sink limitation', in which enhanced photosynthesis exceeds the capacity for carbohydrate synthesis or incorporation in sinks. This leads to a down-regulation of carbon fixation through decreased rubisco levels that restores the balance between carbon production and carbon utilization within the plant; and (ii) a progressive decrease in tissue nitrogen and/or protein contents can develop during long-term exposure to elevated CO$_2$. Such decreases are associated with a lower rubisco content and/or activity, together with a concomitant decrease in overall photosynthetic capacity. However, the leaves become more nitrogen use efficient and can operate high rates of photosynthesis at lower leaf N requirements.

6.3 Elevated CO$_2$ Effects on Stomatal Closure and Patterning

The closure of the stomatal pores on leaves is regulated by atmospheric CO$_2$ concentrations, opening when CO$_2$ is low and closing at high CO$_2$. A doubling of the atmospheric CO$_2$ will thus cause significant stomatal closure, with predicted values of between 20 and 40% in different plant species. Stomata also open in response to light, with rapid and large shifts in the CO$_2$ concentration of the stomatal cavity caused by illumination ranging from 200 to 650 ppm. Little is known about the CO$_2$ signal transduction mechanisms that function upstream of ion channels in guard cells.

Decreases in stomatal conductance have been observed in a wide range of diverse C$_3$ and C$_4$ species grown under elevated CO$_2$ in individual experiments and in FACE trials (Ainsworth and Rogers, 2007). Moreover, the relationships between stomatal conductance, photosynthesis, humidity and atmospheric CO$_2$ availability at the leaf surface are essentially unaltered by long-term growth under elevated CO$_2$ (Leakey et al., 2006a). The observed high CO$_2$-dependent decreases in stomatal conductance are entirely due to the direct effect of CO$_2$ concentration on the stomatal aperture, and there appears to be no long-term acclimatization of the stomatal conductance process that is independent of photosynthetic acclimatization (Leakey et al., 2006a). However, there is now a large body of evidence documenting the effects of high CO$_2$ on stomatal density and patterning, which may be the result of long-distance signals (Beerling and Woodward, 1995; Hetherington and Woodward, 2003). For example, in general stomatal densities have declined since the pre-industrial era and thus stomatal numbers and patterning have acclimatized to changes in the atmospheric CO$_2$ environment that have occurred across geological timescales (Beerling and Woodward, 1995; Hetherington and Woodward, 2003). The effect of high atmospheric CO$_2$ levels on stomatal densities has also been observed under controlled environment conditions (Oberbauer et al., 1985; Woodward, 1987; Woodward et al., 2002). The acclimatization process that controls stomatal density and stomatal function involves long-distance signalling of information concerning CO$_2$ concentration from mature to developing leaves (Lake et al., 2001, 2002; Woodward et al., 2002). Unfortunately, relatively few components of the CO$_2$ signalling pathways have been identified to date. A high-carbon dioxide gene, which encodes a putative 3-keto acyl coenzyme A synthase, was identified in Arabidopsis and shown to be a negative regulator of stomatal density (Gray et al., 2000). While CO$_2$ signalling in plants remains poorly understood, it is possible that the CO$_2$-dependent changes in calcium signalling that regulate stomatal closure may also participate in the longer-term responses that lead to regulation of stomatal density, sizes and patterning on leaves.

The signals that lead to stomatal closure in response to high CO$_2$ have been intensively studied for many years. An early hypothesis that was formulated by Raschke (1977) to explain the action of CO_2 on stomata suggested that a finite amount of CO_2 was needed for the production of malate in the guard cells that control the stomatal aperture. Malate is an important source of the protons that drive the guard cell plasmalemma proton pump, and it is also utilized as a counter-ion for the potassium that is taken up during stomatal opening. Malate is also an important cellular osmoticum per se. Under CO_2 concentrations over the order of 100 µmol/mol, malate production exceeds demand in the guard cells and the pH of the cytoplasm declines. This allows an increased efflux of ions from the guard cells, with a concomitant reduction in stomatal aperture. In the Raschke (1977) hypothesis it was considered that, below a critical atmospheric CO_2 concentration (and hence intercellular CO_2 concentration or C_i), insufficient malate was produced to fully open the stomata, and thus the aperture was less than maximal.

A second hypothesis that has been used to explain the action of CO_2 on stomata considers that CO_2 may act directly upon the H^+ pump of the guard cell plasma membrane (Edwards and Bowling, 1985). However, a detailed analysis of the changes in guard cell membrane potential and conductance suggested that this type of mechanism was not operating (Blatt, 1987). Rather, CO_2 appeared to exert its effects via changes in the pH of the guard cell/apoplastic cell wall space (Blatt, 1987). Consequently, several interacting mechanisms, namely malate production, oxidative photophosphorylation, guard cell apoplast pH and the guard cell plasmalemma proton pumps, can interact in controlling stomatal aperture. In addition, there are calcium-sensitive and calcium-insensitive phases in the response of guard cells to variations in CO_2 concentrations (Young et al., 2006). Some studies have proposed that the signal that controls stomatal aperture in relation to varying atmospheric CO_2 concentrations is not related to changes in mesophyll or guard cell photosynthesis (von Caemmerer et al., 2004). However, more recent studies have shown

that part of the stomatal response to C_i involves the balance between photosynthetic electron transport and carbon reduction, in either the mesophyll or guard cell chloroplasts (Messinger et al., 2006).

High CO_2 causes depolarization of the guard cells. Elevated CO_2 enhances the potassium efflux channel and S-type anion channel activities that mediate extrusion of ions during stomatal closure. Moreover, chloride release (Hanstein and Felle, 2002) from guard cells is triggered by increases in CO_2. CO_2 activation of R-type anion channel currents has also been reported (Raschke et al., 2003; Young et al., 2006 and references therein).

In summary, stomatal closure is a rapid response to elevated CO_2. The signal that modulates the closure of the stomata at high CO_2 is now considered not to be primarily photosynthetic in origin, but much more work is required to fully elucidate the nature of the signal transduction network that controls this process. Growth with elevated CO_2 has immediate effects on stomatal conductance, which decreases as a result of guard cell action to close the stomatal aperture at high CO_2. Other, perhaps interacting, signalling systems modulate stomata density in response to long-term growth at high CO_2. This type of control of stomatal patterning involves long-distance signals that are systemic and arise from mature leaves to control stomatal densities in developing leaves.

6.4 Regulation of Respiration in Response to Elevated CO_2

The balance between carbon gain through photosynthesis and carbon loss through respiration determines the magnitude and direction of net carbon flux throughout the plant. These factors also influence the impact of elevated CO_2 on water use efficiency. On average, a doubling in the atmospheric CO_2 concentration is considered to cause a rapid decrease in leaf dark respiration rates, measured by CO_2 efflux. Some authors consider that such findings are artefactual and related to the way that respiration is often measured. No CO$_2$-dependent decreases in respiratory O$_2$ uptake related to increased CO$_2$ concentrations were detected using a high-resolution, dual-channel oxygen analyser in an open gas exchange system (Davey *et al.*, 2004). In 600 separate measurements this system failed to reveal any effect of CO$_2$ concentration on leaf respiration rates, suggesting that previous observations of a CO$_2$-dependent inhibition of respiration were probably experimental artefacts (Davey *et al.*, 2004). Hence, dark respiration in leaves probably does not decrease rapidly as a result of elevated CO$_2$ concentrations. However, more recently under field conditions, respiration at night in a soybean crop grown under elevated CO$_2$ concentrations was enhanced by 37% (Leakey *et al.*, 2009). This increase was linked to an enhanced abundance of transcripts encoding components of the respiratory pathway (Leakey *et al.*, 2009). It has long been known that the number of mitochondria is increased in the leaves of plants grown at high CO$_2$. The greater respiratory quotient is probably linked to the enhanced leaf carbohydrate contents that occur in plants grown under elevated CO$_2$ (Leakey *et al.*, 2009). High leaf carbohydrate levels stimulate respiration, a response that is consistent across many species. The stimulation of basal maintenance respiration rates may serve to offset carbon gain and thus limit the increases in net primary productivity that might be predicted in response to future increases in atmospheric CO$_2$ availability. Enhanced basal respiration rates at elevated CO$_2$ are probably part of the acclimatizatory mechanisms that plants induce to restore overall carbon balance under conditions of high carbon gain. However, high respiration rates in source and sink tissues contribute to high crop yields through enhanced photo-assimilate export to sink tissues and high sink activities (Leakey *et al.*, 2009).

6.5 Effects of Elevated CO$_2$ on Plant C–N Interactions and Nutrient Status

As discussed above, growth under elevated CO$_2$ has a strong effect on whole-plant nitrogen metabolism and C–N interactions. Such effects have major implications for agriculture – and also for human and animal nutrition. For example, the protein content of cereal grains may decrease and patterns of nitrogen acquisition from the soil may be altered with the absorption of different N sources under elevated CO$_2$. The effects of elevated CO$_2$ on food nutrient biology are complex and poorly understood. For example, growth under CO$_2$ enrichment caused a net decrease in all essential elements except phosphorus and iron in wheat leaves, while nitrogen, calcium, sulfur and iron were decreased in the grains (Fangmeier *et al.*, 1999). While the total nitrogen uptake of the wheat crop was unaffected by the CO$_2$ growth conditions, the N content of grains and straw was significantly lower in plants grown under elevated atmospheric CO$_2$ (Fangmeier *et al.*, 2000). Thus, cereal grain and flour quality could be decreased in nutrient value under high CO$_2$ growth conditions (Fangmeier *et al.*, 1999). In rice plants leaf area index was unaffected by high CO$_2$ (Anten *et al.*, 2004). Other studies have suggested that these effects could be offset by the provision of ample nitrogen fertilizer (Wall *et al.*, 2000; Kimball *et al.*, 2001; Asseng *et al.*, 2004).

Grain protein levels are adversely affected by high CO$_2$ growth conditions (Fangmeier *et al.*, 1999; Wieser *et al.*, 2008). In addition to the effects on total wheat grain N, lysine contents and total protein levels, growth under high CO$_2$ also tended to deplete other mineral nutrients (N, P, K and Zn) in the grains. If growth under high CO$_2$ causes a general increase in the ratio of C to essential elements, then this would intensify the already acute problems of micronutrient malnutrition in some parts of the world (Loladze, 2002). An analysis of data from 228 independent experiments in crops such as barley, wheat, rice, soybean and potato revealed that overall protein levels were lower when plants were grown under high CO$_2$ compared with ambient atmospheric CO$_2$ growth conditions (Taub *et al.*, 2008). In contrast, high CO$_2$-dependent effects on tissue N concentrations showed inter-species variations, with greater effects on C$_3$

compared with C_4 plants and legumes with symbiotic N_2 fixation (Cotrufo *et al.*, 2002). Moreover, the form of N available (as either ammonium or nitrate) has an effect on the responses of plants, such as wheat to high CO_2. When nitrate was used as the sole N source, biomass production in wheat in terms of shoot, stem, root and leaves was similar at high CO_2 and ambient atmospheric CO_2 growth conditions (Bloom *et al.*, 2002). However, when ammonium was provided as the preferred N source, biomass production of all organs (shoot, root, stem, leaf area) was increased under high CO_2 (Bloom *et al.*, 2002). This occurs because reductant cycling in the photorespiratory pathway is an essential driver of primary NO_3 photo-assimilation, and this pathway is depleted of essential reducing power when photorespiration is inhibited under high CO_2 (Bloom *et al.*, 2002). Such results strongly indicate that the major impacts of high CO_2 depend on the source of nitrogen available to plants.

In summary, grain quality in wheat and other C_3 cereals might be adversely affected in a future high-CO_2 world, particularly if CO_2 enrichment is accompanied by increases in global temperature. However, any potential negative impacts of elevated CO_2 might be mitigated, at least in part, by a change in agricultural practices that incorporates utilization of different N sources.

6.6 Interactions of Elevated CO_2 with Other Stresses

Models that aim to predict plant responses to elevated CO_2 have to take into account other features of climate change such as drought, temperature rise and the presence of other greenhouse gases that may have a negative impact on plant productivity. Modelling studies using data obtained from both C_3 (wheat and soybean) and C_4 (maize) species grown in regions of the central USA have shown that in summer an unlimited increase in atmospheric CO_2 did not have desirable effects, even when the positive effects of CO_2 were taken into account

(Okamoto *et al.*, 1991). Other authors have predicted that moderate increases in the temperature of the earth may enhance the production of some major arable grain and legume crops in the temperate zones, but not in the tropics (Long *et al.*, 2005). Overall, the general consensus of opinion is that global food supply may show little change as a result of elevated atmospheric CO_2 levels. This positive view of future food security is possible because the direct effects of rising CO_2 levels in enhancing photosynthesis, while decreasing photorespiration and transpiration, will stimulate yield in all C_3 crops (Long *et al.*, 2005). Furthermore, no negative effects of elevated atmospheric CO_2 levels have been observed in C_4 crops, which benefit only from higher growth temperatures and rising CO_2 levels.

6.6.1 Interactions between elevated CO_2 and drought stress

The effects of water stress are attenuated by elevated CO_2 in both C_3 and C_4 plants because stomatal closure is favoured at high CO_2 (Wullschleger *et al.*, 2002; Wu *et al.*, 2004). For example, well-watered wheat plants exhibited a grain weight increase of 14% under CO_2 enrichment relative to controls under ambient atmospheric CO_2 conditions, when grown in FACE field experiments. However, the relative increase was much higher (24%) in wheat plants subjected to CO_2 enrichment and water stress (Li *et al.*, 2000). The wheat plants were able to perform better under water stress at elevated CO_2 because the leaves could maintain a higher (less negative) leaf water potential compared with plants under ambient atmospheric CO_2 conditions (Wall, 2001). Similar results have also been obtained in studies on plants grown in pots (Schutz and Fangmeier, 2001). A simulation model using field data from wheat FACE experiments showed that response was complex, the high-CO_2 effect being greater under drought because of lower transpiration rates and higher root biomass, together with the non-linear functional dependence of net assimilation rate on leaf internal CO_2

concentration (Grossman-Clarke *et al.*, 2001). Moreover, the relative contributions of each of these different mechanisms changed in significance over the growing season, and these effects were dependent on the degree of soil water limitation (Grossman-Clarke *et al.*, 2001). The effect of high CO$_2$ on water use efficiency (WUE, defined as both the ratio of net CO$_2$ assimilation to transpiration and the ratio of biomass produced to water used) is therefore a key feature of the impact of variations in atmospheric CO$_2$ on the ability of plants to tolerate water deficits. When photosynthesis increases in C$_3$ plants in response to elevated CO$_2$ concentrations, stomatal conductance values decrease. As a consequence WUE is increased.

The optimization theory of transpiration and assimilation was developed in order to explain stomatal responses to vapour pressure deficits (Farquhar *et al.*, 1980; Eamus and Shanahan, 2002). This theory considers that stomatal behaviour is optimal when water loss is minimal for a given amount of carbon assimilated, over a period of time (Farquhar *et al.*, 1980). While stomatal behaviour appears to be optimized with respect to water loss and carbon gain (Farquhar *et al.*, 1980), not all WUE increases can be attributed to a decline in stomatal conductance because increases in WUE exceed decreases in transpiration. Hence, at least part of increased WUE values observed at elevated CO$_2$ can be attributed to enhanced photosynthesis and growth (Morison and Gifford, 1984).

In C$_4$ crops, where exposure to high CO$_2$ has no direct effect on photosynthesis, studies in FACE experiments have revealed that drought tolerance is also enhanced by the reduced stomatal conductance that occurs at elevated CO$_2$ concentrations (Conley *et al.*, 2001). When the high CO$_2$-dependent responses of two tropical maize hybrids to moderate and severe water stress were compared (Fig. 6.1), it was found that both genotypes exhibited better biomass production on both fresh weight and dry weight basis under elevated CO$_2$ concentrations compared with ambient air levels of CO$_2$. Moreover, the degree of leaf

wilting caused by water stress (particularly at SS) was greatest under ambient air CO$_2$ concentrations and occurred well in advance of wilting symptoms in plants grown under CO$_2$ enrichment (Fig. 6.1).

In agreement with these results, biomass production was increased in water-stressed maize plants grown under elevated CO$_2$ concentrations relative to ambient CO$_2$ environment (Samarakoon and Gifford, 1996). Moreover, similar results have been reported in sorghum in FACE studies (Ottman *et al.*, 2001). The better performance of maize under drought under high CO$_2$ may be related to several factors, including enhanced root proliferation providing a better root architecture for whole-plant water uptake, enhanced WUE, lower stomatal conductance rates together with impacts on leaf water potential, accumulation of solutes, osmotic adjustment

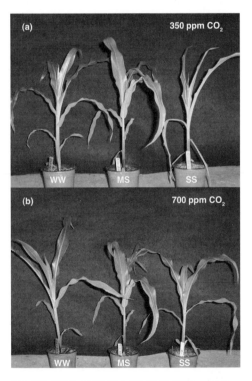

Fig. 6.1. Water depletion effects in maize plants grown under (a) 350 µl/l and (b) 700 µl/l CO$_2$ in different water regimes (WW, well watered; MS, moderate stress; SS, severe stress) during a 7-day period (Lopes and Foyer, unpublished results).

and leaf dehydration tolerance (Wullschleger *et al.*, 2002). Growth under high CO_2 favours carbon allocation to roots, enabling enhanced growth and osmotic adjustment. This may ameliorate the negative impacts of water stress by improving the capacity to extract soil water. High CO_2-induced reductions in stomatal conductance may also ameliorate drought tolerance by increasing leaf and whole-plant WUE values, enabling a better exploitation of the limited amounts of water in the environment. In summary, the reductions in stomatal conductance and transpiration rates that result from growth under high CO_2 favour improved leaf water potentials when plants are subjected to water stress. Higher photosynthetic rates and associated carbohydrate production and transport to growing tissues can be sustained for longer periods at high CO_2. Thus, the negative effects of a restricted carbohydrate supply on growth and yield may therefore be minimized under high CO_2. Such responses are well documented, but the precise underlying mechanisms remain to be fully elucidated. Moreover, the signals that lead to CO_2-dependent stomatal closure and alterations in stomatal conductance are largely unknown. A more precise understanding of the pathways of CO_2 signalling with regard to plant morphology and function, as well as photosynthesis and growth, is required in order to use these pathways to advantage in future plant breeding programmes, particularly with regard to enhancing drought tolerance.

6.6.2 Interactions between elevated CO_2 and heat stress

Anthropogenic contributions to atmospheric CO_2 levels and other greenhouse gases are considered to be largely responsible for recent increases in global temperatures, which rose by 0.6°C from 1990 to 2000. Global temperatures are projected to increase by between a further 1.4°C and 5°C by 2100 (IPCC, 2007). While growth under elevated CO_2 can partially ameliorate the negative effects of drought in both C_3 and C_4

species, it is unlikely that high CO_2 will offer any protection from heat stress (Wang *et al.*, 2008). Rather, high CO_2 may exacerbate the negative effects of high temperatures. Plants grown under elevated CO_2 tend to exhibit higher leaf temperatures because transpiration rates are decreased and thus less latent heat is lost (Lloyd and Farquhar, 2008). Any analysis of the predicted effects of elevated CO_2 on thermo-tolerance thus has to take this feature into account. Photosynthesis may be more sensitive to acute heat stress under elevated CO_2. Photosynthesis is sensitive to inhibition by moderate heat stress, a phenomenon that is often attributed to inactivation of membrane-associated proteins, particularly the oxygen-evolving complex of photosystem (PS) II. Although the light reactions of photosynthesis are disrupted at very high temperatures, the photosynthetic electron transport system is much less sensitive to high temperatures than carbon assimilation. Thus, photosynthetic electron transport can function efficiently at temperatures that inhibit CO_2 fixation. One of most temperature-sensitive reactions of carbon assimilation is rubisco activase (Crafts-Brandner and Salvucci, 2000). Rubisco activase activity is exceptionally sensitive to thermal denaturation. Hence, inhibition of the activase at high temperatures prevents activation of rubisco in leaves suffering heat stress and this inhibition is responsible, at least in part, for deactivation of rubisco and temperature-dependent inhibition of CO_2 fixation (Salvucci *et al.*, 2001). It is possible that photosynthesis could benefit from high CO_2 levels in situations where the efficiency of excitation energy capture by open PSII reaction centres (Fv/Fm) could be increased (Faria *et al.*, 1996; Taub *et al.*, 2000). This type of enhanced PSII thermo-tolerance has been reported in both woody and herbaceous species (Faria *et al.*, 1996; Taub *et al.*, 2000). However, such effects do not protect CO_2 fixation or prevent the effects of heat stress on plant reproductive development (Prasad *et al.*, 2008).

High CO_2 concentrations can undoubtedly provide additional protection for the photosynthetic machinery under certain environmental stresses (Faria *et al.*, 1996; Taub *et al.*, 2000; Donnelly *et al.*, 2001). However, heat stress-dependent inhibition of CO$_2$ fixation may be an increasing problem in the future. Higher global temperatures may have a negative impact on plant growth and development, decreasing crop and ecosystem productivity (Ciais *et al.*, 2005) and possibly also biodiversity (Thomas *et al.*, 2004; Wang *et al.*, 2008). No beneficial interactions between CO$_2$ and temperature have been observed in kidney bean (Prasad *et al.*, 2002). Heat-induced problems with seed set and related decreases in productivity were similar under elevated CO$_2$ and ambient air CO$_2$ conditions (Prasad *et al.*, 2002). Similarly, the damaging effects of high temperature and UV-B irradiation on soybean pollen morphology, pollen production, germination and pollen tube lengths were not ameliorated by high CO$_2$ (Koti *et al.*, 2005). Heat stress exerts a strong negative effect on quality in most cereal grains (Corbellini *et al.*, 1998; Gibson and Paulsen, 1999; Qin *et al.*, 2008). This negative action of heat will probably be exacerbated by high CO$_2$, which as discussed above can also decrease grain quality.

6.6.3 Interactions between elevated CO$_2$ and ozone

Annual average global tropospheric ozone concentrations currently range from 20 to 45 ppb, values that are roughly double those that were present before the Industrial Revolution. Earth's ozone concentrations are predicted to continue increasing by 0.5–2.0% per year over the next century, mainly due to increases in precursor emissions from anthropogenic sources. Ozone diffuses into the leaf apoplast via the stomata, where it is rapidly converted into other reactive oxygen species (ROS) that signal a diverse metabolic response (Kangasjarvi *et al.*, 2005). The ozone-induced ROS burst is similar to that which occurs in the regulation of stomatal closure in response to abscisic acid (ABA). Mutants lacking the 'slow or S-type anion channel' such as 'slow anion channel 1' (SLAC1) cannot close their stomatal pores in response to ozone, CO$_2$ or ABA. Such observations suggest that there is extensive crosstalk between CO$_2$ signalling and ozone signalling with regard to stomatal closure and abiotic stress responses. While much regarding the signalling pathways that control the aperture of stomatal pores remains to be elucidated, it appears that ozone and high CO$_2$ affect the anion channel regulation of stomatal movement in a concerted manner.

Plants grown at high CO$_2$ often have decreased levels of antioxidant enzymes (Pritchard *et al.*, 2000), suggesting a decrease in perceived oxidative stress. In C$_3$ plants this may be linked to increased carboxylation rates and associated lower rates of photorespiration and ROS production (Bowes, 1991). Growth under high CO$_2$ restricts stomatal opening and the entry of air pollutants like ozone. Moreover, plants grown at high CO$_2$ are not impaired in their antioxidant responses and show a large stimulation of antioxidant metabolism in response to acute ozone exposures.

6.7 Conclusions and Perspectives

Plants are the foundations of the human food chain and thus it is crucial to understand the complex network of plant responses to enhanced atmospheric CO$_2$ concentrations in order to avoid any negative impacts that may adversely affect food production and quality. Current models of global climate predict a gradual increase in the atmospheric concentrations of greenhouse gases over the future and associated increases in global temperature (IPCC, 2007). However, the predictions of these changes in global climate on plant productivity are far less certain. Predictability of yield is crucial to the farmer, because much uncertainty remains concerning the effects of predicted increases in atmospheric CO$_2$ concentrations on crop production in different parts of the world. Thus, there is an urgent need for more and better data on plant responses to high CO$_2$ as well as other components of the prediction models. As discussed above, the predicted increases in atmospheric CO$_2$

concentrations can benefit plant growth and development. However, different current models concluded that this will have little impact on crop yields and they may even be decreased in some cases.

Major research efforts are currently involved in the evaluation of future climate change effects, particularly with regard to atmospheric CO_2 concentrations. In order to find suitable solutions to mitigate the negative effects of projected future environmental changes on food security, current plant breeding programmes for the selection of new varieties have to take the impacts of elevated CO_2 into account, particularly in major cereal crops such as wheat. Wheat is currently the major crop worldwide, with some 217 million ha producing approximately 620 million t of grain annually for the period 2004–2006 (FAO, 2007). Wheat provides about one-fifth of the total calorific input of the world's population (FAO, 2007). It is therefore crucial to understand how the nutritional value of cereal crops such as wheat will be affected by the predicted increases in atmospheric CO_2 levels. Currently, crop yields and plant biomass production are tightly linked to N fertilization because of the tight regulation of whole-plant C–N interactions. Studies comparing the NUE (nitrogen use efficiency) of large numbers of *Arabidopsis* genotypes have revealed that there is a large degree of genotypic variation in shoot biomass production in response to limiting N availability. The aspiration is that, once the genetic components and processes that control plant growth and biomass production have been defined more clearly in model species, then this information can be transferred to crop plants by marker-assisted selection coupled to classic plant breeding approaches. Thus, it is feasible to manipulate the C–N relationship of wheat and other major crops at the genetic level, particularly with regard to future increases in atmospheric CO_2. Thus, it is possible to mitigate the more negative effects of high atmospheric CO_2 on wheat quality and prevent any negative impacts on human nutrition. In addition, enhanced productivity may be achieved through improved stress tolerance.

There is considerable potential for enhancement of plant NUE, particularly under high CO_2, as well as the removal of current limitations on carbohydrate production capacity in leaves and sink tissues. Current knowledge of the genome–environment interaction and how growth and biomass production are influenced by the responses of the metabolic networks and C–N signalling pathways under high CO_2 can be used in current breeding programmes to prepare crop plants to face a future world that has an elevated CO_2 environment.

Our future world will suffer more abundant periods of drought as rainfall becomes less predictable. Thus, any high-CO_2-related increases in plant water use efficiency would be beneficial. Drought and heat are currently the main causes of yield losses wordwide. The effects of high CO_2 on stomatal closure may provide a means towards better water use efficiency in both C_3 and C_4 plants. Stomatal closure is a central mechanism in leaves that contributes to plant stress tolerance. The high-CO_2-dependent increase in stomatal closure may be useful as it will help to protect plants against the harmful effects of other greenhouse gases, which are predicted to increase in the atmosphere.

It is generally accepted that high temperatures have a severe negative effect on grain quality, and this will probably be exacerbated by elevated CO_2 levels. The predicted increases in the temperature of the earth, together with the increased likelihood of heat stress, may serve to reverse the positive effects of elevated CO_2 on photosynthesis and plant growth, particularly in C_3 species. Plant breeding strategies must therefore incorporate marker-assisted selection with genes that confer improved thermo-tolerance in order to address the growing problem of heat stress and its likely negative impacts on food security and grain quality.

References

Ainsworth, E.A. and Long, S.P. (2005) What have we learned from 15 years of free-air CO$_2$ enrichment (FACE)? A meta-analytic review of the responses of photosynthesis, canopy properties and plant production to rising CO$_2$. *New Phytologist* 165, 351–372.

Ainsworth, E.A. and Rogers, A. (2007) The response of photosynthesis and stomatal conductance to rising [CO$_2$]: mechanisms and environmental interactions. *Plant Cell and Environment* 30, 258–270.

Ainsworth, E.A., Davey, P.A., Bernacchi, C.J., Dermody, O.C., Heaton, E.A., Moore, D.J. et al. (2002) A meta-analysis of elevated [CO$_2$] effects on soybean (*Glycine max*) physiology, growth and yield. *Global Change Biology* 8, 695–709.

Ainsworth, E.A., Davey, P.A., Hymus, G.J., Osborne, C.P., Rogers, A., Blum, H. et al. (2003) Is stimulation of leaf photosynthesis by elevated carbon dioxide concentration maintained in the long term? A test with *Lolium perenne* grown for 10 years at two nitrogen fertilization levels under free air CO$_2$ enrichment (FACE). *Plant Cell and Environment* 26, 705–714.

Amthor, J.S. (2001) Effects of atmospheric CO$_2$ concentration on wheat yield: review of results from experiments using various approaches to control CO$_2$ concentration. *Field Crops Research* 73, 1–34.

Anten, N.P.R., Hirose, T., Onoda, Y., Kinugasa, T., Kim, H.Y., Okada, M. et al. (2004) Elevated CO$_2$ and nitrogen availability have interactive effects on canopy carbon gain in rice. *New Phytologist* 161, 459–471.

Asseng, S., Jamieson, P.D., Kimball, B., Pinter, P., Sayre, K., Bowden, J.W. et al. (2004) Simulated wheat growth affected by rising temperature, increased water deficit and elevated atmospheric CO$_2$. *Field Crops Research* 85, 85–102.

Beerling, D.J. and Woodward, F.I. (1995) Stomatal responses of variegated leaves to CO$_2$ enrichment. *Annals of Botany* 75, 507–511.

Blatt, M.R. (1987) Electrical characteristics of stomatal guard cells: the ionic basis of the membrane potential and the consequence of potassium chloride leakage from microelectrodes. *Planta* 170, 272–287.

Bloom, A.J., Smart, D.R., Nguyen, D.T. and Searles, P.S. (2002) Nitrogen assimilation and growth of wheat under elevated carbon dioxide. *Proceedings of the National Academy of Sciences USA* 99, 1730–1735.

Bowes, G. (1991) Growth at elevated CO$_2$: photosynthesis responses mediated through Rubisco. *Plant Cell and Environment* 14, 795–806.

Carter, D.R. and Peterson, K.M. (1983) Effects of a CO$_2$-enriched atmosphere on the growth and competitive interaction of a C3 and a C4 grass. *Oecologia* 58, 188–193.

Chaudhuri, U.N., Burnett, R.B., Kirkham, M.B. and Kanemasu, E.T. (1986) Effect of carbon dioxide on sorghum yield, root growth, and water use. *Agricultural and Forest Meteorology* 37, 109–122.

Chen, G.Y., Yong, Z.H., Liao, Y., Zhang, D.Y., Chen, Y., Zhang, H.B. et al. (2005) Photosynthetic acclimation in rice leaves to free-air CO$_2$ enrichment related to both Ribulose-1,5-bisphosphate carboxylation limitation and Ribulose-1,5-bisphosphate regeneration limitation. *Plant and Cell Physiology* 46, 1036–1045.

Ciais, P.H., Reichstein, M. and Viovy, N. (2005) Europe-wide reduction in primary productivity caused by the heat and drought in 2003. *Nature* 437, 529–533.

Conley, M.M., Kimball, B.A., Brooks, T.J., Pinter, P.J. Jr., Hunsaker, D.J., Wall, G.W. et al. (2001) CO$_2$ enrichment increases water use efficiency in sorghum. *New Phytologist* 151, 407–412.

Corbellini, M., Mazza, L., Ciaffi, M., Lafiandra, D. and Borghi, B. (1998) Effect of heat shock during grain filling on protein composition and technological quality of wheats. *Euphytica* 100, 147–154.

Cotrufo, M.F., Ineson, P. and Scott, A. (2002) Elevated CO$_2$ reduces the nitrogen concentration of plant tissues. *Global Change Biology* 4, 43–54.

Cousins, A.B., Adam, N.R., Wall, G.W., Kimball, B.A., Pinter, P.J. Jr, Ottman, M.J. et al. (2003) Development of C4 photosynthesis in sorghum leaves grown under free-air CO$_2$ enrichment (FACE). *Journal of Experimental Botany* 54, 1969–1975.

Crafts-Brandner, S.J. and Salvucci, M.E. (2000) Rubisco activase constrains the photosynthetic potential of leaves at high temperature and CO$_2$. *Proceedings of the National Academy of Sciences USA* 97, 13430–13435.

Cure, J.D. and Acock, B. (1986) Crop responses to carbon dioxide doubling: a literature survey. *Agricultural and Forest Meteorology* 38, 127–145.

Davey, P.A., Parsons, A.J., Atkinson, L., Wadge, K. and Long, S.P. (1999) Does photosynthetic acclimation to elevated CO$_2$ increase photosynthetic nitrogen-use efficiency? A study of three native UK grassland species in open top chambers. *Functional Ecology* 13, 21–28.

Davey, P.A., Hunt, S., Hymus, G.J., DeLucia, E.H., Drake, B.G., Karnosky, D.F. *et al.* (2004) Respiratory oxygen uptake is not decreased by an instantaneous elevation of [CO_2], but is increased with long-term growth in the field at elevated [CO_2]. *Plant Physiology* 134, 1–8.

De Lucia, E.H., Sasek, T.W. and Strain, B.R. (1985) Photosynthetic inhibition after long term exposure to elevated levels of atmospheric carbon dioxide. *Photosynthesis Research* 7, 175–184.

Donnelly, A., Craigon, J., Black, C.R., Colls, J.J. and Landon, G. (2001) Elevated CO_2 increases biomass and tuber yield in potato even at high ozone concentrations. *New Phytologist* 149, 265–274.

Driscoll, S.P., Prins, A., Olmos, E., Kunert, K.J. and Foyer, C.H. (2006) Specification of adaxial and abaxial stomata, epidermal structure and photosynthesis to CO_2 enrichment in maize leaves. *Journal of Experimental Botany* 57, 381–390.

Eamus, D. and Shanahan, S.T. (2002) A rate equation model of stomatal responses to vapour pressure deficit and drought. *BMC Ecology* 2, 8.

Edwards, A. and Bowling, D.J.F. (1985) Evidence for a CO_2-inhibited proton extrusion pump in the stomatal cells of *Tradescantia virginiana*. *Journal of Experimental Botany* 26, 91–98.

Fangmeier, A., De Temmerman, L., Mortensen, L., Kemp, K., Burke, J., Mitchell, R. *et al.* (1999) Effects on nutrients and on grain quality in spring wheat crops under elevated CO_2 concentrations and stress conditions in the European, multiple-site experiment 'ESPACE-wheat'. *European Journal of Agronomy* 10, 215–229.

Fangmeier, A., Chrost, B., Hogy, P. and Krupinska, K. (2000) CO_2 enrichment enhances flag leaf senescence in barley due to greater grain nitrogen sink capacity. *Environmental and Experimental Botany* 44, 151–164.

FAO (2007) FAO production statistics. Published online at http://faostat.fao.org/site/567/Desktop Default.aspx?PageID=567 (accessed 17 September 2010).

Faria, T., Wilkins, D., Besford, R.T., Vaz, M., Pereira, J.S. and Chaves, M.M. (1996) Growth at elevated CO_2 leads to down regulation of photosynthesis and altered response to high temperature in *Quercus suber* L. seedlings. *Journal of Experimental Botany* 47, 1755–1761.

Farquhar, D.G., Schulze, E.D. and Kuppers, M. (1980) Responses to humidity by stomata of *Nicotiana glauca* L. and *Corylus avellana* L. are consistent with the optimization of carbon dioxide with respect to water loss. *Australian Journal of Plant Physiology* 7, 315–327.

Ford, M.A. and Thorne, G.N. (1967) Effect of CO_2 concentration on growth of sugar-beet, barley, kale and maize. *Annals of Botany* 31, 629–644.

Foyer, C.H., Bloom, A., Queval, G. and Noctor, G. (2009) Photorespiratory metabolism: genes, mutants, energetics, and redox signaling. *Annual Review of Plant Biology* 60, 455–484.

Ghannoum, O. (2009) C_4 photosynthesis and water stress. *Annals of Botany* 103, 635–644.

Gibson, L.R. and Paulsen, G.M. (1999) Yield components of wheat grown under high temperature stress during reproductive growth. *Crop Science* 39, 1841–1846.

Gray, J.E., Holroyd, G.H., Van Der Lee, F.M., Bahrami, A.R., Sijmons, P.C., Woodward, F.I. *et al.* (2000) The HIC signalling pathway links CO_2 perception to stomatal development. *Nature* 408, 713–716.

Grossman-Clarke, S., Pinter, P.J. Jr, Kartschall, T., Kimball, B.A., Hunsaker, D.J., Wall, G.W. *et al.* (2001) Modelling a spring wheat crop under elevated CO_2 and drought. *New Phytologist* 150, 315–335.

Gunderson, C.A., Norby, R.J. and Wullschleger, S.D. (1993) Foliar gas exchange of two deciduous hardwoods during three years of growth in elevated CO_2: no loss of photosynthetic enhancement. *Plant Cell and Environment* 16, 797–807.

Hanstein, S.M. and Felle, H.H. (2002) CO_2-triggered chloride release from guard cells in intact fava bean leaves. Kinetics of the onset of stomatal closure. *Plant Physiology* 120, 940–950.

Hetherington, A.M. and Woodward, F.I. (2003) The role of stomata in sensing and driving environmental change. *Nature* 424, 901–908.

Hungate, B.A., Holland, E.A., Jackson, R.B., Chapin III, F.S., Mooney, H.A. and Field, C.B. (1997) The fate of carbon in grasslands under carbon dioxide enrichment. *Nature* 388, 576–579.

Idso, S.B. and Kimball, B.A. (1992) Effects of atmospheric CO_2 enrichment on photosynthesis, respiration and growth of sour orange trees. *Plant Physiology* 99, 341–343.

Intergovernmental Panel on Climate Change (IPCC) (2007) *Climate Change 2007: The Physical Science Basis. Summary for Policy Makers.* WMO, Geneva, Switzerland, pp. 1–21.

Jordan, D.B. and Ogren, W.L. (1984) The CO_2/O_2 specificity of ribulose 1,5-bisphosphate carboxylase/oxygenase. *Planta* 161, 308–313.

Kandel, R.S. (1981) Surface temperature sensitivity to increased atmospheric CO_2. *Nature* 293, 634–636.

Kangasjarvi, J., Jaspers, P. and Kollist, H. (2005) Signalling and cell death in ozone-exposed plants. *Plant Cell and Environment* 28, 1021–1036.

Khatiwala, S., Primeau, F. and Hall, T. (2009) Reconstruction of the history of anthropogenic CO$_2$ concentrations in the ocean. *Nature* 462, 346–349.

Kimball, B.A. (1983) Carbon dioxide and agricultural yield: an assemblage and analysis of 430 prior observations. *Agronomy Journal* 75, 779–788.

Kimball, B.A., Morris, C.F., Pinter, P.J. Jr, Wall, G.W., Hunsaker, D.J., Adamsen, F.J. *et al.* (2001) Elevated CO$_2$, drought and soil nitrogen effects on wheat grain quality. *New Phytologist* 150, 295–303.

Kimball, B.B., Kobayashi, K. and Bindi, M. (2002) Responses of agricultural crops to free-air CO$_2$ enrichment. *Advances in Agronomy* 77, 293–368.

Koti, S., Reddy, K.R., Reddy, V.R., Kakani, V.G. and Zhao, D. (2005) Interactive effects of carbon dioxide, temperature, and ultraviolet-B radiation on soybean (*Glycine max* L.) flower and pollen morphology, pollen production, germination, and tube lengths. *Journal of Experimental Botany* 56, 725–736.

Kramer, P.J. (1981) Carbon dioxide concentration, photosynthesis and dry matter production. *BioScience* 31, 29–33.

Krapp, A., Quick, W.P. and Stitt, M. (1991) Ribulose-1,5-bisphsphate carboxylase-oxygenase, other photosynthetic enzymes and chlorophyll decrease when glucose is supplied to mature spinach leaves via the transpiration stream. *Planta* 186, 58–69.

Lake, J.A., Quick, W.P., Beerling, D.J. and Woodward, F.I. (2001) Plant development: signals from mature to new leaves. *Nature* 411, 154.

Lake, J.A., Woodward, F.I. and Quick, W.P. (2002) Long-distance CO$_2$ signalling in plants. *Journal of Experimental Botany* 53, 183–193.

Leakey, A.D.B., Bernacchi, C.J., Dohleman, F.G., Ort, D.R. and Long, S.P. (2004) Will photosynthesis of maize (*Zea mays*) in the US Corn Belt increase in future [CO$_2$] rich atmospheres? An analysis of diurnal courses of CO$_2$ uptake under free-air concentration enrichment (FACE). *Global Change Biology* 10, 951–962.

Leakey, A.D.B., Nernacchi, C.J., Ort, D.R. and Long, S.P. (2006a) Long-term growth of soybean at elevated [CO$_2$] does not cause acclimation of stomatal conductance under fully open-air conditions. *Plant Cell and Environment* 29, 1794–1800.

Leakey, A.D.B., Uribelarrea, M., Ainsworth, E.A., Naidu, S.L., Rogers, A., Ort, D.R. *et al.* (2006b) Photosynthesis, productivity and yield of maize are not affected by open-air elevation of CO$_2$ concentration in the absence of drought. *Plant Physiology* 140, 779–790.

Leakey, A.D.B., Xu, F., Gillespie, K.M., McGrath, J.M., Ainsworth, E.A. and Ort, D.R. (2009) Genomic basis for stimulated respiration by plants growing under elevated carbon dioxide. *Proceedings of the National Academy of Sciences USA* 106, 3597–3602.

Li, A.G., Hou, Y.S., Wall, G.W., Trent, A., Kimball, B.A. and Pinter, P.J. Jr (2000) Free-air CO$_2$ enrichment and drought stress effects on grain filling rate and duration in spring wheat. *Crop Science* 40, 1263–1270.

Lloyd, J. and Farquhar, G.D. (2008) Effects of rising temperatures and [CO$_2$] on the physiology of tropical forest trees. *Philosophical Transactions of the Royal Society B – Biological Sciences* 363, 1811–1817.

Loladze, I. (2002) Rising atmospheric CO$_2$ and human nutrition: toward globally imbalanced plant stoichiometry? *Trends in Ecology and Evolution* 17, 457–461.

Long, S.P., Ainsworth, E.A., Leakey, A.D.B. and Morgan, P.B. (2005) Global food insecurity. Treatment of major food crops with elevated carbon dioxide or ozone under large-scale fully open-air conditions suggests recent models may have overestimated future yields. *Philosophical Transactions of the Royal Society B – Biological Sciences* 360, 2011–2020.

Makino, A. and Mae, T. (1999) Photosynthesis and plant growth at elevated levels of CO$_2$. *Plant and Cell Physiology* 40, 999–1006.

Maroco, J.P., Edwards, G.E. and Ku, M.S.B. (1999) Photosynthetic acclimation of maize to growth under elevated levels of carbon dioxide. *Planta* 210, 115–125.

Mauney, J.R., Kimball, B.A., Pinter, P.J. Jr., LaMorte, R.L., Lewinn, K.F., Nagy, J. *et al.* (1994) Growth and yield of cotton in response to a free air carbon dioxide enrichment. *Agricultural and Forest Meteorology* 70, 49–67.

Messinger, S.M., Buckley, T.N. and Mott, K.A. (2006) Evidence for involvement of photosynthetic processes in the stomatal response to CO$_2$. *Plant Physiology* 140, 771–778.

Morison, J.I.L. and Gifford, R.M. (1984) Plant growth and water use with limited water supply in high CO$_2$ concentrations. II. Plant dry weight, partitioning and water use efficiency. *Australian Journal of Plant Physiology* 11, 375–384.

Norby, R.J. (1994) Issues and perspectives for

investigating root responses to elevated atmospheric carbon dioxide. *Plant and Soil* 165, 9–20.

Norby, R.J., Gunderson, C.A., Wullschleger, S.D., O'Neill, E.G. and McCraken, M.K. (1992) Productivity and compensatory responses of yellow poplar trees in elevated CO_2. *Nature* 354, 322–324.

Oberbauer, S.F., Strain, B.R. and Fetcher, N. (1985) Effect of CO_2 enrichment on seedling physiology and growth of two tropical tree species. *Physiologia Plantarum* 65, 352–356.

Okamoto, K., Ogiwara, T., Yoshizumi, T. and Watanabe, Y. (1991) Influence of the greenhouse effect on yields of wheat, soybean and corn in the United States for different energy scenarios. *Climate Change* 18, 397–424.

Ottman, M.J., Kimball, B.A., Pinter, P.J., Wall, G.W., Vanderlip, R.L., Leavitt, S.W. *et al.* (2001) Elevated CO_2 increases sorghum biomass under drought conditions. *New Phytologist* 150, 261–273.

Patterson, D.T. and Flint, E.P. (1980) Potential effects of global atmospheric CO_2 enrichment on the growth and competitiveness of C3 and C4 weeds and crop plants. *Weed Science* 28, 71–75.

Paul, M.J. and Foyer, C.H. (2001) Sink regulation of photosynthesis. *Journal of Experimental Botany* 52, 1383–1400.

Peet, M.M., Huber, S.C. and Patterson, D.T. (1986) Acclimation to high CO_2 in monoecious cucumbers. II. Carbon exchange rate, enzyme activities, starch and nutrient concentrations. *Plant Physiology* 80, 63–67.

Poorter, H. (1993) Interspecific variation in the growth response of plant to an elevated ambient CO_2 concentration. *Vegetatio* 104/105, 77–97.

Poorter, H., Van Berkel, Y., Baxter, R., Den Hertog, J., Dijkstra, P., Gifford, R.M. *et al.* (1997) The effect of elevated CO_2 on the chemical composition and construction costs of leaves of 27 C3 leaves. *Plant Cell and Environment* 20, 472–482.

Prasad, P.V.V., Boote, K.J., Allen, L.H. Jr and Thomas, J.M.G. (2002) Effects of elevated temperature and carbon dioxide on seed-set and yield of kidney bean (*Phaseolus vulgaris* L.). *Global Change Biology* 8, 710–721.

Prasad, P.V.V., Pisipati, S.R., Mutava, R.N. and Tuinstra, M.R. (2008) Sensitivity of grain sorghum to high temperature stress during reproductive development. *Crop Science* 48, 1911–1917.

Prins, A., Kunert, K.J. and Foyer, C.H. (2009) Interactions between CO_2 signalling, cellular redox state and ageing in plants. In: Foyer, C.H., Thornalley, P. and Faragher, R. (eds) *Redox Metabolism and Longevity Relationships in Animals and Plants..* SEB Experimental Biology series, Vol. 62. Taylor and Francis Publishers, Oxford, UK, pp. 203–226.

Prins, A., Muchwesi, M.J., Pellny, T.K., Verrier, P., Beyene, G., Lopes, M.S. *et al.* (2011) Acclimation to high CO_2 in maize is related to water status and dependent on leaf rank. *Plant Cell and Environment* 34, 314–331.

Prior, S.A., Rogers, H.H., Runion, G.B. and Mauney, J.R. (1993) Effects of free-air CO_2 enrichment on cotton root growth. *Agricultural and Forest Meteorology* 70, 69–86.

Pritchard, S.G., Ju, Z.L., van Santen, E., Qiu, J.S., Weaver, D.B., Prior, S.A. *et al.* (2000) The influence of elevated CO2 on the activities of antioxidative enzymes in two soybean genotypes. *Australian Journal of Plant Physiology* 27, 1061–1068.

Pritchard, S.G., Prior, S.A., Rogers, H.H., Davis, M.A., Runion, G.B. and Popham, T.W. (2006) Effects of elevated atmospheric CO2 on root dynamics and productivity of sorghum grown under conventional and conservation agricultural management practices. *Agriculture, Ecosystems and Environment* 113, 175–183.

Qin, D., Wu, H., Peng, H., Yao, Y., Ni, Z., Li, Z. *et al.* (2008) Heat stress-responsive transcriptome analysis in heat susceptible and tolerant wheat (*Triticum aestivum* L.) by using wheat genome array. *BMC Genomics* 9, 432, 1–19.

Raper, D.C. and Peedin, G.F. (1978) Photosynthetic rate during steady-state growth as influenced by carbon dioxide concentration. *Botanical Gazette* 139, 147–149.

Raschke, K. (1977) The stomatal turgor mechanism and its response to CO_2 and abscisic acid: observations and a hypothesis. In: Mare, E. and Cifferi, O. (eds) *Regulation of Cell Membrane Activities in Plants*. Elsevier, North Holland Biomedical Press, Amsterdam, pp. 173–183.

Raschke, K., Shabahang, M. and Wolf, R. (2003) The slow and the quick anion conductance in whole guard cells: their voltage-dependent alternation, and the modulation of their activities by abscisic acid and CO_2. *Planta* 217, 639–650.

Rogers, G.S., Milham, P.J., Gillings, M. and Conroy, J.P. (1996) Sink strength may be the key to growth and nitrogen responses in N-deficient wheat at elevated CO_2. *Australian Journal of Plant Physiology* 23, 253–264.

Rogers, H.H., Peterson, C.M. and McCrimmon, C.J.D. (1992) Response of plant roots to elevated atmospheric carbon dioxide. *Plant Cell and Environment* 15, 749–752.

Rogers, H.H., Runion, G.B. and Krupa, S.V. (1994) Plant responses to atmospheric CO$_2$ enrichment with emphasis on roots and the rhizosphere. *Environmental Pollution* 83, 155–189.

Rowland-Bamford, A.J., Baker, J.T., Allen, Jr L.H. and Bowes, G. (2006) Acclimation of rice to changing atmospheric carbon dioxide concentration. *Plant Cell and Environment* 14, 577–583.

Sage, R.F., Sharkey, T.D. and Seemann, J.R. (1988) Acclimatization of photosynthesis to elevated CO$_2$ in five C3 species. *Plant Physiology* 89, 590–596.

Salvucci, M.E., Osteryoung, K.W., Crafts-Brandner, S.J. and Vierling, E. (2001) Exceptional sensitivity of rubisco activase to thermal denaturation in vitro and in vivo. *Plant Physiology* 127, 1053–1064.

Samarakoon, A.B. and Gifford, R.M. (1996) Elevated CO$_2$ effects on water use and growth of maize in wet and drying soil. *Australian Journal of Plant Physiology* 23, 53–62.

Sasek, T.W., Delucia, E.H. and Strain, B.R. (1985) Reversibility of photosynthetic inhibition in cotton after long term exposure to elevated CO$_2$ concentrations. *Plant Physiology* 78, 619–622.

Schutz, M. and Fangmeier, A. (2001) Growth and yield responses of spring wheat (*Triticum aestivum* L. cv. Minaret) to elevated CO$_2$ and water limitation. *Environmental Pollution* 114, 187–194.

Sicher, R.C. and Bunce, J.A. (1997) Relationship of photosynthetic acclimation to changes of Rubisco activity in field-grown winter wheat and barley during growth in elevated carbon dioxide. *Photosynthesis Research* 52, 27–38.

Soares, A.S., Driscoll, S.P., Olmos, E., Harbinson, J., Arrabaça, M.C. and Foyer, C.H. (2007) Adaxial/abaxial specification in the regulation of photosynthesis with respect to light orientation and growth with CO$_2$ enrichment in the C$_4$ species *Paspalum dilatatum*. *New Phytologist* 177, 186–198.

Stitt, M. (1991) Rising CO$_2$ levels and their potential significance for carbon flow in photosynthetic cells. *Plant Cell and Environment* 14, 741–762.

Stulen, I. and den Hertog, J. (1993) Root growth and functioning under atmospheric CO$_2$ enrichment. *Vegetatio* 104/105, 99–115.

Taub, D.R., Seeman, J.R. and Coleman, J.S. (2000) Growth in elevated CO$_2$ protects photosynthesis against high temperature damage. *Plant Cell and Environment* 23, 649–656.

Taub, D.R., Miller, B. and Allen, H. (2008) Effects of elevated CO$_2$ on the protein concentration of food crops: a meta-analysis. *Global Change Biology* 14, 565–575.

Thomas, C.D., Cameron, A. and Green, R.E. (2004) Extinction risk from climate change. *Nature* 427, 145–158.

von Caemmerer, S., Lawson, T., Oxborough, K., Baker, N.R., Andrews, T.J. and Raines, C.A. (2004) Stomatal conductance does not correlate with photosynthetic capacity in transgenic tobacco with reduced amounts of Rubisco. *Journal of Experimental Botany* 55, 1157–1166.

Wall, G.W. (2001) Elevated atmospheric CO$_2$ alleviates drought stress in wheat. *Agriculture, Ecosystems and Environment* 87, 261–271.

Wall, G.W., Adam, N.R., Broks, T.J., Kimball, B.A., Pinter, Jr. P.J., LaMorte, R.L., Adamsen, F.J., Hunsaker, D.J., Wechsung, G., Wechsung, F., Grossman-Clarke, S., Leavitt, S.W. *et al.* (2000) Acclimation response of spring wheat in a free-air CO$_2$ enrichment (FACE) atmosphere with variable soil nitrogen regimes. 2. Net assimilation and stomatal conductance of leaves. *Photosynthesis Research* 66, 79–95.

Wand, S.J.E., Midgley, G.F. and Stock, W. (2001) Growth responses to elevated CO$_2$ in NADP-ME, NAD-ME and PCK C4 grasses and a C3 grass from South Africa. *Australian Journal of Plant Physiology* 28, 13–25.

Wang, D., Heckathorn, S.A., Barua, D., Joshi, P., Hamilton, E.W. and LaCroix, J.J. (2008) Effects of elevated CO$_2$ on the tolerance of photosynthesis to acute heat stress in C3, C4, and CAM species. *American Journal of Botany* 95, 165–176.

Watling, J.R., Press, M.C. and Quick, W.P. (2000) Elevated CO$_2$ induces biochemical and ultrastructural changes in leaves of the C4 cereal sorghum. *Plant Physiology* 123, 1143–1152.

Wechsung, G., Wechsung, F., Wall, G.W., Adamsen, F.J., Kimball, B.A., Pinter, P.J. Jr *et al.* (1999) The effects of free-air CO$_2$ enrichment and soil water availability on spatial and seasonal patterns of wheat root growth. *Global Change Biology* 5, 519–529.

Whipps, J.M. (1985) Effect of CO$_2$ concentration on growth, carbon distribution and loss of carbon from the roots of maize. *Journal of Experimental Botany* 36, 649–651.

Wieser, H., Manderscheid, R., Erbs, M. and Weigel, H.J. (2008) Effects of elevated atmospheric CO$_2$ concentrations on the quantitative protein composition of wheat grain. *Journal of Agricultural and Food Chemistry* 56, 6531–6535.

Wingler, A., Lea, J.P., Quick, P.W. and Leegood, C.R. (2000) Photorespiration: metabolic pathways and their role in stress protection. *Philosophical Transactions of the Royal Society B – Biological Sciences* 355, 1517–1529.

Witter, S.H. (1978) Carbon dioxide fertilization of crop plants. In: Gupta, U.S. (ed.) *Problems in Crop Physiology*. Haryana Agricultural University, Hissar, India, pp. 310–333.

Witter, S.H. and Robb, W.M. (1964) Carbon dioxide enrichment of greenhouse atmospheres for food crop production. *Economic Botany* 18, 34–56.

Woodward, F.I. (1987) Stomatal numbers are sensitive to increases in CO_2 from pre-industrial levels. *Nature* 327, 617–618.

Woodward, F.I., Lake, J.A. and Quick, W.P. (2002) Stomatal development and CO_2: ecological consequences. *New Phytologist* 153, 477–484.

Wu, D.X., Wang, G.X., Bai, Y.F. and Liao, J.X. (2004) Effects of elevated CO_2 concentration on growth, water use, yield, and grain quality of wheat under two soil water levels. *Agriculture Ecosystems and Environment* 104, 493–507.

Wullschleger, S.D., Tschaplinski, T.J. and Norby, R.J. (2002) Plant water relations at elevated CO_2 – implications for water-limited environments. *Plant Cell and Environment* 25, 319–331.

Yelle, S., Beeson, R.C. Jr, Trudel, M.J. and Gosselin, A. (1989a) Acclimation of two tomato species to high atmospheric CO_2. I. Sugar and starch concentrations. *Plant Physiology* 90, 1465–1472.

Yelle, S., Beeson, R.C. Jr, Trudel, M.J. and Gosselin, A. (1989b) Acclimation of two tomato species to high atmospheric CO_2. II. Ribulose-1,5-bisphosphate carboxylase/oxygenase and phosphor-enolpyruvate carboxylase. *Plant Physiology* 90, 1473–1472.

Young, J.J., Mehta, S., Israelsson, M., Godoski, J., Grill, E. and Schroeder, J.I. (2006) CO_2 signalling in guard cell: calcium sensitivity response modulation, a Ca^{2+} independent phase, and CO_2 insensitivity of the gca2 mutant. *Proceedings of the National Academy of Sciences USA* 103, 7506–7511.

Ziska, L.H. and Bunce, J.A. (1999) Effect of elevated carbon dioxide concentration at night on the growth and gas exchange of selected C4 species. *Australian Journal of Plant Physiology* 26, 71–77.

7 Breeding to Improve Grain Yield in Water-limited Environments: the CSIRO Experience with Wheat

R.A. Richards, G.J. Rebetzke, A.G. (Tony) Condon and M. Watt

7.1 Introduction

Timely rainfall is the universal need of dryland farmers. Where crops are reliant on rain and have no supporting irrigation, as is common in semi-arid environments world-wide, a single rainfall event around the optimal planting time and, more importantly, around the time of flowering, can result in a 'boom or a bust' year for farmers. Extreme seasonal rainfall variability, which is common in semi-arid environments such as in Australia, means that the economic well-being of most dryland farmers is generally determined by the fortunes or misfortunes of the rainfall lottery averaged over, say, a 7-year period. This vulnerability of crops, and of the farmers that grow them, to the timing of rainfall events draws attention to taking advantage of limited resources (soil water) when they are available. It also draws attention to the expected large genotype–environment (year) interactions and to the difficulty in making genetic progress in dry environments when selection is primarily for grain yield.

The Commonwealth Scientific and Industrial Research Organisation (CSIRO) was established as a government-funded research agency in Australia to investigate scientific issues of importance to Australia. Temperate cereals such as wheat have been the most profitable rainfed crops in southern Australia for centuries, and the physiology and nutrition of cereals have been important research endeavours at CSIRO for many years. Breeding was not a discipline that

CSIRO, as a national agency, undertook until recently, as this was traditionally the domain of universities and of the regional (state) agricultural agencies that were also responsible for agricultural extension in their region. A consequence of this was that much of the cereal research at CSIRO was not immediately relevant to farmers, but was more fundamental with a strong focus on plant physiology and biochemistry. Accordingly, there was little connection between farmers and CSIRO research until the early 1980s.

During the 1970s considerable research interest in plant responses to drought was developing internationally, nurtured by the availability of new instrumentation to measure plant–water relations and leaf gas exchange. This interest stemmed from agriculturalists as well as ecophysiologists, and a terminology developed that was common to both disciplines concerning drought tolerance and drought resistance (Levitt, 1972). Interest in plant responses to drought rapidly resulted in a strong understanding of the morphological and physiological factors associated with plant survival in both natural and cultivated plant communities, and some speculation on ways to improve the drought resistance of crops were also proposed (e.g. Levitt, 1972). Many studies of drought response were conducted with pot-grown plants. However descriptions of crop growth, water relations and gas exchange in the field were almost non-existent. At CSIRO in the 1970s an interest in plant–water relations was also emerging but focusing on crop communities (Begg and Turner, 1976) and supported by a research interest in the crop physiology of water-limited crops (Passioura, 1976; Fischer and Turner, 1978; Fischer, 1979). This changed focus led to questions about whether the factors that determine crop productivity in water-limited conditions were the same as those for drought resistance. It soon became evident that many of the strategies plants use to survive extended periods of drought to produce seed were generally unrelated to the most important factors associated with determining crop productivity (Fischer and Turner, 1978).

Associated with this interest in drought physiology was an awareness that yield improvements were being made through both empirical plant breeding and changed crop management practices such as altered row spacing and earlier sowing, and so tentative advice was being given to plant breeders seeking improvement in dry environments. Drought in Australia commonly occurs during grain-filling at the end of the season (i.e. terminal drought). Hence the main advice to plant breeders was to select for earlier flowering so that crops could escape the effects of drought. Since the beginning of wheat improvement from the mid-1800s in Australia, breeders, in selecting for greater grain yield, were in fact selecting for earlier maturity, although the importance of phenology was probably not evident at the time. A close association between the time of floral initiation and grain yield in a historical set of varieties grown in Western Australia is shown in Fig. 7.1. Other advice to breeders was to select for plant survival, which was then expected to be important in an agricultural context (Levitt, 1972). Also beginning to emerge was the idea that some traits specifically contributing to yield under drought may not be present in breeding programmes targeting water-limited environments, a good example being the axial resistance to water flow in the seminal roots of wheat, which will be discussed shortly. There were also questions being asked about whether genetic variation in root depth was present in breeding programmes (Hurd, 1974) and whether morphological characters under simple genetic control, such as awns in cereals, contribute to yield under drought (Evans *et al.*, 1972). Separate to this, the

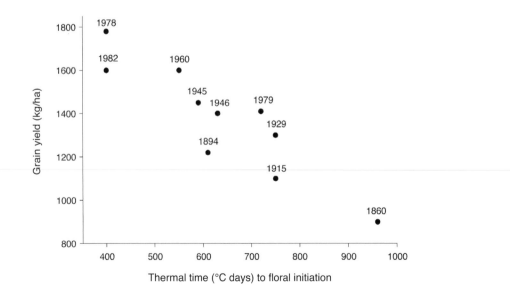

Fig. 7.1. Relationship between grain yield and thermal time to floral initiation (°C days) for wheat cultivars released in Western Australia. The year of release of each cultivar is indicated above each data point (adapted from Richards, 1991).

question was being asked as to whether more attention be given to genotypic differences in traits related to crop productivity under drought so that breeding programmes could further enhance or accelerate genetic progress.

At CSIRO, Dr John Passioura, inspired by the recognition by Nix and Fitzpatrick (1968) of the correlation between grain yield and soil water at anthesis in crops growing largely on stored water, was investigating water use by crops over time and the importance of slowing water use to improve harvest index and grain yield in crops experiencing terminal drought (Passioura, 1972). In a radical departure from the thinking at that time, he proposed that, for crops growing on stored soil moisture, grain yield is a function of the proportion of water used by the crop post-flowering. Passioura argued that if somehow plants were able to ration their water use so that more becomes available after flowering then grain yield should be increased. For temperate cereals, Passioura proposed an extremely novel idea that might achieve this based on their root architecture and root geometry (see below).

Around this time Passioura also proposed a novel and more general framework within which to consider yield in water-limited environments (Passioura, 1977). This was not just for environments where crops are grown on stored soil moisture but for all water-limited environments. That is, the framework is relevant for environments where rainfall is winter-dominant (e.g. Mediterranean), summer-dominant or may occur at any time during the year. It is relevant to all grain crops. The Passioura identity states that:

Grain yield = Water use × Water-use efficiency × Harvest index.

Passioura suggested that an increase in any one of these three determinants should increase grain yield in water-limited environments. Furthermore, he suggested that, unlike the yield determinants (grain number, grain size, etc.), each component is likely to be largely independent of the other, enabling breeders to focus on selection for one or all determinants.

7.2 Major Traits Studied by CSIRO

This framework was a radical departure from earlier thinking on ways to improve the growth and yield of water-limited crops. Importantly, it focused on grain yield and not on drought resistance, and therefore turned our attention to longer-term processes associated with crop production and to resource limitations. This framework laid the foundations for germplasm improvement at CSIRO for water-limited environments. With some knowledge of genetic variation available in wheat for morphological and physiological characteristics and some thought into the likely traits that may influence any one of these yield determinants, a number of traits were identified for further investigation at CSIRO. These traits were expected directly to influence water use, water-use efficiency and/or harvest index for use in breeding. The traits are shown in Table 7.1, together with the components of the Passioura identity that are likely to be increased so that grain yield may be increased. The intention of our research at CSIRO has since been to search for important genetic variation and then determine the underlying genetic control for these traits, develop appropriate crossing and selection strategies, and then develop suitable lines/populations for validation and germplasm development so that breeders can confidently use them in crop improvement.

This trait approach to breeding is intended to complement existing breeding programmes, and is not dissimilar to approaches taken to improve specific resistances/tolerances to diseases, soil chemical constraint or for components of grain quality. Possible advantages of this trait-based approach to breeding have previously been enunciated (Richards et al., 2002). Briefly, these are:

1. Genetic variability for important traits can be enhanced, leading to faster and greater genetic gain.
2. The trait may have a higher heritability than grain yield, and this may lead to faster genetic gain in yield.

Table 7.1. Priority list of traits identified in wheat that are expected to increase grain yield and where there is known genetic variation. Traits that are listed against the determinants of the Passioura identity and expected positive or negative impacts are indicated. Parentheses show possible additional positive or negative impacts on the determinants in certain environments.

Trait	Water use	Water-use efficiency	Harvest index
Xylem vessel diameter of seminal roots			+
Tiller inhibition	(+)		+
Early seedling vigour	+	+	(–)
GA-responsive dwarfing genes	+	+	+
Extended vegetative growth period	+	+	(–)
Root growth and architecture	+		(+)
Waxiness/glaucousness	+	+	
Transpiration efficiency (carbon isotope discrimination)	(–)	+	(+)
Stem carbohydrate storage and remobilization			+
Leaf-rolling/stay-green		(+)	+

GA, gibberellic acid.

3. Selection for the trait may be more cost-effective than selection for yield.
4. Out-of-season selection or selection in controlled environments may be possible, resulting in more than one cycle of selection per year.
5. The trait may be amenable to marker-assisted selection, whereas grain yield is not.
6. Multiple yield-enhancing traits may be pyramided.

For this approach to be successful a necessary prerequisite is that water is the major limiting factor and that other factors, such as disease or soil chemical and physical constraints, do not limit crop yield. Conventional breeding continues to make progress in water-limited environments and it will remain the backbone of further yield progress as it simultaneously combines improved yield with the required disease resistances, grain quality characteristics and tolerances to other abiotic stresses required in varieties for the target set of environments and needed by farmers and consumers. A major challenge in breeding is seamlessly and efficiently to incorporate this trait-based physiological approach into conventional breeding programmes. Means of doing this are described in Richards *et al.* (2010).

In the following sections the traits studied by CSIRO will be described, together with an outline of the environments where they are likely to be most effective, ways to select for them, results of validation, proof-of-concept tests and whether they are currently incorporated in breeding programmes.

7.2.1 Xylem vessel diameter of seminal roots

Temperate cereals have a dual root system, composed of seminal roots that develop from the seed and nodal roots that develop later from nodes above the seed. In wheat, there are typically three seminal axes that grow from each seed. Because the seminal roots initiate growth well before the nodal roots they grow deepest into the subsoil. When this is dry, crops are largely reliant on subsoil water and this water must pass through the single main xylem vessel in each axis. If plants use this subsoil water too fast during the vegetative period then little will be available for grain-filling. Use of subsoil water is slowed if there is a high hydraulic resistance in the seminal roots. Passioura (1972) proposed that genetically reducing the xylem vessel in the seminal roots of wheat should increase hydraulic resistance and force plants to use the subsoil water more slowly. This idea was based on Poiseuille's law, which states that the resistance to water flow in a tube is proportional to the fourth power of the diameter. Thus, if the diameter of the main xylem vessel is halved then resistance to water flow is increased 16-fold. A major attraction of this idea is that if the season is wet then grain yield is unlikely to be affected, as the nodal roots are shallower and will use the available water in the topsoil. On the other hand, if the season is dry and crops are reliant on water stored in the soil, then the increased axial resistance will reduce leaf area growth and thereby crop water use and save more water for use during grain-filling.

A breeding programme in wheat was initiated after developing a screening protocol for xylem vessel diameter in the seminal roots and following identification of suitable genetic variation. An understanding of the genetic control and of the environmental factors associated with xylem vessel diameter was also investigated to ensure that a breeding programme would be successful (Richards and Passioura, 1981a, b). This breeding programme reduced the xylem vessel diameter of two Australian commercial wheat varieties from 65 to <55 μm. In field trials in eastern Australia, narrow-vessel selections, averaged over both genetic backgrounds, yielded 8% more than the unselected controls in the driest environments, whereas yield differences in the wetter environments were largely not significant (Richards and Passioura, 1989).

7.2.2 Tiller inhibition

Selection for xylem vessel diameter is not expected to be readily adopted by breeding programmes as it is not under simple genetic control, and this raises questions about the incorporation of complex traits into breeding programmes, issues discussed in Richards *et al.* (2010). As an alternative strategy a reduction in the tiller number per plant, which is more easily selected, may also reduce leaf area development and thereby ration soil water over the duration of the crop so that more water is available for grain-filling. A reduction in tillering may not penalize the yield of temperate cereals,

as these produce a large number of unproductive tillers. If these unproductive tillers could be eliminated then more water could be saved for grain-filling. A landrace wheat from North Africa with restricted tillering was first reported by Atsmon and Jacobs (1977). We used a derivative of this to alter leaf area development in spring wheat, as this had been validated in pot trials (Richards and Townley-Smith, 1987). Tillering from this donor line was regulated by a single recessive gene (tiller inhibition gene, *tin*), although the phenotypic expression of *tin* can be highly variable depending on background genotype and the environment (Richards, 1988). This major gene for tiller regulation has a dramatic effect on tillering, leaf area development and ear size in spaced plants, although its effect is modified when plants are grown in the field as a canopy. A tightly linked molecular marker on chromosome 1AS has been developed for this gene to assist selection (Spielmeyer and Richards, 2004).

Near-isogenic wheat lines with and without the *tin* gene have been developed and tested over a range of dry environments. Lines with strongly reduced tillering (uniculm) had substantially less biomass and yield than the commercial parent when tested in a range of environments in Western Australia (Condon and Giunta, 2003). A different set of reduced-tillering lines, with less extreme tillering reduction and unselected for yield, had on average the same grain yield as commercial near-isogenic counterparts that had been selected for yield (Duggan *et al.*, 2005a). The final spike number for lines containing the *tin* gene was lower and, even though above-ground biomass was a little lower, this was compensated by the higher harvest index (Duggan *et al.*, 2005a). Light interception and leaf area index of reduced-tillering lines were also slightly smaller (Duggan *et al.*, 2005b). Interestingly, early root growth of the reduced-tillering lines has been found to be enhanced in both controlled (Duggan *et al.*, 2005b) and field studies (Richards *et al.*, 2007). Little yield penalty under favourable conditions has been noted in lines with the *tin* gene as long as the tillering reduction is not extreme. Lines in commercial back-grounds are currently in extensive field trials with Australian breeding programmes.

7.2.3 Early seedling vigour

Where crops are largely reliant on in-season rainfall rather than stored soil moisture, a large proportion of the soil water evaporates from the soil surface and is therefore unavailable to crops. A common estimate for Mediterranean-type environments is that about half of the rainfall is lost as evaporation from the soil surface (Cooper *et al.*, 1987). Crops with greater vigour are expected to achieve canopy closure faster and hence increase the proportion of water transpired by the crop, leading to a higher biomass but with a similar total water use (Condon *et al.*, 1993). Additional benefits are likely to accrue from the higher intrinsic transpiration efficiency associated with more growth when air temperatures are cooler and evaporative demand lower (Richards, 1991).

Studies comparing winter cereals have shown that barley has a much faster early leaf area growth than wheat, and that in regions where early in-season rainfall is important, yet low, the yield of barley is greater than that of wheat (Lòpez-Castañeda and Richards, 1994a). We have shown that this greater early leaf area growth of barley is due to a larger embryo and a high specific leaf area (Lòpez-Castañeda *et al.*, 1996). Deliberately pyramiding these traits from different wheat sources has also resulted in progeny with a greatly enhanced leaf area (Richards and Lukacs, 2002). One of these progeny, Vigour 18, and further progeny derived from it have also been shown to have a larger root system and an enhanced nitrogen uptake (Liao *et al.*, 2006), faster root system growth (Watt *et al.*, 2005; Richards *et al.*, 2007) and a greater competitive ability in the presence of weeds (Coleman *et al.*, 2001). Fast early growth is also expected to be beneficial where growing seasons are short, because of the much greater opportunity for light capture from rapid leaf area development. Breeding lines contrasting for early vigour, selected from a

cross of Vigour 18 and a commercial parent, were tested in field environments in Western Australia (Botwright *et al.*, 2002). 'High-vigour' lines out-yielded 'low-vigour' lines when growing-season rainfall was > 250 mm, but there was no difference in yield when rainfall was < 250 mm.

Selection for embryo size and specific leaf area is difficult and slow, and unsuitable when screening large numbers of lines in a breeding programme. We therefore proposed that selection for the breadth of young seedling leaves should integrate these two traits and result in greater vigour (Lòpez-Castañeda *et al.*, 1996). This method of selection has proved very effective (Rebetzke and Richards, 1999; Richards and Lukacs, 2002), and successive cycles of recurrent selection for breadth of seedling leaves continue to result in significant gains in early vigour (Rebetzke, unpublished data). Progeny from the recurrent selection are now being used as donor lines in breeding for enhanced vigour.

Whereas an increase in tillering induced by favourable soil conditions, such as high nutrition, generally results in an increased leaf area index, a genetic increase or decrease in the production of main stem tillers does not typically alter leaf area index (Rebetzke and Richards, 1999; Richards and Lukacs, 2002; Duggan *et al.*, 2005b). An exception to this is the coleoptile tiller, which arises from the coleoptilar node when there are about two main stem leaves. This tiller, whose formation is under strong genetic control, forms a new branch separate from the main stem and can also be important for increasing early leaf vigour (Liang and Richards, 1994: Rebetzke *et al.*, 2008c).

7.2.4 GA-responsive dwarfing genes

A prerequisite to developing wheat germplasm with substantially increased early vigour for breeding is the deployment of new dwarfing genes (Richards, 1992). The green revolution wheat dwarfing genes *Rht-B1b* and *Rht-D1b* are associated with reduced cell length in above-ground plant parts, especially leaves and coleoptiles (Keyes *et al.*,

1989) and thereby reduced early growth (Richards, 1992). *Rht-B1b* and *Rht-D1b* have been very successful globally in lowering plant height, to reduce lodging and increase harvest index without compromising biomass in favourable environments. However, for some dry environments where vigour is important for biomass production, new dwarfing genes that are not associated with reduced cell length are required to maximize the expression of vigour (Richards, 1992). Also, in any environment where soil moisture may not be available in the top 5 cm at sowing, gibberellic acid (GA)-responsive dwarfing genes that allow expression of long coleoptiles are required to ensure good crop establishment and early growth (Richards, 1992: Rebetzke *et al.*, 2007a). The GA-responsive dwarfing genes alone may not be enough to provide an optimal coleoptile length; additional chromosomal regions independent of the dwarfing genes are also going to be important if coleoptile length is to be maximized (Rebetzke *et al.*, 2007b). GA-responsive dwarfing genes may be especially important in conservation farming systems where seedlings also need to

emerge through stubble (Rebetzke *et al.*, 2005) and when sowing occurs in warm soils, as this results in shorter coleoptiles (Rebetzke *et al.*, 1999).

A number of GA-responsive dwarfing genes have been identified that are good candidates to replace *Rht-B1b* and *Rht-D1b* (Rebetzke and Richards, 2000; Ellis *et al.*, 2004; Rebetzke *et al.*, 2011a), and these have been mapped to wheat chromosomes (Ellis *et al.*, 2005; Fig. 7.2). These genes will therefore be important in facilitating breeding to improve the proportion of water used by the crop, as they will improve the timeliness of sowing and facilitate the expression of the fast early vigour trait so as to maximize early growth and water-use efficiency. *Rht4*, *Rht5*, *Rht8*, *Rht13* and *Rht18* have been identified as having the most likely potential for commercial use. We have found that *Rht13* has the added attraction in that it is a reduction in the length of the peduncle that contributes most to reduction in plant height (Rebetzke *et al.*, 2011b). This is likely to be important because this internode expands at a time when carbon resources for floret development are most needed to increase grain

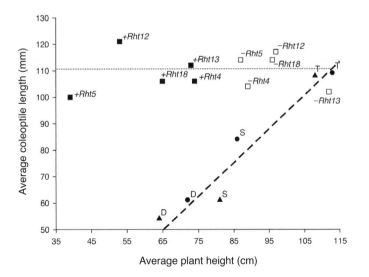

Fig. 7.2. Gibberellic acid (GA)-responsive dwarfing genes identified where coleoptile length is independent of plant height. Dotted line, open symbols indicate the tall wild-type allele whereas the closed symbols indicate the dwarf allele; this line shows the relationship between coleoptile length and plant height for wheats with Rht-B1b and Rht-D1b (D, double dwarf; S, single dwarf; T, tall).

number (Fischer and Stockman, 1986). A reduced peduncle expansion may increase the carbon available for floret development.

Work is now under way to introduce GA-responsive dwarfing genes, shown in Fig. 7.2, into commercial Australian wheats. We are aiming for improved crop establishment and early growth in commercial wheats by combining these new *Rht* genes with additional genes for coleoptile length, and also for much greater early vigour. Advanced breeding lines containing *Rht8* are near commercial release in some of the Australian breeding programmes.

7.2.5 Extended vegetative growth

Selection to alter the timing of phenological events such as the time of sowing, flowering and harvest has had a greater impact on crop yield than any other trait – with the possible exception of crop height. In Australia, the time of floral initiation or that of flowering accounts for most of the variation in grain yield in historical data sets (e.g. Fig. 7.1). Flowering is optimized to minimize the combined risks from later spring frost if flowering is too early, and from spring/summer drought and high temperature if flowering is too late. Selective breeding has probably optimized flowering time in recent decades for wheat sown at the conventional time, but important opportunities remain to develop varieties that can be planted earlier but still flower at the optimal time. Earlier sowing can have a dramatic impact on the efficiency of both water use (Gomez-Macpherson and Richards, 1995) and phosphorus use (Batten *et al.*, 1999), and on root depth (Kirkegaard *et al.*, 2007). It can also result in wheat having a dual use – forage for grazing prior to grain production. Thus, longer-season wheats in Australia, or the US Great Plains, are now often planted in late summer to early autumn. The early forage produced is grazed by animals in winter, the animals are then removed in spring and grain is harvested in summer. The economic advantages to farmers with both livestock and grain can be large (Kelman and Dove, 2009). CSIRO and partners have released numerous longer-season winter wheats suitable for grazing and grain (e.g. cvs Brennan, Rudd, Mackellar, Revenue), as well as longer-season spring wheats (cvs Forrest, Preston) that can be planted several weeks earlier than most current spring wheats. The former can be grazed heavily whereas the latter may be grazed sparingly. Breeding programmes continue at CSIRO in partnership with other entities with dedicated programmes for spring- and winter-milling wheats (e.g. HRZ Wheats P/L) and for dual-purpose feed wheats.

The improved water-use efficiency for biomass of early-sown wheats (Fig. 7.3) comes from faster growth after planting when temperatures are warmer, so that by winter there is often full canopy cover resulting in more crop transpiration and less soil evaporation. Intrinsic transpiration efficiency is also greater, as there is more growth in winter when vapour pressure deficit is low. However, water-use efficiency for grain production may not be greater. This is because earlier-sown crops, while they produce more biomass, generally do not convert this biomass into extra grain (Penrose, 1993; Gomez-Macpherson and Richards, 1995). One explanation given for this is that, to sow earlier, crops must have genes that delay flowering, e.g. photoperiod and/or vernalization. This increases leaf numbers on the main stem and also the number of internodes, resulting in taller crops with a low harvest index (Gomez-Macpherson and Richards, 1995). Another explanation could be that increased crop duration may exhaust soil moisture, resulting in limited water for grain-filling. However, this was considered unlikely as reduced grain size and lower leaf water potential were not observed (Gomez-Macpherson and Richards, 1995). One possible way to increase harvest index and yield of early-sown crops is to incorporate additional dwarfing genes to reduce final plant height. Intense grazing pressure during winter also appears to reduce plant height and, by cutting back leaf area and crop water use, increases water supply for grain-filling and thereby yield and harvest index (Virgona *et al.*, 2006).

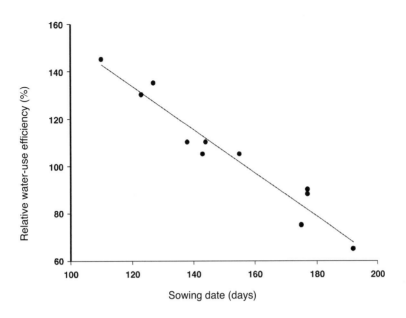

Fig. 7.3. Relationship between sowing date and water-use efficiency of wheat among near-isogenic populations differing in flowering time and grown at different locations and years in eastern Australia (adapted from Gomez-Macpherson and Richards, 1995).

7.2.6 Root growth and architecture

The importance of a more extensive root system in water-limited environments has been a commonly stated goal in the literature. However, there are few reports to suggest it has been incorporated into new varieties (e.g. Hurd, 1974). Contrary to popular belief. a deeper root system will not usually be beneficial in very dry seasons or dry environments. In the very dry seasons/ environments the available water will not be very deep and roots of most cereal varieties should be able to access it (Lilley and Kirkegaard, 2007). Exceptions to this are where roots fail to penetrate a hard pan. However, soil water will be deeper in wetter seasons/environments where water from earlier rains is stored at depth. To access this deeper soil water late in the season roots require time to grow into it and then to use it, but when time is not available, i.e. in annual crops, a faster-growing, deeper and more branched root system is required.

We have found extensive variation in root system growth (Fig. 7.4). Some of this is

associated with variation in above-ground traits such as early leaf growth and tillering, whereas other variation in root systems appears unrelated to shoot characteristics (Liao *et al.*, 2006; Richards *et al.*, 2007). Duggan *et al.* (2005b) showed in pot experiments that the *tin* gene, which inhibits tillering, allocates more carbon for root growth at about the time tillering commences. This was confirmed in field plots (Richards *et al.*, 2007), where greater branching of the seminal roots was associated with the *tin* gene. Selection for greater above-ground vigour has also resulted in a deeper root system in pot studies, as well as in the field (Richards *et al.*, 2007), and in more vigorous root growth to overcome the physical and biological soil constraints associated with conservation farming systems (Watt *et al.*, 2005).

Selection methods for the number and length of seminal roots at the early seedling stages on absorbent paper sheets have been developed in controlled conditions, whereas long tubes containing sandy soil are used to grow plants to a later stage outside. These

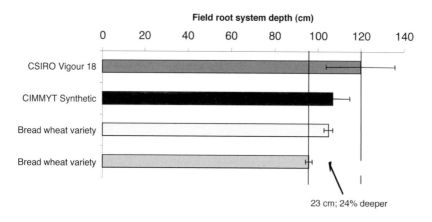

Fig. 7.4. Depths of wheat root systems in the field around the time of flowering and grain-filling in four different wheat lines. The CSIRO Vigour line has roots up to 23 cm deeper than the varieties currently available to farmers; this extra depth is sufficient to increase yield by at least 0.5 t/ha if water is available at depth.

methods are being validated in the field (Watt *et al.*, in press). Measurements of root depth in the field can be fast using hydraulic coring methods, and this has demonstrated significant variation in root depth (Fig. 7.4). In addition, indirect selection methods in the field using canopy temperature depression, leaf senescence and carbon isotope discrimination are also being tested (Richards *et al.*, 2010).

7.2.7 Waxiness/glaucousness

We have advocated selection for enhanced glaucousness in wheat in water-limited environments (Johnson *et al.*, 1983; Richards *et al.*, 1986). The expression of glaucousness is greatest on the abaxial (lower surface) of later-formed leaves around ear emergence, as well as on the leaf sheath and the ear. These photosynthetic surfaces are most exposed to solar radiation when conditions are dry and temperatures increasing. Glaucousness acts as a highly reflective and protective covering on photosynthetic surfaces, and the timing of its greatest expression coincides with determination of the yield components, grain number and grain size (Richards *et al.*, 1986). Because it reflects radiation, we have

found that glaucousness reduces leaf and ear temperature as it reflects radiation, thereby reducing leaf senescence (Richards *et al.*, 1986) and increasing water-use efficiency and yield (Johnston *et al.*, 1983). Glaucousness is largely controlled by major genes on chromosomes 2A and 2B in wheat; however, there are additional modifiers that alter its expression. It has been actively selected for in the Roseworthy wheat breeding programme in Australia (Gil Hollamby, personal communication), and could partly be responsible for the recent outstanding performance of the variety Gladius and related lines in dry conditions in Australia. Glaucousness is a visual trait and therefore easily selected in breeding programmes, provided the appropriate high expression of glaucousness is present in breeding gene pools.

7.2.8 Transpiration efficiency (carbon isotope discrimination)

An improvement in transpiration efficiency (TE), i.e. the ratio of the rate of photosynthesis to transpiration, will be important in all water-limited environments provided it is not negatively associated with factors that increase water use or harvest index.

Farquhar *et al.* (1982) proposed that, for C_3 species, the stable-isotopic composition of plant carbon should reflect TE. Farquhar and Richards (1984) then went on to demonstrate that the degree of discrimination against ^{13}C was indeed related to TE in wheat and that there were genetic differences. An understanding of how carbon isotope discrimination ($\Delta^{13}C$) varies with season, genotype, growth conditions and the tissue used for its measurement has also been described (Condon *et al.*, 1992). A breeding programme was commenced to backcross low $\Delta^{13}C$ (high TE) from the donor parent Quarrion into commercially acceptable wheats. Studies demonstrated that in South-eastern Australia lines selected for low $\Delta^{13}C$ resulted in a 2–15% yield advantage at yield levels between 1 and 5 t/ha when compared with high-$\Delta^{13}C$ sister lines (Rebetzke *et al.*, 2002). Subsequently the varieties Drysdale and Rees were released commercially; these varieties combined high TE with broad-spectrum disease resistance and high grain quality suitable for international markets.

The spring wheats Drysdale and Rees were developed through a limited backcrossing breeding programme targeting the northern wheat-growing region of Australia. This is the region where cultivars with high TE are expected to perform best, as crops are mostly reliant on metering-out water stored in the soil at sowing from monsoonal summer rains. Because low $\Delta^{13}C$ can be associated with lower stomatal conductance, there may be an extra benefit for low $\Delta^{13}C$ in water-limited environments where there is a terminal drought, such as in Australia's northern region, as a lower conductance may conserve soil moisture for use during grain filling, which is likely to increase harvest index. There is evidence that this may contribute to increased yield in varieties such as Drysdale and Rees, as a large grain size was also associated with increased yield (Rebetzke *et al.*, 2002, 2005). In more Mediterranean-type environments, where rainfall is frequent during the winter period, lines with low $\Delta^{13}C$ may be at a disadvantage due to a possible negative association between early growth and low $\Delta^{13}C$ (Condon *et al.*, 2002).

In breeding Drysdale and Rees, the old Australian winter wheat Quarrion was used as the source of high TE (low $\Delta^{13}C$), and Hartog was chosen as the recurrent parent due to its good grain quality, generally robust disease resistance and its broad adaptation in the northern Australian wheat belt. $BC_2F_{4.6}$ lines were developed from selections for low $\Delta^{13}C$ made in the field in the F_3 generation, and again at BC_2. Limited backcrossing was done to retain as much variation as possible in agronomic and grain quality traits so that selection for these traits could also be carried out. Drysdale and Rees were selected in New South Wales and Queensland, respectively, following extensive testing for grain yield, disease resistance and grain quality by the agricultural agencies in those states. Unfortunately, soon after their release, a new exotic strain of stripe rust entered Australia that was virulent on Drysdale and Rees, and this has limited the adoption of these varieties. A more recent spring wheat variety released in Australia derived from parents with low $\Delta^{13}C$ is LPB Scout.

In breeding for high TE, selection for low $\Delta^{13}C$ is made on plant tissue collected at mid-tillering from plants grown under favourable water conditions, in order to maximize the genetic component of $\Delta^{13}C$ variation. However, a low $\Delta^{13}C$ is not always associated with a high yield. There are numerous studies where high $\Delta^{13}C$ was shown to be associated with high yield (e.g. Condon *et al.*, 1987, 2002; Fischer *et al.*, 1998; Voltas *et al.*, 1999; Araus *et al.*, 2003). This may be expected under favourable conditions where more open stomata, and hence a high leaf $\Delta^{13}C$, will increase water use, growth and yield and where the increased water use is replaced, so does not incur a yield penalty. A high $\Delta^{13}C$ in grain may also be related to high yield if that is associated with plants that can access more soil water. This is often found with plants that flower early or in those that have a more effective root system. In addition, high grain yield and high grain $\Delta^{13}C$ may be associated with more effective re-translocation of stored assimilates (see following section).

Typically, such stored assimilates would be expected to have a higher $\Delta^{13}C$ than photosynthate acquired during grain-filling. Clearly, $\Delta^{13}C$ is a complex trait and, while quantitative trait loci (QTL) for $\Delta^{13}C$ have also been identified in several wheat populations, each of these QTL has had a marginal effect and is therefore unlikely to be useful in breeding (Rebetzke *et al.*, 2008b).

7.2.9 Stem carbohydrate storage and remobilization

Temperate cereals accumulate surprisingly large amounts of water-soluble carbohydrates (WSC) in leaf sheaths, and especially stems around anthesis and shortly after (Ruuska *et al.*, 2006). This is a temporary storage only, as most of the carbohydrate is remobilized to the developing grains during grain-filling. This remobilization, together with post-anthesis photosynthesis, is responsible for final grain yield. When dry the stored carbohydrate and its subsequent remobilization make a major

contribution to final grain yield and grain size (van Herwaarden *et al.*, 1998).

Pot studies where water supply can be carefully regulated demonstrated that carbohydrate reserves at anthesis can contribute up to 60% of the final yield under extreme water stress conditions (Richards and Townley-Smith, 1987). The importance of carbohydrate storage and remobilization for increases in grain yield in dry-field environments was demonstrated in mixed-cereal trials involving bread wheat, durum wheat, barley, triticale and oat cultivars (Lòpez-Castañeda and Richards, 1994b). In this study, barley, which is the preferred species in extreme dry conditions, remobilized more stem dry matter between anthesis and maturity, and had the highest yield followed by triticale, bread wheat, durum wheat and then oats (Fig. 7.5).

In wheat, very significant genetic variation in WSC has been reported (Ruuska *et al.*, 2006; Rebetzke *et al.*, 2008a), which is heritable, although multiple loci are involved. Rebetzke *et al.* (2008a) identified between 8 and 16 QTL across three separate wheat populations grown in mainly favourable

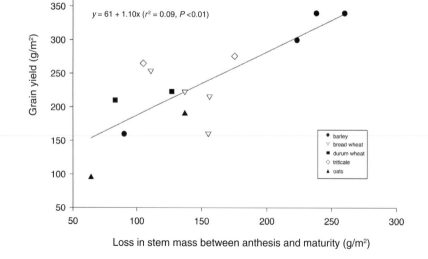

Fig. 7.5. Relationship between grain yield and apparent remobilization of stem and leaf sheath reserves for bread wheat, durum wheat, barley, triticale and oats (adapted from Lòpez-Castañeda and Richards, 1994b).

environments. These QTL often co-located with QTL for plant height, flowering time and spike number. Due to the small contribution of each locus and the influence of genetic background, it is unlikely that these QTL will be used in breeding.

Numerous methods have been proposed for rapid measurement of WSC in crops. Blum (1988) suggested the use of chemical desiccants in large field trials, followed by selection for maintenance of grain size. Although this has significant appeal, due to its potential to screen large numbers in breeding, it is likely to be confounded with variation in flowering time. Other methods include difference in mass between anthesis and final maturity, chemical analyses, near-infrared reflectance of powdered stem samples (van Herwaarden *et al.*, 1998) and, more recently, dry weight per unit of fresh weight (Xue *et al.*, 2009).

There are some issues in relation to WSC accumulation that still need to be addressed before it is used in breeding. These are, first, determining the importance of WSC to grain yield in water-limited environments. Although there is some circumstantial evidence that suggests it is important, it is still confounded with other traits. Secondly, there is the measuring of WSC on a per-unit area basis in field-grown plots. Rather than reporting WSC content, many studies report WSC concentration and this is unlikely to be important. Thirdly, there is the need to develop more effective screening methods for large populations evaluated in rows and/ or canopies.

7.2.10 Leaf-rolling

In certain cereal genotypes the leaves immediately below the reproductive ear often roll longitudinally, exposing the waxy abaxial surface. This is very common in rice (O'Toole and Cruz, 1979) and sorghum (Begg, 1980), but less common in temperate cereals such as wheat. It is considered that the capacity to roll the final formed leaves is a protective mechanism when available soil moisture is depleted, as it can temporarily reduce leaf area and thereby reduce transpiration. It can also reduce the impact of high temperatures and high radiation on otherwise water-stressed leaves, and this is further assisted by the glaucousness expressed on the abaxial surface. This is expected to protect leaf area on days of extreme weather conditions and maintain leaf area so as to complete grain-filling if further rainfall events are likely. The propensity for upper leaves to roll depends on the transverse shape of flag leaves. The transverse shape can be readily observed under well-watered conditions (Sirault, 2007) and is easily selected in breeding programmes.

7.3 Concluding Remarks

Retrospectively, it is evident that the approach enunciated by Passioura (1977) to increase the yield of water-limited crops has been enlightening and has provided clear guidelines for both breeders and agronomists (see also Passioura and Angus, 2010). It has been successful because it proposed a resource-driven approach linked to crop productivity rather than associating yield with drought resistance. A further extension of these ideas, developed by French and Schultz (1984), identified a practical upper limit to the yield of field-grown crops in water-limited environments. This upper limit, linearly related to water supply, was adopted as a benchmark by agronomists and farmers, and has been particularly important in improving the management of water-limited crops.

The approach taken by CSIRO has benefited from a number of factors, the most important of which are summarized thus:

1. A clear physiological framework complemented by a rigorous understanding of the target environment.
2. A strong focus on wheat improvement for a target set of environments.
3. An integrated stable team with skills in agronomy, physiology, molecular biology, genetics and breeding and who are mainly located together and have daily dialogue.

4. A focus on quality phenotyping.
5. A commitment to field research and field validation using appropriate populations fixed for height and maturity but varying for the target trait(s).
6. Stability in funding and a long-term commitment to maintaining a broad skills base.
7. A commitment to the application of results and germplasm to commercial plant breeders, combined with a regular dialogue with breeders.
8. An interaction with farmers and knowledge of the broader cereal industry.

The main bottleneck remains at the interface between the pre-breeding research described here and the commercial breeders who must release new varieties to farmers. Plant breeding is a very integrative science and requires input from multiple disciplines if successful varieties are to be released. Wheat in Australia, for example, is largely sold into discriminating international markets that demand grain with a particular protein content as well as varieties known to be highly regarded for their ability to produce quality end products such as bread, noodles, etc. This generally means they are required to have a flour with suitable mixing times, strength, elasticity, etc. These markets also require varieties that have a high milling yield. Farmers also are very discriminating; not only do they select varieties to grow that have yield and adaptation to their region, but they also want varieties with inbuilt genetic resistance to diseases for dry environments, which are typically low yielding, since additional inputs such as fungicides or herbicides can dramatically increase costs and reduce profits. Farmers who sell grain into the marketplace will also want varieties that yield well in the favourable years, as these years are most profitable. On the other hand, subsistence farmers are more interested in growing varieties that are productive in the poor years but which may not express the yield potential available in the favourable years. Thus successful breeding programmes must be able to integrate yield improvement with disease resistance, with breeding for end product requirements and with farmers' needs.

Weaving a physiological approach into this milieu is not easy. Some of the approaches and pitfalls of integrating pre-breeding research with trait validation and varietal release have recently been outlined by Richards *et al.* (2010). However, a scientific understanding of factors underpinning adaptation to water-limited environments, coupled with good genetics and breeding, will deliver potential varieties and/or parents with potential for improved performance under drought.

7.4 Acknowledgement

John Passioura continues to challenge us and to contribute to our research and to the grains industry.

References

Araus, J.L., Villegas, D., Aparicio, N., García del Moral, L.F., El Hani, S., Rharrabti, Y. *et al.* (2003) Environmental factors determining carbon isotope discrimination and yield in durum wheat under Mediterranean conditions. *Crop Science* 43, 170–180.

Atsmon, D. and Jacobs, E. (1977) A newly bred 'gigas' form of bread wheat (*Triticum aestivum* L.); morphological features and thermo-periodic responses. *Crop Science* 17, 31–35.

Batten, G.D., Fettell, N.A., Mead, J.A. and Khan, M.A. (1999) Effect of sowing date on the uptake and utilisation of phosphorus by wheat (cv. Osprey) grown in central New South Wales. *Australian Journal of Experimental Agriculture* 39, 161–170.

Begg, J.E. (1980) Morphological adaptations of leaves to water stress. In: Turner, N.C. and Kramer, P.J. (eds) *Adaptation of Plants to Water and High Temperature Stresses*. John Wiley and Sons, Inc., New York, pp. 33–42.

Begg, J.E. and Turner, N.C. (1976) Crop water deficits. *Advances in Agronomy* 28, 161–217.

Blum, A. (1988) Improving wheat grain filling under stress by stem reserve mobilization. *Euphytica* 100, 77–83.

Botwright, T.L., Condon, A.G., Rebetzke, G.J. and Richards, R.A. (2002) Field evaluation of early vigour for genetic improvement of grain yield in wheat. *Australian Journal of Agricultural Research* 53, 1137–1145.

Coleman, R.K., Gill, G.S. and Rebetzke, G.J. (2001) Identification of quantitative trait loci for traits conferring weed competitiveness in wheat (*Triticum aestivum* L.). *Australian Journal of Agricultural Research* 52, 1235–1246.

Condon, A.G. and Giunta, F. (2003) Yield response of restricted-tillering wheat to transient waterlogging on duplex soils. *Australian Journal of Agricultural Research* 54, 957–967.

Condon, A.G., Richards, R.A. and Farquhar, G.D. (1987) Carbon isotope discrimination is positively correlated with grain yield and dry matter production in field grown wheat. *Crop Science* 27, 996–1001.

Condon, A.G., Richards, R.A. and Farquhar, G.D. (1992) The effect of variation in soil water availability, vapour pressure deficit and nitrogen nutrition on carbon isotope discrimination in wheat. *Australian Journal of Agricultural Research* 43, 935–947.

Condon, A.G., Richards, R.A. and Farquhar, G.D. (1993) Relationships between carbon isotope discrimination, water use efficiency and transpiration efficiency for dryland wheat. *Australian Journal of Agricultural Research* 44, 1693–1711.

Condon, A.G., Richards, R.A., Rebetzke, G.R. and Farquhar, G.D. (2002) Improving intrinsic water-use efficiency and crop yield. *Crop Science* 42, 122–131.

Cooper, P.J.M., Gregory, P.J., Keatinge, J.D.H. and Brown, S.C. (1987) Effects of fertilizer, variety and location on barley production under rainfed conditions in northern Syria. 2. Soil water dynamics and crop water use. *Field Crops Research* 16, 67–84.

Duggan, B.L., Richards, R.A., van Herwaarden, A.F. and Nettell, N.A. (2005a) Agronomic evaluation of a tiller inhibition gene (*tin*) in wheat. I. Effect on yield, yield components, and grain protein. *Australian Journal of Agricultural Research* 56, 169–178.

Duggan, B.L., Richards, R.A. and van Herwaarden, A.F. (2005b) Agronomic evaluation of a tiller inhibition gene (*tin*) in wheat. II. Growth and partitioning of assimilate. *Australian Journal of Agricultural Research* 56, 179–186.

Ellis, M.H., Rebetzke, G.J., Chandler, P., Bonnett, D., Spielmeyer, W.I. and Richards, R.A. (2004) The effect of different height reducing genes on the early growth of wheat. *Functional Plant Biology* 31, 583–589.

Ellis, M.H., Rebetzke, G.J., Spielmeyer, W., Richards, R.A. and Bonnett, D.B. (2005) Molecular mapping of gibberellin-responsive dwarfing genes in bread wheat (*Triticum aestivum* L.). *Theoretical and Applied Genetics* 111, 423–430.

Evans, L.T., Bingham, J., Jackson, P. and Sutherland, J. (1972) Effect of awns and drought on the supply of photosynthate and its distribution within wheat ears. *Annals of Applied Biology* 70, 67–76.

Farquhar, G.D., O'Leary, M.H. and Berry, J.A. (1982) On the relationship between carbon isotope discrimination and the intercellular carbon dioxide concentration in leaves. *Australian Journal of Plant Physiology* 9, 121–137.

Farquhar, G.D. and Richards, R.A. (1984) Isotopic composition of plant carbon correlates with water-use efficiency of wheat genotypes. *Australian Journal of Plant Physiology* 11, 539–552.

Fischer, R.A., (1979) Growth and water limitation to dryland wheat yield in Australia: a physiological framework. *Journal of the Australian Institute of Agricultural Science* 45, 83–94.

Fischer, R.A. and Stockman, Y.M. (1986) Increased kernel number in Norin 10-derived dwarf wheat: evaluation of the cause. *Australian Journal of Plant Physiology* 13, 767–784.

Fischer, R.A. and Turner, N.C. (1978) Plant productivity in the arid and semi-arid zones. *Annual Review of Plant Physiology* 29, 277–317.

Fischer, R.A., Rees, D., Sayre, K.D., Lu, Z., Condon, A.G. and Saavendra, A.L. (1998) Wheat yield progress is associated with higher stomatal conductance, higher photosynthetic rate and cooler canopies. *Crop Science* 38, 1467–1475.

French, R.J. and Schultz, J.E. (1984) Water use efficiency of wheat in a Mediterranean-type environment. I. The relation between yield, water use and climate. *Australian Journal of Agricultural Research* 35, 743–764.

Gomez-Macpherson, H. and Richards, R.A. (1995) Effect of sowing time on yield and agronomic characteristics of wheat in south-eastern Australia. *Australian Journal of Agricultural Research* 46, 1381–1399.

Hurd, E.A. (1974) Phenotype and drought tolerance in wheat. *Agricultural Meteorology* 14, 39–55.

Johnson, D.A., Richards, R.A. and Turner, N.C. (1983) Yield, water relations, gas exchange, and surface reflectances of near-isogenic wheat lines differing in glaucousness. *Crop Science* 23, 318–325.

Kelman, W.M. and Dove, H. (2009) Growth and phenology of winter wheat and oats in a dual-purpose management system. *Crop and Pasture Science* 60, 921–932.

Keyes, G.J., Paolillo, D.J. and Sorrells, M.E. (1989) The effects of dwarfing genes *Rht1* and *Rht2* on cellular dimensions and rate of leaf elongation in wheat. *Annals of Botany* 64, 683–690.

Kirkegaard, J.A., Lilley, J.M., Howe, G.N. and Graham, J.M. (2007) Impact of subsoil water use on wheat yield. *Australian Journal of Agricultural Research* 58, 303–315.

Levitt, J. (1972) *Responses of Plants to Environmental Stresses*. Academic Press, New York.

Liang, Y.L. and Richards, R.A. (1994) Coleoptile tiller development is associated with fast early vigour in wheat. *Euphytica* 80, 119–124.

Liao, M., Palta, J.A. and Fillery, I.R.P. (2006) Root characteristics of vigorous wheat improve early nitrogen uptake. *Australian Journal of Agricultural Research* 57, 1097–1107.

Lilley, J.M. and Kirkegaard, J.A. (2007) Seasonal variation in the value of subsoil water to wheat: simulation studies in southern New South Wales. *Australian Journal of Agricultural Research* 58, 1115–1128.

Lòpez-Castañeda, C. and Richards, R.A. (1994a) Variation in temperate cereals in rainfed environments. I. Grain yield, biomass and agronomic characteristics. *Field Crops Research* 37, 51–62.

Lòpez-Castañeda, C. and Richards, R.A. (1994b) Variation in temperate cereals in rainfed environments. II. Phasic development and growth. *Field Crops Research* 37, 63–75.

Lòpez-Castañeda, C. and Richards, R.A. (1994c) Variation in temperate cereals in rainfed environments. III Water use and water-use efficiency. *Field Crops Research* 39, 85–98.

Lòpez-Castañeda, C., Richards, R.A., Farquhar, G.D. and Williamson, R.E. (1996) Seed and seedling characteristics contributing to variation in early vigour among temperate cereals. *Crop Science* 36, 1257–1266.

Nix, H.A. and Fitzpatrick, E.A. (1968) An index of crop water stress related to wheat and grain sorghum yields. *Agricultural Meteorology* 6, 321–337.

O'Toole, J.C. and Cruz, R.T. (1979) Leaf rolling and transpiration. *Plant Science Letters* 16, 111–114.

Passioura, J.B. (1972) The effect of root geometry on the yield of wheat growing on stored water. *Australian Journal of Agricultural Research* 23, 745–752.

Passioura, J.B. (1976) Physiology of Grain Yield in Wheat Growing on Stored Water. *Australian Journal of Plant Physiology* 3, 559–565.

Passioura, J.B. (1977) Grain yield, harvest index, and water use of wheat. *Journal of the Australian Institute of Agricultural Science* 43, 117–121.

Passioura, J.B. and Angus, J.F. (2010) Improving productivity of crops in water-limited environments. *Advances in Agronomy* 106, 37–75.

Penrose, L.D.J. (1993) Yield of early dryland sowing of wheat with winter and spring habit in southern and central New South Wales. *Australian Journal of Experimental Agriculture* 33, 601–608.

Rebetzke, G.J. and Richards, R.A. (1999) Genetic improvement of early vigour in wheat. *Australian Journal of Agricultural Research* 50, 291–301.

Rebetzke, G.J. and Richards, R.A. (2000) Gibberellic acid-sensitive dwarfing genes reduce plant height to increase kernel number and grain yield of wheat. *Australian Journal of Agricultural Research* 51, 235–245.

Rebetzke, G.J., Richards, R.A., Fischer, V.M. and Micklelson, B.J. (1999) Breeding long coleoptile, reduced height wheats. *Euphytica* 106, 158–168.

Rebetzke, G.J., Condon, A.G., Richards, R.A. and Farquhar, G.D. (2002) Selection for reduced carbon isotope discrimination increases aerial biomass and grain yield of rainfed bread wheat. *Crop Science* 42, 739–745.

Rebetzke, G.J., Bruce, S. and Kirkegaard, J.A. (2005) Genotypic increases in coleoptile length improves emergence and early vigour with crop residues. *Plant and Soil* 270, 87–100.

Rebetzke, G.J., Richards, R.A., Fettell, A., Long, M., Condon, A.G. and Botwright, T.L. (2007a) Genotypic increases in coleoptile length improve wheat establishment, early vigour and grain yield with deep sowing. *Field Crops Research* 100, 10–23.

Rebetzke, G.J., Ellis, M.H., Bonnett, D.G. and Richards, R.A. (2007b) Molecular mapping of genes for coleoptile growth in bread wheat (*Triticum aestivum* L.). *Theoretical and Applied Genetics* 114, 1173–1183.

Rebetzke, G.J., van Herwaarden, A., Jenkins, C., Ruuska, S., Tabe, L., Lewis, D. *et al.* (2008a) Quantitative trait loci for water soluble carbohydrates and associations with agronomic traits in wheat. *Australian Journal of Agricultural Research* 59, 891–905.

Rebetzke, G.J., Condon, A.G., Richards, R.A., Appels, R. and Farquhar, G.D. (2008b) Quantitative trait loci for carbon isotope discrimination are repeatable across environments and wheat mapping populations. *Theoretical and Applied Genetics* 118, 123–137.

Rebetzke, G.J., Lòpez-Casteñeda, C., Botwright-Acuna, T., Condon, A.G. and Richards, R.A. (2008c) Inheritance of coleoptile tiller appearance and size in wheat. *Australian Journal of Agricultural Research* 59, 863–873.

Rebetzke, G.J., Ellis, M.H., Bonnett, D.G., Mickelson, B., Condon, A.G. and Richards, R.A.

(2011a) Utility of gibberellin-responsive dwarfing genes for genetic improvement of bread wheat. *Field Crops Research* (in press).

Rebetzke, G.J., Ellis, M.H., Bonnett, D.G., Condon, A.G., Falk, D. and Richards, R.A. (2011b) The Rht13 dwarfing gene reduces peduncle length and plant height to increase gain number and yield of wheat. *Field Crops Research* (in press).

Richards, R.A. (1988) A tiller inhibitor gene in wheat and its effect on plant growth. *Australian Journal of Agricultural Research* 39, 749–757.

Richards, R.A. (1991) Crop improvement for temperate Australia: Future opportunities. *Field Crops Research* 26, 141–169.

Richards, R.A. (1992) The effect of dwarfing genes in spring wheat in dry environments. II. Growth, water use and water use efficiency. *Australian Journal of Agricultural Research* 43, 529–539.

Richards, R.A. and Lukacs, Z. (2002) Seedling vigour in wheat – sources of variation for genetic and agronomic improvement. *Australian Journal of Agricultural Research* 53, 41–50.

Richards, R.A. and Passioura, J.B. (1981a) Seminal Root Morphology and Water Use of Wheat I. Environmental Effects. *Crop Science* 21, 249–252.

Richards, R.A. and Passioura, J.B. (1981b) Seminal Root Morphology and Water Use of Wheat II. Genetic Variation. *Crop Science* 21, 253–255.

Richards, R.A. and Passioura, J.B. (1989) A breeding program to reduce the diameter of the major xylem vessel in the seminal roots of wheat and its effect on grain yield in rain-fed environments. *Australian Journal of Agricultural Research* 40, 943–950.

Richards, R.A. and Townley-Smith, T.F. (1987) Variation in leaf development and its effect on water use yield and harvest index of droughted wheat. *Australian Journal of Agricultural Research* 38, 983–992.

Richards, R.A., Rawson, H.M. and Johnson, D.A. (1986) Glaucousness in wheat: Its development and effect on water-use efficiency, gas exchange and photosynthetic tissue temperatures. *Australian Journal of Plant Physiology* 13, 465–473.

Richards, R.A., Rebetzke, G.J., Condon, A.G. and van Herwaarden, A.F. (2002) Breeding opportunities for increasing the efficiency of water use and crop yield in temperate cereals. *Crop Science* 42, 111–121.

Richards, R.A., Watt, M. and Rebetzke, G.J. (2007) Physiological traits and cereal germplasm for sustainable agricultural systems. *Euphytica* 154, 409–425.

Richards, R.A., Rebetzke, G.J., Watt, M., Condon, A.G., Spielmeyer, W. and Dolferus, R. (2010) Breeding for improved water productivity in temperate cereals: phenotyping, quantitative trait loci, markers and the selection environment. *Functional Plant Biology* 37, 85–97.

Ruuska, S., Rebetzke, G.J., van Herwaarden, A., Richards, R.A., Fettell, N., Tabe, L. *et al.* (2006) Genotypic variation in water-soluble carbohydrate accumulation in wheat. *Functional Plant Biology* 33, 799–809.

Sirault, W.R.R. (2007) Leaf rolling in wheat. PhD thesis, The Australian National University, Canberra.

Spielmeyer, W. and Richards, R.A. (2004) Comparative mapping of wheat chromosome 1AS which contains the tiller inhibition gene (tin) with rice chromosome 5S. *Theoretical and Applied Genetics* 109, 1303–1310.

van Herwaarden, A.F., Angus, J.F., Richards, R.A. and Farquhar, G.D. (1998) 'Haying-off', the negative grain yield response of dryland wheat to nitrogen fertiliser. II. Carbohydrate and protein dynamics. *Australian Journal of Agricultural Research* 49, 1083–1093.

Virgona, J.M., Gummer, F.A.J. and Angus, J.F. (2006) Effects of grazing on wheat growth, yield, development, water use, and nitrogen use. *Australian Journal of Agricultural Research* 57, 1307–1319.

Voltas, J., Romagosa, I., Lafarga, A., Armesto, A.P., Sombrero, A. and Araus, J.L. (1999) Genotype by environment interaction for grain yield and carbon isotope discrimination of barley in Mediterranean Spain. *Australian Journal of Agricultural Research* 50, 1263–1271.

Watt, M., Kirkegaard, J.A. and Rebetzke, G.J. (2005) A wheat genotype developed for rapid leaf growth copes well with the physical and biological constraints of unploughed soil. *Functional Plant Biology* 32, 695–706.

Xue, G., McIntyre, C.L., Rattey, A.R., van Herwaarden, A.F. and Shorter, R. (2009) Use of dry matter content as a rapid and low-cost estimate for ranking genotypic differences in water-soluble carbohydrate concentrations in the stem and leaf sheath of *Triticum aestivum*. *Crop and Pasture Science* 60, 51–59.

8

Molecular Breeding for a Changing Climate: Bridging Ecophysiology and Molecular Biology

R. Tuberosa, M. Maccaferri, C. Colalongo and S. Salvi

8.1 Introduction

The recent changes in weather patterns brought about by global warming have for the most part negatively impacted crop productivity and food security, thus threatening the livelihood of the billions living in less-developed countries where the consequences of a changing climate have been more pronounced. During the past decade, the frequency, duration and intensity of extreme weather conditions have increased, thus exacerbating the reduction in crop yield due to abiotic constraints (Schmidhuber and Tubiello, 2007; Pennisi, 2008; Ainsworth and Ort, 2010; Challinor et al., 2010; Parry and Hawkesford, 2010; Langridge and Fleury, 2011). Among all abiotic stresses most affected by climate change, drought is the major one curtailing crop productivity worldwide, thus contributing to socio-economic unrest as underlined by the food riots sparked by the sharp increase in rice and wheat price consequent to the reduction in cereal production due to the severe droughts that occurred in 2008 in several developing countries (Archer et al., 2008; Xiong et al., 2009) and in 2010 in Russia, in combination with excessive heat.

Science and technology contribute strategies to mitigate this daunting scenario, even more worrying in view of the expected increase in human population and parallel decrease in natural resources that will be required for a more profitable and environmentally sustainable agriculture (Borlaug and Dowswell, 2005). Until recently, conventional breeding and agronomic approaches have allowed for a constant increase in crop productivity across a broad range of environmental constraints (Blum, 1988; Ludlow and Muchow, 1990; Boyer, 1996; Duvick and Cassman, 1999; Tollenaar and Wu, 1999; Duvick, 2005; Passioura, 2006; Richards, 2006; Delgado and Berry, 2008; James et al., 2008; Farooq et al., 2009; Li et al., 2009; Serraj et al., 2009; Ella et al., 2010; Passioura and Angus, 2010; Verulkar et al., 2010; Zhang et al., 2010). This remarkable progress has mainly been achieved through an empirical approach in which yield has been the preferred target of the selection process, often pursued with no or rather limited knowledge of the factors involved in the adaptive response to the prevailing environmental constraints (Campos et al., 2004; Borlaug and Dowswell, 2005; Borlaug, 2007; Reynolds et al., 2009b).

None the less, the present rate of increase in crop production will be insufficient to meet the future need of a rapidly increasing mankind while meeting the requirements for a more sustainable agriculture (Tester and Langridge, 2010). Genomics-assisted crop improvement (Varshney and Tuberosa, 2007) and other biotechnological approaches (e.g. double-haploid production, genetic engineering, etc.) provide a means to enhance the yearly rate of increase in crop yield while advancing our knowledge of the genetic basis of the adaptive response of crops to abiotic stress. Irrigation and other agronomic interventions will also be instrumental in mitigating the negative effects of climate change (Kang et al., 2009;

Srivastava *et al.*, 2010). In rice, one of the crops most exposed to the negative effects of climate change, the buffer effect of irrigation against climate change impacts will depend on the nature of the respective irrigation system (Wassmann *et al.*, 2009). In this respect, the value of interdisciplinary approaches should be underlined, particularly when dealing with multiple stresses (Parry and Lea, 2009). Surprisingly, the co-occurrence of different stresses has rarely been addressed in molecular studies investigating plant acclimatization (Mittler, 2006). The response of plants to a combination of two different abiotic stresses is usually unique and cannot be directly extrapolated from the response of plants to each of the different stresses applied individually.

Clearly, a better understanding of the genetic and molecular framework underlying crop performance under environmentally constrained conditions is an essential, albeit insufficient, prerequisite for more effective exploitation of the novel approaches ushered in by the genomics era (Pakniyat *et al.*, 1997; Tuberosa *et al.*, 2002a, 2011a; Nguyen and Blum, 2004; Pelleschi *et al.*, 2006; Valliyodan and Nguyen, 2006; Maccaferri *et al.*, 2008b; Xu and Crouch, 2008; Habash *et al.*, 2009; Ruan *et al.*, 2010; Tester and Langridge, 2010). Genomics approaches offer ample opportunities to: (i) dissect the genetic and functional basis of yield under environmentally constrained conditions (Tuberosa *et al.*, 2002a; Leung, 2008; Colmer and Voesenek, 2009; Manavalan *et al.*, 2009; Ashraf, 2010; Langridge and Fleury, 2011); and (ii) deploy this information to implement marker-assisted selection (MAS) to improve crop performance under abiotic stress. Although remarkable progress has been achieved in identifying and in some cases cloning the loci regulating adaptation to abiotic stress (Salvi *et al.*, 2007; Collins *et al.*, 2008; Fleury *et al.*, 2010; Xue-Xuan *et al.*, 2010), only in a limited number of cases has this knowledge contributed towards the release of improved cultivars obtained via MAS (Sinclair *et al.*, 2004; Passioura, 2007; Collins *et al.*, 2008; Xu and Crouch, 2008; Herve and Serraj, 2009), particularly for the improvement of crop yield under water-limited conditions, one of the least heritable and most difficult traits to select for.

This chapter surveys how genomics approaches have helped to elucidate the genetic basis of crop performance under abiotic constraints. Additionally, the review shows how this information has been used in breeding programmes to improve crop performance and how it might help agriculture better to withstand the negative effects of climate change.

8.2 Dissecting the Genetic Basis of Abiotic Stress Tolerance

The dissection of the genetic basis of any trait can be carried out via forward-genetics or reverse-genetics approaches. While the forward-genetics approach capitalizes on the observation of the phenotype as its starting point, the reverse-genetics approach focuses on sequence and functional information of candidate sequences (e.g. expressed sequence tags: ESTs) which are postulated to play a role in the expression of the target trait (Tuberosa and Salvi, 2006; Leung, 2008; Chen *et al.*, 2009; Raju *et al.*, 2010). Under the latter approach, the use of *Arabidopsis* and other model species (e.g. rice, *Brachypodium*, etc.) has been instrumental in elucidating the metabolic and signalling pathways regulating the adaptive response to environmental constraints (Shinozaki and Yamaguchi-Shinozaki, 2007; Fujii and Zhu, 2009; Moore *et al.*, 2009; Park *et al.*, 2009; Kuromori *et al.*, 2010; Yoshida *et al.*, 2010). As sequence information becomes more readily available, the reverse-genetics approach becomes more feasible. None the less, the majority of the results obtained so far in the dissection of the genetic basis of crop performance have been obtained via forward-genetics. In particular, in view of the quantitative nature of the genetic control of yield and the morpho-physiological features that influence crop performance under abiotic stress conditions, extensive work has been carried out to identify the quantitative trait loci (QTLs) that underscore such traits.

8.2.1 QTL discovery via biparental linkage mapping

Following the introduction in the 1980s and 1990s of DNA-based markers capable of reporting the effects of functional polymorphisms, genome-wide search for loci controlling the adaptive response of crops to abiotic stress became a reality. Until the introduction in the past decade of association mapping based on the evaluation of large sets (>150–200) unrelated accessions, QTL identification has been pursued via linkage mapping based on the evaluation of biparental populations of recombinant inbred lines (RILs, usually varying from 100 up to 300) derived from the cross of parental lines differing for the target trait(s) (Tanksley, 1993; Hospital and Charcosset, 1997; Price and Tomos, 1997; Beavis, 1998; Melchinger *et al.*, 1998; Kao *et al.*, 1999; Specht *et al.*, 2001; Bernardo *et al.*, 2006; Bernardo and Yu, 2007; Ragot and Lee, 2007; Varshney and Tuberosa, 2007; Xu and Crouch, 2008; Varshney and Dubey, 2009; Xu, 2010).

To date, hundreds of studies have reported on the identification of QTLs for yield and other agronomically relevant traits. Notwithstanding these remarkable results, a number of factors hinder a more effective deployment of this information in breeding programmes. One of such factors is the limited accuracy in defining the most likely position of the QTL and the boundaries of its support interval along the chromosome. Notably, when a number of studies investigate similar traits in the same species and share common markers, it is possible to apply meta-analysis in order to confine the QTL more precisely (Salvi *et al.*, 2010), hence improving the effectiveness of MAS and facilitating the identification of the most feasible candidate genes. Importantly, the availability for a particular species of maps obtained with different crosses and sharing common polymorphisms allows for the construction of a consensus map that in turn enables an even more accurate comparative analysis (e.g. meta-analysis) of QTL positions, an important prerequisite for the identification of the most interesting QTLs

from a selection standpoint (Khavkin and Coe, 1997; Goffinet and Gerber, 2000; Tuberosa *et al.*, 2002b; Zheng *et al.*, 2003; Chardon *et al.*, 2004; Sawkins *et al.*, 2004; Li, X. *et al.*, 2005; Khowaja *et al.*, 2009; Salvi *et al.*, 2010; Hao *et al.*, 2010).

A recent meta-analysis of drought-related QTLs in the Bala × Azucena mapping population allowed Khowaja *et al.* (2009) to compile data from 13 experiments and 25 independent screens providing 1650 individual QTLs separated into five trait categories: drought avoidance, plant height, plant biomass, leaf morphology and root traits. The confidence intervals along the chromosomes for the meta-QTLs identified by Khowaja *et al.* (2009) ranged from 5.1 to 14.5 cM with an average of 9.4 cM, an interval that on average contains ~180 genes in rice. In maize a recent meta-analysis, based on a total of 239 QTLs detected under water-stressed conditions and 160 QTLs detected under control conditions from 12 populations tested in 22 experiments, allowed Hao *et al.* (2010) to identify 39 consensus QTLs under water stress and 36 under control conditions. A valuable feature of consensus maps is that they enable us to compare the map position of QTLs with that of mutants that might represent feasible candidate genes influencing the investigated trait (Dubey *et al.*, 2009). Accordingly, one mutant at the *ERECTA* locus capable of influencing transpiration efficiency in *Arabidopsis* was shown to co-localize with a naturally occurring QTL for the same trait (Masle *et al.*, 2005). In maize, Salvi *et al.* (2011b) have confined a major QTL for number of seminal roots to a 5 cM interval where the *rtcs* (*rootless for crown and lateral seminal root*) mutant has been assigned (Taramino *et al.*, 2007). Hence, the information acquired with the evaluation of mutants can have great value for unravelling the nature of QTLs, even more so in view of the availability of methods such as EcoTILLING (Comai *et al.*, 2004; Till *et al.*, 2007) that enable us to evaluate the effect of allelic diversity at candidate loci.

The first study attempting the identification of QTLs for crop adaptation to an abiotic stress (drought) was published (on tomato) over two decades ago (Martin *et al.*, 1989). In the mid-1990s, the introduction of new molecular markers such as simple sequence repeats (SSRs; Taramino and Tingey, 1996) and amplified fragment length polymorphisms (AFLPs; Vuylsteke *et al.*, 1999) facilitated a more systematic quest for QTLs for abiotic stress tolerance in several crops (Lebreton *et al.*, 1995; Quarrie *et al.*, 1997; Tuberosa *et al.*, 1998; Sanguineti *et al.*, 1999; Flowers *et al.*, 2000; Price *et al.*, 2000a, b, 2002; Herve *et al.*, 2001). The construction of maps reporting the position of ESTs ushered in the candidate gene approach for the discovery of the sequence that underpinned target QTLs (Pelleschi *et al.*, 1999; Pflieger *et al.*, 2001; Nguyen *et al.*, 2004; Chao *et al.*, 2006; Tondelli *et al.*, 2006; Varshney *et al.*, 2009a). Accordingly, linkage maps enriched with function-specific genes have repeatedly been utilized for QTL analysis (Andersen and Lubberstedt, 2003; Diab *et al.*, 2004; Gardiner *et al.*, 2004; Nguyen *et al.*, 2004; Marino *et al.*, 2009). In barley, a number of differentially expressed sequence tags (dESTs) and candidate genes for drought response identified by Ozturk *et al.* (2002) were mapped in Tadmor × Er/Apm (Diab *et al.*, 2004). A survey of 100 sequenced probes from two cDNA libraries previously constructed from drought-stressed barley (Ozturk *et al.*, 2002) and 12 candidate genes allowed for the addition of 33 loci to a previously published map (Diab *et al.*, 2004). Two candidate genes and ten dESTs were found associated with one or more of the 68 QTLs for drought tolerance traits that were mapped on the same population.

8.2.2 QTL discovery via association mapping

During the past decade, association mapping based on the evaluation of sufficiently large panels of unrelated accessions (~150 or more) has provided an additional option to identify the loci (genes and/or QTLs) for target traits (Flint-Garcia *et al.*, 2005; Gupta *et al.*, 2005; Maccaferri *et al.*, 2005, 2011b; Buckler *et al.*, 2006; Burke *et al.*, 2007; Ersoz *et al.*, 2007; Zhu *et al.*, 2008; Rafalski,

2010). Although most of the association mapping studies published so far have targeted traits (e.g. resistance to biotic stress and quality traits) with a genetic basis less complex than abiotic stress tolerance, applications of the association approach to investigate traits providing drought adaptation have been reported in loblolly pine (Gonzalez-Martinez *et al.*, 2008), maize (Krill *et al.*, 2010; Setter *et al.*, 2011) and durum wheat (Sanguineti *et al.*, 2007; Maccaferri *et al.*, 2011b).

Important factors requiring careful consideration for best deploying association mapping are the level of linkage disequilibrium (LD, namely the level of non-random assortment of alleles at different loci) among the investigated accessions and the presence of population structure that could greatly increase false-discovery rate (i.e. Type-I error; Ersoz *et al.*, 2007). Populations characterized by high LD (>1 cM) are well suited for a genome-wide search (Maccaferri *et al.*, 2005, 2011b; Breseghello and Sorrells, 2006; Rostoks *et al.*, 2006; Bagge *et al.*, 2007; Crossa *et al.*, 2007; Somers *et al.*, 2007; Royo *et al.*, 2010). Conversely, validating the role of a candidate sequence requires the utilization of panels with much lower LD (<10 kb, i.e. a small fraction of 1 cM, depending on the ratio of the genetic and physical distance), hence a much higher level of genetic resolution, a condition that is typical of allogamous species like maize (Buckler *et al.*, 2009). An interesting example of the high level of genetic resolution made possible through association mapping is shown by the fine mapping and, in at least one case, cloning of QTLs for flowering time in maize (Salvi *et al.*, 2007; Ducrocq *et al.*, 2008; Buckler *et al.*, 2009). In particular, association mapping revealed that the most important QTL for flowering time per se in maize (*Vgt1*: *Vegetative to generative transition 1*) is controlled by a ~2.3 kb, non-coding, long-distance enhancer region that regulates the expression of *ZmRap2.7*, a gene that encodes for a transcription factor known to regulate flowering time also in *Arabidopsis* (Salvi *et al.*, 2007). Another remarkable example in which the functional polymorphism responsible for phenotypic variability was located in a non-coding region far (~5 kb) from the structural gene is provided by the work of Kochian and colleagues for the cloning of a major QTL for aluminium tolerance in sorghum (Magalhaes *et al.*, 2007; Magalhaes, 2010). Clearly, only a positional cloning approach is able to provide this kind of result and unequivocally to highlight the role of non-coding regions in controlling the level of expression of a particular gene. Recently, association mapping has been used in maize to investigate the role of several hundred candidate genes in the adaptive response to drought (Setter *et al.*, 2011).

Notwithstanding the clear advantages of association mapping over biparental linkage mapping (e.g. multi-allelism, higher genetic variability and genetic resolution, no need to assemble mapping populations, shorter time required to identify relevant loci, etc.), a major limitation to a more widespread utilization of association mapping is represented by the high rate of false positives (i.e. Type-I error rate) and hence spurious association, due to the presence of a hidden population structure (Ersoz *et al.*, 2007). An additional constraint to a more widespread utilization of association mapping for QTL mapping and/or candidate gene validation, particularly for yield under water-limited conditions, may well relate to factors other than computational and statistical issues. For highly integrative and functionally complex traits such as yield, particularly under adverse conditions, association mapping may fall short of expectations due to the fact that similar phenotypic values in different genotypes can result from: (i) the action of different gene networks; and/or (ii) trait compensation (e.g. yield components), thus inevitably undermining the identification of significant marker–trait association. Although the number of studies in this direction is clearly too limited to draw more certain conclusions on the validity of association mapping to identify QTLs for yield, recent results in durum wheat suggest that the identification of yield QTLs under different water regimes might be more effectively pursued via biparental mapping (Maccaferri *et al.*, 2008a, 2011b).

Although similar limitations also pertain to a mapping population developed from the cross of two lines, their relevance in the case of association mapping for complex traits is increased by the wider functional variability explored with association mapping. This is particularly the case whenever the investigated trait (e.g. yield under drought conditions) is strongly influenced by flowering time or plant height; in this case, the overwhelming effects on yield of these phenological covariates will overshadow the effects due to the action of yield per se loci, certainly more valuable in order to boost yield potential. Therefore, the high functional variation typical of association mapping panels, actually perceived advantageous when compared with biparental mapping, might partially undermine the effectiveness in identifying QTLs for yield or other physiologically complex traits. Clearly, more case studies are required before drawing conclusions on the validity of association mapping for the discovery of QTLs for abiotic stress tolerance.

A valuable evolution of association mapping is represented by nested association mapping (NAM; Buckler *et al.*, 2009). Maize is the first crop in which NAM has been applied to identify QTLs for traits (e.g. flowering time, root architecture, leaf angle, etc.) relevant for adaptation to drought-prone environments. By combining association mapping with QTL linkage mapping based on 25 biparental RIL populations developed from crosses of 25 founders with a common tester, Buckler and co-workers have assembled the most powerful QTL dissection platform ever devised in plant species (Yu *et al.*, 2008). The first results obtained using this NAM platform for flowering time have validated the value and the power of this approach (Buckler *et al.*, 2009; McMullen *et al.*, 2009), the main limiting factor being the phenotyping requirements (~5000 lines) to take full advantage of the superior genetic power made possible by the NAM plaform. The high diversity of the 25 maize founders used in the NAM scheme coupled with the rapid LD decay (~2 kb) allow for a very high mapping resolution, a valuable feature for an effective validation of the role and function of candidate genes. With the NAM approach, the improvement in resolution due to the rapid decay of LD is thus fully exploited while avoiding its drawbacks (e.g. the need for a large number of markers) by projecting the genomic information from the founders to their RILs (Yu *et al.*, 2008). Presently, NAM schemes are being implemented in several other crops and will provide new sets of QTL data and a better understanding of how the QTL lanscape is shaped and influences tolerance to abiotic stresses.

8.2.3 QTL discovery and cloning with introgression libraries

An additional option in QTL mapping is provided by the use of congenic strains obtained through the introgression, via backcrossing, of portions (~20–30 cM) of a donor genome of a line with valuable features for the target trait(s) into a common, agronomically valid recurrent background (Zamir, 2001). The final objective is to assemble a collection of introgression library lines (ILLs; at least 70–80 or more lines for each cross), basically a collection of near-isogenic lines (NILs) each one differing for the introgressed chromosome portion and collectively representing most of the donor genome (Eshed and Zamir, 1995; Zamir, 2001; Li, Z. *et al.*, 2005; Tan *et al.*, 2007; Varshney and Dubey, 2009). A major advantage of ILLs is the rapid progress that they allow for the fine mapping and positional cloning of major QTLs (Eshed and Zamir, 1995; Paran and Zamir, 2003). Besides the well-documented effectiveness of ILLs for the mapping and cloning of QTLs in tomato (Eshed *et al.*, 1992; Paran and Zamir, 2003; Gur *et al.*, 2004; Xu *et al.*, 2008), ILLs have been used for mapping drought-adaptive QTLs in rice (Moncada *et al.*, 2001; Li, Z. *et al.*, 2005; Zhang *et al.*, 2006; Takai *et al.*, 2009) and maize (Szalma *et al.*, 2007; Hao *et al.*, 2009; Li *et al.*, 2009; Salvi *et al.*, 2011a).

8.3 QTLs for Abiotic Stress Tolerance

Among all abiotic stresses, drought has been and still is the one most widely investigated. Dedicated reviews and volumes have surveyed the eco-physiology of crop resistance to drought (Levitt, 1972; Boyer, 1982; Morgan, 1984; O'Toole and Bland, 1987; Blum, 1988, 1996, 2009; Ludlow and Muchow, 1990; Quarrie, 1991; Passioura, 1996, 2007; Turner, 1997; Sinclair and Muchow, 2001; Richards *et al.*, 2002; Bartels and Sunkar, 2005; Ribaut, 2006; Serraj *et al.*, 2009) and how a better knowledge of the molecular and biochemical factors controlling yield under such constrained conditions can improve selection strategies (Blum, 1988; Passioura, 1996, 2002; Quarrie *et al.*, 1999; Richards, 2000; Araus *et al.*, 2002, 2008; Richards *et al.*, 2002, 2010; Boyer and Westgate, 2004; Campos *et al.*, 2004; Chaves and Oliveira, 2004; Reynolds *et al.*, 2005; Alpert, 2006; Ribaut, 2006; Tuberosa and Salvi, 2006; Jenks *et al.*, 2007; Kumar *et al.*, 2008; Morison *et al.*, 2008; Reynolds and Tuberosa, 2008; Manavalan *et al.*, 2009; Serraj *et al.*, 2009; Wassmann *et al.*, 2009).

8.3.1 QTLs for drought-adaptive traits

The main mechanisms that contribute to maintaining yielding ability under water-limited conditions are dehydration avoidance and dehydration tolerance (Levitt, 1972). Morpho-physiological features such as deep rooting and osmotic adjustment – classified under dehydration avoidance – enable the plant to maintain better hydration, while other biochemical and physiological features (e.g. accumulation of molecular protectants, remobilization of stem water-soluble carbohydrates, etc.) – classified under dehydration tolerance – enable the plant to sustain metabolism in a severely dehydrated state. Carefully planned experiments carried out under controlled conditions provide important clues on the prevailing mode of action (e.g. avoidance versus tolerance) of the relevant loci (Yue *et al.*, 2006). Notably, drought episodes that jeopardize crop survival rarely occur in farmers' fields. Therefore, any breeding strategy aimed at increasing plant survival under such conditions will need to consider how frequently severe drought occurs in the target environments and, more importantly, the metabolic costs of features (e.g. excessive root mass) that might have a negative trade-off on yield under more favourable conditions (Collins *et al.*, 2008; Blum, 2009). Many of the genes that are induced under extreme dehydration have been shown to belong to metabolic pathways with doubtful functional significance under the water-limited field conditions to which field-grown crops are most commonly exposed (Passioura, 2007). Conversely, exploitation of naturally occurring variation for yield and/or drought-adaptive traits has allowed for slow but unequivocal progress in crop performance under drought conditions (Duvick, 2005; Reynolds and Tuberosa, 2008).

An important aspect to be considered for a successful breeding programme, whether on a conventional or a non-conventional basis, is to define the population of environments to be targeted (TPE, targeted population of environments). The identification and characterization of TPE is facilitated by the use of crop simulation models based on historical records of weather data (Chapman *et al.*, 2003; Chapman, 2008). Simulation can describe a TPE by the frequency of occurrence of specific abiotic stresses and based on the soil moisture profile along the crop cycle (Chapman *et al.*, 2003; Cooper *et al.*, 2009). In Mediterranean environments, barley and wheat usually undergo terminal drought associated with high temperatures during the grain-filling period (Araus *et al.*, 2008). This notwithstanding, within each TPE, genotype–year interactions are frequently observed consequent to yearly fluctuations in the severity of abiotic (e.g. high temperature, low nutrients, etc.) and biotic (e.g. foliar disease, nematodes, insects, etc.) factors that curtail yield potential (Sadras and Angus, 2006).

Multi-environment trials have been pivotal to increasing yield potential and yield stability of crops under environmentally constrained conditions (Casanoves *et al.*, 2005; Duvick, 2005; Tollenaar and Lee, 2006; Crossa *et al.*, 2007; Lafitte *et al.*, 2007; Acuna *et al.*, 2008). Given the quantitative nature of abiotic stress tolerance, QTLs have been the main target of studies attempting to identify the loci regulating the adaptive response of crops to environmentally constrained conditions. In a few cases, clearly the exception, major QTLs affecting yield and other drought-adaptive traits across a broad range of soil moisture conditions have been identified (Quarrie *et al.*, 2005, 2006; Ejeta and Knoll, 2007; Maccaferri *et al.*, 2008a; Venuprasad *et al.*, 2009). Along this line, one of the main reasons accounting for the modest impact of genomics-assisted breeding on the release of drought-tolerant cultivars is that screening conditions adopted under controlled conditions (e.g. growth chamber) usually provide a rather poor surrogate of the dynamics and timing of the drought episodes to which crops are exposed under field conditions (Passioura *et al.*, 2007; Herve and Serraj, 2009).

QTLs for root architecture and size

A vast number of morpho-physiological traits affect the water balance of the plant. Particular attention to the identification of QTLs has been devoted to root features, most likely due to the difficulty and cost required for their phenotyping and the key role that roots play in conditions of limited soil moisture (O'Toole and Bland, 1987; Richards, 2008). Accordingly, among a number of traits and their order of priority for improving drought resistance in four different scenarios identified by the combination of either intermittent or terminal stress environments with either subsistence or modern agriculture practices, Ludlow and Muchow (1990) identified rooting depth and density as traits of primary importance in three of these four scenarios.

Roots show a high level of morphological and developmental plasticity, a peculiarity that allows plants more properly to respond and promptly adapt to the main soil characteristics, particularly under moisture-limited conditions (Sharp and Davies, 1985; O'Toole and Bland, 1987; Liu et al., 2004; Davies, 2007; de Dorlodot et al., 2007; Richards, 2008; Den Herder et al., 2010) or other soil constraints. Other authors have further underlined the pivotal role of root features in regulating adaptation to scarce soil moisture, and have advocated for a better knowledge of the genetic control of such features and the application of MAS to tailor roots in order to optimize yield (Nguyen et al., 1997; Jackson et al., 2000; Maggio et al., 2001; Price et al., 2002a, b; Tuberosa et al., 2002c, 2007b, 2011b; de Dorlodot et al., 2007; Gregory et al., 2009; Hammer et al., 2009; Hochholdinger and Tuberosa, 2009).

The merits of a deep root system under drought conditions have been underlined in common bean (Mohamed et al., 2002), coffee (Pinheiro et al., 2005), lettuce (Johnson et al., 2000), maize (Lorens et al., 1987; Landi et al., 2007; Tuberosa et al., 2007b; Hammer et al., 2009; Hund et al., 2009a; Ruta et al., 2010b), barley (Forster et al., 2005), wheat (Lopes and Reynolds, 2010) and, particularly, rainfed rice (Nguyen et al., 1997; Price et al., 1997, 2000; Babu et al., 2003; Courtois et al., 2003; Zheng et al., 2003; Li, Z. et al., 2005; Steele et al., 2006, 2007; Kamoshita et al., 2008; Liu et al., 2008; Witcombe et al., 2008; Bernier et al., 2009; Cairns et al., 2009; Serraj et al., 2009; Coudert et al., 2010). However, it should be noted that other studies in rice have reported a weak or even absence of correlation between root features and drought resistance (Pantuwan et al., 2002; Subashri et al., 2009). The final effects of root architecture and size on yield will depend on the distribution of soil moisture and the level of competition for water resources within the plant community (King et al., 2009). Accordingly, when additional moisture is available in deeper soil layers, selection for faster-growing and deeper roots could improve water use and mitigate the negative effects of drought on yield, particularly in those crops that are characterized by a low capacity to adjust osmotically.

A most challenging aspect for those

willing to select for root features is to define the most desirable root ideotype able to optimize yield according to the prevailing dynamic of soil moisture profile, while also accounting for the concurrent presence of gradients in the soil profile for other abiotic factors (e.g. salinity, toxic elements, high pH, etc.) that may impair plant growth. Therefore, each root ideotype should be established based upon the prevailing soil features in the target environment, a good understanding of the root architectural features that limit water uptake and the metabolic cost required to develop and functionally sustain the root system. Along this line, loci that affect root growth under particular abiotic (e.g. boron toxicity: Langridge, 2005; McDonald et al., 2010; Reid, 2010) and biotic (e.g. nematode resistance: Williams et al., 2006; Barloy et al., 2007) constraints are interesting targets for MAS aimed at improving drought resistance through a more vigorous root system of crops grown in problematic soils.

Rice is the species most extensively investigated for root QTLs, in view of the fact that it is cultivated across a broad range of soil conditions that range from waterlogged paddies to the hard and compact soils experienced by upland rice. Bernier et al. (2009) have suggested improved root architecture as the possible cause of the increased water uptake of rice under water-limited conditions associated with the high-yielding allele at qtl12.1. The rice mapping population most extensively investigated for QTLs governing root traits has been derived from the cross between Bala and Azucena, two cultivars that differ for root features and other morpho-physiological traits that influnce water balance such as stomatal conductance, osmotic adjustment, leaf rolling, etc. Although several QTLs were identified across a broad range of environments (Price et al., 1997, 2000, 2002a, b; MacMillan et al., 2006a, b; Emrich et al., 2008; Norton and Price, 2009), only a limited overlap was observed between QTLs for root features and QTLs for other drought-avoidance traits, a finding that was related to the difficulty of collecting precise data from field trials because of variability in soils and rainfall (Cairns et al., 2009). Other studies have reported QTLs for root architecture in rice and discussed their value in improvement of drought tolerance (Nguyen et al., 1997; Mackill et al., 1999; Courtois et al., 2000, 2003; Tripathy et al., 2000; Zhang, J. et al., 2001; Zhang, W. et al., 2001; Lafitte et al., 2002; Mei et al., 2003; Zou et al., 2005; Yue et al., 2006; Ikeda et al., 2007; Khowaja et al., 2009). Collectively, these studies have highlighted a number of QTLs (on chrs 1, 2, 3, 7, 9 and 11) with a more substantial and consistent effect on root features (Coudert et al., 2010). A meta-analysis for root architectural feature was recently conducted (Courtois et al., 2009). Information was extracted from 24 studies carried out on 12 mapping populations that revealed 675 QTLs for 29 root parameters including root number, maximum root length, root thickness, root/shoot ratio and root penetration index. A web-accessible database that includes the QTLs for root features and also all QTLs for drought resistance traits in rice published between 1995 and 2007 was constructed. An overview of the number of root QTLs in 5-Mb segments covering the whole genome revealed the existence of 'hot spots' that represented 10% of the genome and carried 30% of the total number of QTLs. In particular, two regions on chrs 1 and 9 were those most frequently affecting root features in the 12 populations. In several cases, the meta-analysis considerably reduced the number of QTLs. The most striking examples were the regions on chrs 1 and 9, with a reduction from 15 to 3 QTLs and from 14 to 1 QTLs, respectively. As for resolution power, the increase in precision of the QTL location was also striking. Following the meta-analysis, the average confidence interval of the QTLs decreased from the original 4.14 to 1.98 Mb. Notably, the meta-QTL confidence interval was generally smaller than the smallest confidence interval of the QTLs of the cluster, particularly when the cluster encompassed a large number of QTLs. Twenty-five meta-QTLs had a confidence interval below 250 kb, small enough to attempt the identification of possible candidate genes from the

Nipponbare sequence. Among those that could be of particular interest in the context of root development, genes involved in auxin signaling and transport, transcription factors and sugar metabolism were identified (Courtois et al., 2009). One of the added values of meta-analysis studies is that they facilitate a comparative analysis with the QTL position in other species based on the known syntenic relationships. Interestingly, the QTL region on rice chr. 3 shown to harbour root QTLs is orthologous to the maize region on chr. bin 1.06 that has been reported to affect root architecture and yield (Tuberosa et al., 2002b, 2003; Landi et al., 2010).

In maize, a major QTL for root architecture was originally mapped on chromosome bin 1.06 in a number of different crosses (Lebreton et al., 1995; Barriere et al., 2001; Tuberosa et al., 2002c, 2003). Importantly, the QTL region on bin 1.06 in Lo964 × Lo1016 also affected grain yield under both well-watered and drought-stressed conditions (Tuberosa et al., 2002c). NILs differing for the parental segment at this QTL region on bin 1.06 have been tested per se and in hybrid combination, more accurately to evaluate the effects of the QTL on root traits, grain yield and other agronomic traits under different soil moisture conditions (Landi et al., 2010). Collectively, the results on this QTL suggest a constitutive mode of action (i.e. independent of the water regime) on root architecture and yield, a feature that makes this QTL an interesting candidate for a positional cloning approach (Salvi and Tuberosa, 2007). Strong effects on root architecture have also been reported for bin 1.03 (Steve Quarrie, personal communication; Tuberosa et al., 2002b) where the rtcs locus cloned by Taramino et al. (2007) has also been mapped (Hetz et al., 1996). Recent work carried out with a collection of introgression lines developed from the cross B73 × Gaspé Flint, where a major QTL for the number of seminal roots has also been identified on bin 1.03, indicates that this QTL overlaps with rtcs (Salvi et al., 2011b). Another major QTL for root architecture, root lodging, leaf ABA concentration (L-ABA) and grain yield has been mapped on bin 2.04 (Lebreton *et al.* 1995; Tuberosa *et al.*, 1998; Sanguineti *et al.*, 1999; Landi *et al.*, 2002, 2005, 2007; Giuliani *et al.*, 2005). The evaluation under well-watered and water-stressed conditions of NILs, both per se (Giuliani *et al.*, 2005; Landi *et al.*, 2005) and in hybrid combination with several testers (Landi *et al.*, 2007), confirmed the effect of the QTL (*root-ABA1*) on root traits, L-ABA and yield, particularly under drought conditions. The fine mapping of *root-ABA1* is now under way (Salvi *et al.*, unpublished results). More recently, further work to identify QTLs for root features in maize has been carried out by Hund and co-workers (Hund *et al.*, 2009a, b; Ruta *et al.*, 2010a, b; Trachsel *et al.*, 2010). These studies, while confirming the effects on root traits of a number of major QTLs that had been previously described (e.g. *root-ABA1* on bin 2.04), have also identified new QTLs and present valuable insights as to the role of different root features in determining maize yield under water-limited conditions. QTLs for root traits and plasticity in maize exposed to stresses other than drought have also been described that may be relevant under concurrent conditions of water shortage. An example is provided by the QTLs for root hair length, taproot length, root thickness and root biomass that were identified using a paper-roll culture system under conditions of high- and low-phosphorus availability (Zhu *et al.*, 2005).

QTLs for carbon isotope discrimination

Carbon isotope discrimination ($\Delta^{13}C$) is expressed as the ratio of stable carbon isotopes (^{13}C /^{12}C) in the plant dry matter compared with the value of the same ratio in the atmosphere (Farquhar *et al.*, 1989). Due to differences in the underlying leaf anatomy and the mechanisms of carbon fixation in C_3 and C_4 species, studies on $\Delta^{13}C$ have broader implications for the former group (Farquhar *et al.*, 1989). Under drought conditions, $\Delta^{13}C$ is a good predictor of stomatal conductance (Condon *et al.*, 2002) and water-use efficiency (WUE) in a number of crops (Araus *et al.*, 1993; Turner, 1997;

Tambussi *et al.*, 2007). However, although increasing WUE would seem a desirable feature for drought-prone environments, increasing evidence indicates otherwise (Blum, 2009), most likely due to the fact that a conservative strategy aiming to increase WUE, hence a lower transpiration, provides reduced yield benefit under moderate stress and can cause a yield penalty under favourable conditions (Condon *et al.*, 2004). When bread wheat was grown under different conditions of soil moisture availability, the correlation values between $\Delta^{13}C$ and final grain yield varied from positive when plenty of water was available to the crop, to negative in drought conditions, with no correlation in intermediate conditions (Condon *et al.*, 1993). Additional complexity is added when considering other physiological traits such as leaf temperature (Turner, 1997; Richards *et al.*, 2002) and growth cycle duration (Araus *et al.*, 1997). Consequently, the relationships between $\Delta^{13}C$ and grain yield will depend on the phenology of the crop, the prevailing environmental conditions and the plant organ (e.g. leaf or grain) from which samples are collected (Araus *et al.*, 1997; Royo *et al.*, 2001; Tambussi *et al.*, 2007).

Although grain $\Delta^{13}C$ has shown high heritability (Merah *et al.*, 2001; Ellis *et al.*, 2002) and also a low genotype–environment interaction (Richards, 1996; Rebetzke *et al.*, 2008a), the high cost required to measure each sample limits its utilization as a selection trait in large populations. Therefore, the identification of major QTLs for $\Delta^{13}C$ would provide the opportunity for applying MAS to select for this trait and WUE. It should be noted that selection for high $\Delta^{13}C$ per se (rather than genetic markers linked to QTLs) in wheat grown in Australian environments, where water must be used conservatively to allow the crop to complete its life cycle, has led to the release of two cultivars ('Drysdale' and 'Rees'; Condon *et al.*, 2004) characterized by yield increases of up to 23% when compared with control cultivars (Passioura, 2004; Richards, 2004). A more recent study has shown the importance of high transpiration efficiency in pearl millet, where leaf water loss in the vegetative phase was negatively associated with grain yield under conditions of terminal drought (Kholova *et al.*, 2010a, b). NILs for a major QTL that affect the concentration of abscisic acid (ABA) in the leaf, leaf water loss and grain yield have been derived, thus providing an interesting opportunity for more refined physiological work and for MAS to test the effects of the QTL allele in different genetic backgrounds (Yadav *et al.*, 2011).

In tomato, a dominant QTL for $\Delta^{13}C$ (*QWUE5.1*) that accounted for 26% of the total phenotypic variance was mapped to a 2.2 cM interval using an introgression library line carrying a *Solanum pennellii* chromosome fragment at that target region (Xu *et al.*, 2008). The markers and genetic stocks that were developed in this study are valuable for MAS of *QWUE5.1* and for cloning of the gene underlying this QTL.

In cotton, backcrossed generations from a cross between *Gossypium hirsutum* and *Gossypium barbadense* showed only incidental association of QTLs for $\Delta^{13}C$ with productivity, indicating that high WUE can be associated with either high or low productivity, a finding that led Saranga *et al.* (2004) to postulate that different cotton species have evolved different alleles related to physiological responses and productivity under water deficit, a feature that may thus enable the selection of genotypes that are better adapted to arid conditions.

In barley, $\Delta^{13}C$ of the shoot tissue has been reported to be more heritable than other seedling traits (Ellis *et al.*, 2002), thus indicating that a substantial fraction of the phenotypic variation in plant $\Delta^{13}C$ can be genetically manipulated. A set of barley RILs derived from Tadmor × Er/Apm were investigated for QTLs for $\Delta^{13}C$ in three field environments differing in water availability (Teulat *et al.*, 2002). Among the ten QTLs that were identified, one was specific to one environment, two showed a significant genotype–environment interaction and six presented main effects in two to three environments. Although seasonal rainfall and the ratio of rainfall to evapotranspiration contributed considerably to the environmental effect, their influence on genotype–

environment interaction was weak. Eight QTLs for $\Delta^{13}C$ were found to co-locate with QTLs for physiological traits related to plant water status or osmotic adjustment, and/or for agronomic traits previously measured on the same population (Teulat *et al.*, 2002).

QTLs for flowering time

Flowering time is the single most important factor for optimizing crop performance as a function of the prevailing climatic conditions. Therefore, flowering time will play a pivotal role in tailoring cultivars able to maintain high yield under changing climate condtions. One of the best-known examples of the importance of flowering time to optimize yield is provided by crops grown in Mediterranean-type environments, where breeders have traditionally selected for an early flowering date that allows the crop to escape, at least partially, the onset of drought as well as the high temperatures and terminal drought that typically occur during late spring and early summer. Under these conditions, grain filling in early flowering genotypes occurs under less constrained conditions (Araus *et al.*, 2003, 2008; Slafer, 2003).

Many QTLs and major genes associated with the control of flowering time have been described in different crops (Chardon *et al.*, 2004; Buckler *et al.*, 2009; Distelfeld *et al.*, 2009; Salvi *et al.*, 2011a), and could be used to further exploit variability for this trait in drought-prone farming conditions and to optimize the choice of parental lines of new crosses. In maize, the meta-analysis conducted by Chardon *et al.* (2004) summarizes the information provided by 313 QTLs identified in 22 studies where flowering time was recorded. This meta-analysis highlighted the presence of 62 consensus QTLs, with six playing a major effect. The 62 consensus QTLs were then compared with the positions of the few flowering time candidate genes that have been mapped in maize, and also with rice candidates, using a synteny conservation approach based on comparative mapping. However, it should be pointed out that the degree of collinearity between such distantly related species as maize and *Arabidopsis* is very low (van Buuren *et al.*, 2002). More recently, Salvi *et al.* (2011a) have extended the meta-analysis of Chardon *et al.* (2004) by including the results reported by Buckler *et al.* (2009) obtained in a large study of the 25 NAM populations. Based on the comparative analysis of these results, Salvi *et al.* (2010) concluded that the main breeding pools of maize have been exhaustively investigated for flowering time QTLs and thus the major QTLs have been appropriately identified. One of these major QTLs for flowering time per se (*Vgt1*) has been positionally cloned (Salvi *et al.*, 2007). It should be noted that under the growing conditions typical of the Pò Valley in northern Italy, one of the highest productive regions for maize, the Gaspé Flint *Vgt1* allele in late-maturing backgrounds (topcrosses) reduces by up to one week the flowering time of the isogenic counterpart, while causing a parallel reduction in the number of internodes (up to one internode), biomass and yield (Salvi *et al.*, unpublished). Notably, although under well-watered conditions the early test crosses obtained with the late *Vgt1* allele usually out-yield their early counterpart, the same test crosses significantly out-yielded their late counterparts during the severe 2003 drought in northern Italy that curtailed maize yield by ~30% (Salvi *et al.*, unpublished). These results provide yet another example of how the action of a QTL allele on yield can drastically change according to the prevailing environmental conditions (Collins *et al.*, 2008).

In rice, five genes have been shown to determine a large portion of the variation for photoperiod response. A major QTL (*Hd1*) that controls the response to photoperiod was cloned by means of a map-based cloning strategy (Yano *et al.*, 2000). *Hd1* and *Hd3a* are flowering promoters, homologues of *A. thaliana CO* and *FT* genes (Yano *et al.*, 2000; Kojima *et al.*, 2002). *Ghd7* and *Hd6* encode a CCT domain protein and a CK2 kinase, respectively, which probably repress Hd3a under long-day photoperiod (Takahashi *et al.*, 2001; Xue *et al.*, 2008). Loss-of-function alleles of both genes produce early flowering, but additional alleles differing in several

Ghd7 missense substitutions have been suggested to contribute further to flowering variation in rice. *EhdD1* encodes a response regulator that independently up-regulates *Hd3a* under short days (Doi *et al.*, 2004). Several *Hd1* loss-of-function alleles reduce the expression of *Hd3a* and produce late flowering under short days, which indicates a conservation of this photoperiod regulatory module in phylogenetically distant short- and long-day flowering species (Alonso-Blanco *et al.*, 2009).

QTLs for carbohydrate accumulation and relocation

The accumulation of carbohydrates and their partitioning to storage organs are key factors for the optimization of yield under adverse environmental conditions that negatively affect final yield, particularly in small-grain cereals (Blum *et al.*, 1994; Blum, 1998; Cui *et al.*, 2003; Trouverie and Prioul, 2006; Cuellar-Ortiz *et al.*, 2008; Luquet *et al.*, 2008). Experimental evidence indicates that sucrose and hexoses are important signal molecules in source–sink regulation and in the integration between signalling pathways responding to phytohormones, nutrients, light and abiotic stress-related stimuli (Roitsch, 1999; Xue-Xuan *et al.*, 2010). In maize grown under drought-stressed conditions, a number of studies have unequivocally shown the important role played by sucrose in reproductive fertility (Boyer, 1996; Zinselmeier *et al.*, 2002; Boyer and Westgate, 2004; McLaughlin and Boyer, 2004a, b; Boyer and McLaughlin, 2007). Early QTL studies have attempted the dissection of the physiological basis of grain yield in maize by measuring the level of activity of some of the key enzymes influencing photosynthetic capacity (Prioul *et al.*, 1997, 1999; Thevenot *et al.*, 2005). Notably, some of the structural genes of the relevant enzymes (e.g. soluble invertase, sucrose synthase and ADP-glucose pyrophosphorylase) mapped to regions with QTLs affecting the activity of the encoded enzymes and/or concentration of their products, and occasionally growth traits as well. In addition to the role played by carbohydrate supply in regulating reproductive fertility in cereals (Saini and Westgate, 2000), other biochemical factors (e.g. ABA) appear to be involved in kernel abortion during the early stages of kernel growth (Setter *et al.*, 2001, 2011; Zinselmeier *et al.*, 2002).

In cereals, the remobilization of water-soluble carbohydrates (WSC) from the stem and leaves is an important mechanism that allows plants to mitigate the negative effects of post-anthesis drought tolerance of other limitations (e.g. foliar diseases, early senescence due to high temperature, etc.) impairing post-anthesis accumulation of photosynthates (Blum, 1988; Turner *et al.*, 2010). In wheat, QTLs for stem-reserve mobilization were mapped on chrs 2D, 5D and 7D (Salem *et al.*, 2007). In bread wheat, QTLs for WSC remobilization and leaf senescence have also been reported across both well-watered and water-stressed conditions (Snape *et al.*, 2007; Rebetzke *et al.*, 2008b). Although these studies showed an important role for WSC in assuring stable yield and grain size, Rebetzke *et al.* (2008b) concluded that the small effects of many independent WSC QTLs may limit their direct use for MAS in breeding programmes.

In perennial ryegrass, WSC is an important factor determining the nutritional value of grass forage. Markers associated with QTLs for WSC (Turner *et al.*, 2006) were used to design narrow-based populations with homozygosity for WSC QTLs (Turner *et al.*, 2010). When the divergent populations produced via MAS were analysed for WSC content in the glasshouse and the field, complex interactions between WSC content and other factors and traits (e.g. the scale of assessment, forage quality parameters, etc.) were observed. Although differences between the divergent pairs of the various populations were small, the roles of the QTL regions in regulating forage WSC content were confirmed, thus setting the scene for marker-assisted breeding strategies in ryegrass (Turner *et al.*, 2010).

QTLs for stay-green trait and chlorophyll fluorescence

Stay-green indicates a delay in senescence that is associated with post-flowering

drought tolerance independently of the maturity of the genotype (Thomas and Howarth, 2000; Ejeta and Knoll, 2007). Because different physiological mechanisms can influence leaf senescence, particularly under abiotic stress conditions, similar stay-green phenotypes may underline opposite functional models in terms of cause–effect relationships. Green leaf-area duration during grain filling appears to be a product of different combinations of three distinct factors: green leaf area at flowering, time of onset of senescence and subsequent rate of senescence (Borrell *et al.*, 2000a, b). The evaluation of a historical series of maize hybrids has clearly shown that delayed leaf senescence was higher yielding in drought-stressed conditions (Duvick, 2005). In sorghum, stay-green improves genotype adaptation to post-flowering drought stress, particularly when crop performance depends prevalently on the amount of moisture stored in the soil profile explored by the roots (Mahalakshmi and Bidinger, 2002).

The early work carried out in sorghum by Tuinstra *et al.* (1996, 1997) identified several QTLs for post-flowering drought tolerance, six of which influenced stay-green (Tuinstra *et al.*, 1997). Three of these QTLs were also positively associated with grain yield under irrigated conditions, suggesting a physiological link between the expression of stay-green under post-flowering drought and grain yield under non-drought conditions. Because of the importance of the stay-green trait, additional studies have been conducted more accurately to characterize and map the relevant QTLs in various populations (Subudhi *et al.*, 2000; Xu *et al.*, 2000), subsequently prompting four major stay-green QTLs to be named as *Stg1* and *Stg2* on chr. 3, *Stg3* on chr. 2 and *Stg4* on chr. 5. These results prompted Ejeta and Knoll (2007) to postulate that different mechanisms contribute to the stay-green phenotype during post-flowering drought stress in sorghum. Additionally, these results suggest the possibility of using MAS to pyramid multiple stay-green QTLs from various sources in order to improve drought tolerance.

Other stay-green-related QTLs have been reported in pearl millet (Yadav *et al.*, 2011), rice (Jiang *et al.*, 2004; Yoo *et al.*, 2007), wheat (Yang *et al.*, 2007a, b) and barley (Guo *et al.*, 2008), where two major QTLs were shown to be involved in the development of functional chloroplasts at the post-flowering stage under drought stress.

QTLs for other traits of interest for the control of water balance

Measurement of traits such as stomatal conductance, canopy temperature and leaf rolling provides indications of water extraction patterns and the water status of the plant. Therefore, measuring these traits together with soil moisture may help in selecting deep-rooted germplasm in environments where water is available at depth (Blum *et al.*, 1989; Reynolds *et al.*, 2005, 2009b).

Stomatal conductance integrates important environmental and metabolic cues and allows the plant to modulate and optimize its transpiration accordingly. Therefore, stomatal conductance plays an important role in determining $\Delta^{13}C$ and WUE (Farquhar *et al.*, 1989; Condon *et al.*, 2002; Brennan *et al.*, 2007). A study conducted on a series of successful bread wheat cultivars released from 1962 to 1988 showed a strong and positive correlation between stomatal conductance and grain yield ($r = 0.94$; Fischer *et al.*, 1998), suggesting that the more modern cultivars extract more water from the soil. Thus, a positive association between stomatal conductance and yield has also been reported in cotton (Lu *et al.*, 1998; Ulloa *et al.*, 2000; Levi *et al.*, 2009a, b), in this case also owing to the cooling effect on leaf temperature due to the water transpired from the stomata. These results indicate the possibility of raising the yield potential, hence water use and productivity, of these two species through an indirect selection for stomatal conductance, and suggest the value of identifying the relevant QTLs in order to implement MAS. The application of MAS would eliminate the time-consuming procedures required properly to measure stomatal conductance, particularly under field conditions, where it is often difficult adequately to account for the fluctuations of the environmental factors known to affect stomatal conductance during the day such as wind, solar radiation, humidity, etc.). Notwithstanding the difficulty in measuring stomatal conductance in large numbers of plants, several studies have searched for QTLs for this trait (Lebreton *et al.*, 1995; Price *et al.*, 1997, 2002a; Sanguineti *et al.*, 1999; Ulloa *et al.*, 2000; Herve *et al.*, 2001; Khowaja and Price, 2008; Takai *et al.*, 2009).

Canopy temperature and leaf rolling are integrative traits that report on the water balance at the leaf and whole-plant level; as such, they provide an indirect assessment of the capacity of the plant to extract soil moisture. In field conditions, genotypes with a cooler canopy temperature under drought stress, or a higher canopy temperature depression (CTD), will extract and use more of the water available in the soil, hence avoiding excessive dehydration and reduction in grain yield (Blum, 1988, 2009; Ludlow and Muchow, 1990; Reynolds and Tuberosa, 2008; Jones *et al.*, 2009). New thermal-imaging technology can now report subtle differences in leaf temperature in both laboratory and field conditions (Jones *et al.*, 2003). CTD is mainly useful in hot and dry environments, with measurements preferably made on recently irrigated crops on cloudless and windless days at high vapour pressure deficits (Blum, 1988; Reynolds and Pfeiffer, 2000; Reynolds *et al.*, 2009b). Under these circumstances, which maximize the heritability of the trait, and provided that data are collected when the canopy is sufficiently expanded to cover the soil, CTD can be a good predictor of grain yield in bread wheat ($r = 0.6–0.8$; Reynolds and Pfeiffer, 2000), where yield progress has been associated with cooler canopies, hence higher transpiration (Fischer *et al.*, 1998; Saint Pierre *et al.*, 2010); additionally, genetic gains in yield have also been reported in response to direct selection for CTD (Reynolds *et al.*, 2009b).

In rice, canopy temperature (CT), leaf water potential (LWP) and spikelet fertility (SF) of a set of RILs were investigated under two soil moisture regimes in a drought-screen facility (Liu *et al.*, 2005). Although the correlations between the investigated traits were very low, several QTLs were shown to influence more than one trait. A total of 44 main-effect QTLs were associated with CT, LWP and SF, with multiple R^2 values for CT, LWP and SF equal to 88, 15 and 79% under well-watered conditions and 73, 88 and 33% under drought-stressed conditions, respectively.

QTLs for yield under different water regimes

As global climate change increases the vagaries of weather patterns and their negative impact (Pennisi, 2008), the identification and selection of loci with a consistent per se effect on yield (i.e. not loci for flowering time) across a broad range of soil moisture regimes becomes increasingly important to raise yield potential (Maccaferri *et al.*, 2008a; Foulkes *et al.*, 2011; Parry *et al.*, 2011; Reynolds *et al.*, 2011). Although these loci are clearly the exception rather than the rule, a few notable cases have been reported. In cereals, markers associated with major QTLs for grain yield and its components across a broad range of soil moisture regimes have been described in rice (Xing and Zhang, 2010), bread wheat (Groos *et al.*, 2003; Quarrie *et al.*, 2005, 2006; Kirigwi *et al.*, 2007; Kuchel *et al.*, 2007a, b; Laperche *et al.*, 2007; Snape *et al.*, 2007), durum wheat (Maccaferri *et al.*, 2008a) and maize (Prasanna *et al.*, 2009). A combined QTL analysis for yield of several wheat DH populations evaluated across different environments and seasons enabled Snape *et al.* (2007) to identify QTLs showing stable and differential expression across irrigated and non-irrigated conditions. Variation for stem water-soluble carbohydrate reserves was associated with the chr. 1RS arm of the 1BL/1RS translocated (from rye to wheat) chromosome, and was positively associated with yield under both irrigated and rainfed conditions, thus contributing to general adaptability (Snape *et al.*, 2007). The beneficial role of this translocation on wheat performance under drought-stressed conditions had already been reported (Ehdaie *et al.*, 2003, 2008). In durum wheat, Maccaferri *et al.* (2008a) searched for QTLs for grain yield in 249 RILs derived from the cross of two elite cvs (Kofa and Svevo) that were

evaluated in ten rainfed and six irrigated environments in the Mediterranean Basin characterized by a broad range of grain yield values (0.56–5.88 t/ha), mainly consequent to different soil moisture availability. Among the 16 QTLs that affected grain yield in one or more of the 16 environments, two major QTLs on chrs 2BL and 3BS (*QYld.idw-2B* and *QYld.idw-3B*, respectively) showed significant and consistent effects in eight and seven environments, respectively. In both cases, an extensive overlap was observed between the LOD profiles for grain yield and plant height, and hence biomass production, but not with those for heading date, thus indicating that the effects of these two QTLs on yield were not due to escape from drought, a well-known factor in determining yield under terminal drought stress conditions that typically characterize Mediterranean environments (Araus *et al.*, 2008). Accordingly, this population was originally chosen because it had shown limited variability in flowering time. For plant height and grain yield, a strong epistasis between *QYld.idw-2B* and *QYld.idw-3B* was detected across several environments, with the parental combinations providing the higher performance. These two QTLs evidenced significant additive and epistatic effects also on ear peduncle length and kernel weight (Graziani *et al.*, 2011). Isogenic lines have been derived for fine mapping of both QTLs (Maccaferri *et al.*, 2011a).

Major QTLs for seed weight and grain yield at more specific water regimes have also been identified in soybean (Specht *et al.*, 2001; Chung *et al.*, 2003; Du *et al.*, 2009), sunflower (Ebrahimi *et al.*, 2008, 2009), rice (Lanceras *et al.*, 2004; Wang *et al.*, 2006; Bernier *et al.*, 2007, 2009; Guan *et al.*, 2010; Liu *et al.*, 2010), maize (Frova *et al.*, 1999; Xiao *et al.*, 2005; LeDeaux *et al.*, 2006) and pearl millet (Bidinger *et al.*, 2007; Kholova *et al.*, 2010a, b; Yadav *et al.*, 2011).

8.3.2 QTLs for salinity tolerance

Under certain conditions, global climate change will impact crop yield also via effects due to changes in soil salinity. One example is when irrigation practices aimed at counteracting an increased frequency and/or intensity of drought episodes also increase soil salinity because saline irrigation water is used (Rengasamy, 2006). At present, it is estimated that salinity affects approximately 830 million ha worldwide and is fast increasing in regions where saline water is used for irrigation. Salt tolerance has often been found, frequently, albeit not always, associated with lower accumulation of sodium (Na^+) in the shoot, with tolerance achieved by sequestration of toxic Na^+ in the vacuole (Munns and Tester, 2008). Many QTLs for Na^+ accumulation and other measures of salt tolerance have been mapped in various crops (Flowers et al., 2000; Flowers and Flowers, 2005; Jenks et al., 2007; Koyama et al., 2001; Munns and Tester, 2008; Genc et al., 2010). Members of the HKT (high-affinity K1 transporter) family of K1 and Na1 transporters (Platten et al., 2006) were shown or postulated to control natural variation in salinity tolerance at a number of loci in rice, wheat and Arabidopsis. Two loci, Nax1 and Nax2, controlling shoot Na^+ accumulation via Na^+ exclusion were identified by QTL mapping in durum wheat (genome AABB; Munns et al., 2003; James et al., 2006). Both exclusion genes represent introgressions from an accession of Triticum monococcum (genome AA). The Na^+ exclusion genes have also been introgressed into backgrounds of five cultivars of bread wheat (genome AABBDD) by using durum × bread wheat interspecific crosses and MABC; the resulting lines are currently under field evaluation in three Australian states (R. James and R. Munns, personal communication). TmHKT7-A2 and TmHKT1;5-A (HKT8) are the candidates for Nax1 (chr. arm 2AL; Huang et al., 2006) and for Nax2 (chr. arm 5AL), while TaHKT1;5-D is the candidate for the Kna1 gene on chr. arm 4DL, which accounts for the superior salt tolerance of bread wheat compared with durum wheat (Byrt et al., 2007). Interestingly, the putative orthologue of the TaHKT1;5 gene in rice (OsHKT1;5) underscores a QTL on chromosome 1 that affects Na accumulation and salt tolerance (Ren et al., 2005; Sunarpi et al., 2005). In Arabidopsis, an HKT1 homologue (AtHKT1) coincided with

the peak of a natural QTL (Rus et al., 2006). Plant HKT transporters reduce shoot Na accumulation by facilitating the unloading of Na from the xylem (Ren et al., 2005; Sunarpi et al., 2005). Nax1 promotes Na retention in the leaf sheath relative to the leaf blade, whereas Nax2 and Kna1 do not affect this relative accumulation, suggesting that the former promotes xylem unloading of Na in the leaf sheath as well as in the roots (James et al., 2006).

Genc et al. (2010) evaluated a bread wheat population in supported hydroponics to identify QTLs associated with salinity tolerance traits (e.g. leaf symptoms, tiller number, seedling biomass, chlorophyll content and shoot Na^+ and K^+ concentrations), to investigate the relationships among these traits and to determine their genetic value for MAS. There was considerable segregation within the population for all traits measured. A total of 40 QTLs were detected and, for the first time in a cereal species, a QTL interval for Na^+ exclusion (wPt-3114-wmc170) was associated with an increase (10%) in seedling biomass. Of the five QTLs identified for Na^+ exclusion, two were co-located with seedling biomass on chrs 2A and 6A. The chr. 2A QTL appears to coincide with the previously reported Na^+ exclusion locus in durum wheat that hosts one active HKT1;4 (Nax1) and one inactive HKT1;4 gene. Using these sequences as a template for primer design enabled Genc et al. (2010) to assign three HKT1;4 genes to chr. 2AL, suggesting that bread wheat carries more HKT1;4 gene family members than durum wheat. However, the combined effects of all Na^+ exclusion loci accounted for only 18% of the variation in seedling biomass under salinity stress, indicating that there were other mechanisms of salinity tolerance operative at the seedling stage in this population. Na^+ and K^+ accumulation appeared to be under separate genetic control. The molecular markers wmc170 (2A) and cfd080 (6A) are expected to facilitate breeding for salinity tolerance via MAS in bread wheat, the latter being associated with seedling vigour (Genc et al., 2010).

The evaluation of a mapping population derived from a cross between a wild barley

(*Hordeum vulgare* ssp. *spontaneum*) accession and cultivated barley allowed Shavrukov *et al.* (2010) to identify a major QTL on the short arm of chr. 7H capable of limiting Na^+ accumulation in the shoots under saline hydroponic conditions. Additional work attributed the control of the Na^+ exclusion trait to a single locus (*HvNax3*) able to reduce shoot Na^+ accumulation by 10–25% in plants grown in 150 mM NaCl. The evaluation of congenic strains with markers generated using colinearity with rice and *Brachypodium* enabled Shavrukov *et al.* (2010) to map *HvNax3* within a 1.3 cM interval. Several plausible candidates for *HvNax3* were identified from the corresponding rice and *Brachypodium* intervals.

8.3.3 QTLs for low- and high-temperature tolerance

Terminal drought occurring during grain filling in cereals is often accompanied by high temperatures, particularly in Mediterranean-type environments. The interaction between drought and heat greatly accelerates leaf senescence and severely curtails yield (Pinto *et al.*, 2010). Additionally, excessive heat disrupts many cellular and developmental processes and directly affects grain production by increasing sterility and decreasing grain quality (She *et al.*, 2010). In rice, a recent study based on 227 intensively managed irrigated farms has shown that higher minimum temperature reduced yield, whereas higher maximum temperature raised it (Welch *et al.*, 2010). These effects imply a net negative impact on yield from moderate warming expected in the next decades. Beyond that, the impact would probably become even more negative, because prior research indicates that higher maximum temperature will impact yield more negatively. The work of Welch *et al.* (2010) in Asia has clearly shown that diurnal temperature variation must be considered when investigating the impacts of climate change on irrigated rice. Conversely, chilling stress prior to anthesis can also impair male gametophytic functions and cause sterility, a condition not uncommon in small-grain

cereals grown in areas where low temperature may occur at flowering. In crops grown in temperate regions and more northern latitudes, even more severe damage can occur when freezing temperatures cause extensive injuries at an early growth stage. Overall, the genetic basis of freezing and chilling tolerance in crops appears better understood than the genetic basis of heat stress tolerance, which remains poorly understood.

Higher temperatures are thought to affect rice yields adversely through two main pathways: (i) high maximum temperatures that cause – in combination with high humidity – spikelet sterility, hence adversely affecting grain quality; and (ii) increased night-time temperatures that may reduce assimilate accumulation (Wassmann *et al.*, 2009). In rice, a RIL population was screened for heat tolerance at anthesis by measuring spikelet fertility at 30 and 38°C (Jagadish *et al.*, 2010). No correlation was observed between spikelet fertility in control and high-temperature conditions and no common QTLs were identified. Two QTLs for spikelet fertility under control conditions were identified on chrs 2 and 4. Eight QTLs for spikelet fertility under high-temperature conditions were identified on chrs 1, 2, 3, 8, 10 and 11. The most significant heat-responsive QTL, accounting for up to 18% of the phenotypic variation, was mapped on chromosome 1. This QTL was also found to influence plant height, explaining 36.6% of the phenotypic variation. A comparison with other studies of abiotic (e.g. drought, cold and salinity) stresses showed QTLs at similar positions on chromosomes 1, 3, 8 and 10, suggesting common underlying stress-responsive regions of the genome.

In wheat, two QTLs were identified that controlled grain-filling duration, a trait thought to be correlated with heat tolerance (Yang *et al.*, 2002). In maize, QTLs were identified that controlled heat tolerance of pollen germinability and pollen tube growth, a factor that influences heat-induced sterility (Frova and Sari-Gorla, 1994). Studies investigating multiple parameters related to heat tolerance in wheat provided evidence for genetic variability and multiple tolerance mechanisms (Dhanda and Munjal, 2006).

The study of Pinto *et al.* (2010), conducted on a population with a restricted range in anthesis, investigated the interaction between high temperature and drought in wheat. Among the 104 QTLs that were identified across a combination of 115 traits under three environments over 2 years, six genomic regions influenced a large number of traits. A yield QTL located on chr. 4A explained 27 and 17% of variation under drought and heat stress, respectively. At the same location, a QTL explained 28% of the variation in canopy temperature under heat, while 14% of canopy temperature variation under drought was explained by a QTL on chr. 3B. Notably, in this particular background the T1BL.1RS (rye) trans-location donated by the Seri parent was associated with decreased yield. Common QTLs for drought and heat stress traits were identified on chrs 1B, 2B, 3B, 4A, 4B and 7A, confirming their generic value across stresses (Pinto *et al.*, 2010). Yield QTLs were shown to be associated with components of other traits, supporting the prospects for dissecting crop performance into its physiological and genetic components in order to facilitate a more strategic approach to breeding (Reynolds and Tuberosa, 2008).

Maize and rice originate from tropical/subtropical regions and, as such, are rather cold-sensitive crops. Chilling temperatures lead to poor seedling establishment and, at booting stage, reduce fertility due mainly to the arrest of microspore development. QTLs for chilling sensitivity have been described in maize seedlings (Hund *et al.*, 2004, 2005; Presterl *et al.*, 2007), in sorghum seedlings (Ejeta and Knoll, 2007) and in rice at the seedling and booting stages (Andaya and Mackill, 2003; Andaya and Tai, 2006; Kuroki *et al.*, 2007, 2009). Recently, Saito *et al.* (2010) positionally cloned *Ctb1* (*cold tolerance at booting stage* on chr. 1), a major QTL for cold tolerance in rice, and were thus able to unravel the molecular mechanism under-lying the action of this QTL. In particular, the target gene encodes for an F-box protein gene that confers cold tolerance. Cold tolerance was associated with greater anther length, and the transgenic plants had longer anthers than non-transformed controls. The F-box protein interacts with a subunit of the E3 ubiquitin ligase, Skp1, suggesting that a ubiquitin-proteasome pathway is involved in cold tolerance at the booting stage (Saito *et al.*, 2010).

Freezing damage derives from dehydration and membrane damage caused by the expansion of ice crystals. Commonly, the expression of frost tolerance at the vegetative stage typically requires exposure to low, non-freezing temperatures. Loci controlling vegetative frost tolerance have been mapped at corresponding positions across the Triticeae species at two locations on the long arms of group 5 chromosomes. The proximal loci (*Fr-1*) are close to or coincident with the *Vrn-1* loci involved in the response of flowering to vernalization. Clusters of *C-repeat Binding Factor* (*CBF*) genes map at the distal (*Fr-2*) loci (Francia *et al.*, 2007). CBFs, also known as dehydration-responsive element-binding (DREB) factors, are cold-responsive transcription factors with probable roles in coordinating cold responses leading to cold/freezing tolerance in plants (Shinozaki and Yamaguchi-Shinozaki, 2000, 2007; James *et al.*, 2008). In many parts of the world, frost at flowering time can also cause sterility or shrivelled grains in cereals and other crops, leading to sporadic episodes of severe losses (Fuller *et al.*, 2007). In barley, QTLs for reproductive frost tolerance have been reported near the *Fr-H1* locus, and on chr. arm 2HL (Reinheimer *et al.*, 2004). This notwithstanding, limited progress has been achieved in improving reproductive frost tolerance, most likely due to low genetic variation and the dependence on frost simulation chambers for efficient pheno-typing. Other species that have been investigated for QTLs for winter hardiness include lentil (Kahraman *et al.*, 2004), rapeseed (Asghari *et al.*, 2007) and ryegrass (Xiong *et al.*, 2007).

8.3.4 QTLs for submergence and anoxia tolerance

Flooding is one of the abiotic stresses whose frequency and intensity is increasing due to global warming and record rainfall. The devastating floods of the early months in 2011 in Queensland and Sri Lanka clearly show how dramatic the consequences of excessive rain can be and the importance of having crops able to withstand the anoxia conditions associated with extended submergence. Among cereals, rice is the crop most heavily damaged by submergence stress that periodically affects approximately 15 million ha of rainfed lowland areas in Asia, where it has been estimated to cause annual losses in excess of US$1 billion (Mackill, 2007). A suite of traits contribute to flooding tolerance through adaptive features at the morpho-physiological and molecular level (Colmer and Voesenek, 2009; Bailey-Serres *et al.*, 2010; Chen *et al.*, 2010; Singh *et al.*, 2010). In rice, Xu and Mackill (1996) identified *Sub1*, a major QTL on chr. 9 that accounted for a major portion of variability for survival under prolonged flooding conditions. Positional cloning of the *Sub1* QTL has revealed a cluster of three putative ethylene response factor genes, namely *Sub1A*, *Sub1B* and *Sub1C*. Further work unequivocally assigned the functional polymorphism to *Sub1A* (Xu *et al.*, 2006; Jung *et al.*, 2010).

In maize, Mano *et al.* (2005a, b, 2007) identified QTLs for adventitious root formation at the soil surface, one of the most important adaptations to soil flooding or waterlogging, a condition that can severely impair root growth at an early stage, thus reducing the capacity of the plant to extract soil moisture at a later stage when water shortage is more likely to occur. Several QTLs for adventitious root formation have been mapped and a major QTL was mapped on chr. 8 (Mano *et al.*, 2005b).

8.3.5 QTLs for tolerance to soil toxicities

Plant growth is severely constrained in soils containing toxic concentrations of particular minerals or insufficient concentrations of specific minerals (Ismail *et al.*, 2007). An example is provided by soil acidity (pH 5.0), a constraint that affects over 50% of the world's arable land mainly through the effect of promoting aluminium (Al) toxicity. Although liming can be used to treat soil acidity, the release of Al-tolerant varieties provides a more profitable option for growing crops in soils with toxic levels of Al (Kochian *et al.*, 2005). Major loci and QTLs controlling Al tolerance have been identified in several crops: lucerne (Narasimhamoorthy *et al.*, 2006), soybean (Bianchi-Hall *et al.*, 2000; Qi *et al.*, 2008), rice (Xue *et al.*, 2007), sorghum (Magalhaes *et al.*, 2007), maize (Ninamango-Cárdenas *et al.*, 2003), barley (Wang *et al.*, 2007), wheat (Raman *et al.*, 2005), oat (Wight *et al.*, 2006) and rye (Matos *et al.*, 2005). Comparative mapping has indicated possible homologies between Al-tolerance loci in different cereal species (Kochian *et al.*, 2005). The most commonly documented mechanism of Al tolerance is the Al-activated extrusion of Al-chelating anions such as malate and citrate from root tips, followed by the formation of non-toxic Al complexes in the apoplast or rhizosphere. The four recently cloned genes that govern Al tolerance all encode organic anion transporters involved in this tolerance mechanism. Malate transporters belonging to the novel plant-specific Al-activated malate transporter 1 (ALMT1) family of proteins are encoded by Al-tolerance genes in Arabidopsis and wheat (Hoekenga *et al.*, 2006; Sasaki *et al.*, 2006), while citrate transporters belonging to the multi-drug and toxic compound extrusion (MATE) family of membrane transporters are encoded by Al-tolerance genes in sorghum and barley (Furukawa *et al.*, 2007; Magalhaes *et al.*, 2007; Maron *et al.*, 2010). A cluster of *ALMT1* genes also controls Al tolerance in rye (Collins *et al.*, 2008). mRNA expression levels of the tolerance genes cloned from wheat, barley and sorghum showed strong positive correlations with the tolerance and organic acid secretion levels in collections of diverse genotypes of these species (Raman *et al.*, 2005; Furukawa *et al.*, 2007; Magalhaes *et al.*, 2007; Wang *et al.*, 2007).

A study conducted in wheat identified a major Al-tolerance QTL on chr. arm 4DL, explaining 31% of the phenotypic variance present in the population (Navakode *et al.*, 2008). Using a different population, a second major Al-tolerance QTL was assigned to chr. arm 3BL. This major QTL ($Qalt_{CS}.ipk-3B$) in 'Chinese Spring' accounted for 49% of the phenotypic variation, suggesting the feasibility of applying marker-assisted selection (MAS) and pyramiding of this new QTL to improve the Al tolerance of wheat cultivars in breeding programmes (Navakode et al., 2008).

Researchers funded by the Generation Challenge Program (GCP) have made significant progress in elucidating the molecular basis for tolerance to both Al toxicity and P deficiency, and how to translate this information for the release of improved cultivars in rice and sorghum via MAS (Kochian et al., 2011).

8.3.6 QTLs for tolerance to low nutrients

One of the major factors contributing to the environmental footprint of modern agriculture and its effects on climate is the large amounts of nitrogen (N) fertilizers that are used to boost biomass production and yield. The increase in crop yields made possible by the Green Revolution during the past century is attributed to the selection of genotypes with a higher yield potential and to the parallel increase in the application of fertilizers, particularly N (Borlaug and Dowswell, 2005). The vast majority of N used in agriculture is synthesized artificially, thus requiring a vast amount of energy and CO_2 production, since most of the synthesis is achieved through burning of fossil oil or coal. The sharp increase in energy cost has also made N fertilizer more expensive. In the long term, an even more worrying picture emerges when considering phosphorus (P) availability, a major factor limiting crop yield, particularly in the low-input agricultural systems of many developing countries. Phosphorus presents a more critical long-term prospect, as it is a non-renewable fertilizer that cannot be synthesized artificially, unlike N. Forecasts indicate that P reserves, given the present rate of deployment, will be completely depleted by the end of the current century.

Another negative factor in terms of the impact of agricultural activities on the environment is due to the leaching of P and N into surface and sea water, causing a number of environmental problems (e.g. algal blooms). Therefore, increasing crop N- and P-use efficiency represents an important objective for ensuring a cost-effective, profitable and more sustainable agriculture.

In maize, a set of QTLs for NUE, grain yield and its components at high and low N levels have been described (Hirel et al., 2001, 2007; Gallais and Hirel, 2004; Coque and Gallais, 2006; Coque et al., 2008). These studies showed that QTLs mapped in clusters, with those identified under low-N conditions being a subset of those identified under high N, except for grain protein content, for which a higher number of QTLs were detected in low N. A number were close to the confidence interval of several QTLs for vegetative development, as well as for grain yield and its components. These QTLs overlapped with genes that encode enzymes involved in N and C metabolism (Gallais and Hirel, 2004) such as glutamine synthetase (GS), glu-ammonia ligase, suc-P synthase, suc synthase and invertase (b-fructofuranosidase). Notably, a major grain-yield QTL on chr. 5 overlapped with the gene encoding cytosolic GS (gln4 locus). Gln1 and gln2 (GS genes) on chr. 1 and gln3 on chr. 4 are other interesting candidate loci encoding enzymes involved in N metabolism that were found to co-locate with NUE QTLs. Based on these results, it has been suggested that the increased productivity in modern maize genotypes under low N may be due to their ability to accumulate nitrate in the leaves during vegetative growth and to remobilize it efficiently at grain filling. In wheat, a QTL meta-analysis and factorial regression showed that major phenological loci controlling phenology (e.g. Ppd-D1, Rht-B1, and B1) have a major impact on N-related QTLs (Laperche et al., 2007). Additionally, QTL clusters for GS activity on wheat chrs 2A and 4A coincided with the location of GS and GSr genes, respectively, and, although QTL alleles for higher GS activity were associated with higher grain N, they did not show a noteworthy effect on grain yield components (Habash *et al.*, 2007).

Lack of P will inevitably impact negatively crop growth and its final biomass and yield. Due to the low mobility of P in the soil, several studies have investigated the effects of root architecture QTLs on P uptake. Relevant QTLs have been identified in common bean (Ochoa *et al.*, 2006), soybean (Kuang *et al.*, 2005; Li, Y.D. *et al.*, 2005), rice (Wissuwa *et al.*, 2002, 2005; Heuer *et al.*, 2009), maize (Kaeppler *et al.*, 2000; Zhu *et al.*, 2005, 2006) and wheat (Su *et al.*, 2006). Notably, a common feature of the above-mentioned studies is that the QTL alleles for high-P efficiency were associated with greater root surface area due to an increase in either root mass or root hair density. Interestingly, QTLs for P uptake in bean were found to influence total acid exudation from the root (Yan *et al.*, 2004), a process capable of mobilizing soil-bound P through soil P desorption or mineralization. In rice, the application of marker-assisted backcrossing (MABC) of the beneficial allele at *Phosphate uptake 1* (*Pup1*), a major QTL for P-uptake efficiency that was mapped to a 3 cM interval on chromosome 12 (Wissuwa *et al.*, 2002), allowed for up to fourfold increase in P uptake (Wissuwa *et al.*, 2005). Fine mapping of *Pup1* is under way to identify and clone the controlling gene (Ismail *et al.*, 2007; Heuer *et al.*, 2009). Additional work has also been undertaken to further validate the effects of *Pup1* in rice (Chin *et al.*, 2010). Based on the *Pup1* genomic sequence of the tolerant donor variety (Kasalath) that recently became available, markers were designed that target: (i) putative genes that are partially conserved in the Nipponbare reference genome; and (ii) Kasalath-specific genes that are located in a large insertion-deletion (INDEL) region that is absent in Nipponbare. Testing these markers in 159 diverse rice accessions confirmed their diagnostic value across genotypes and showed that *Pup1* is present in more than 50% of rice accessions adapted to stress-prone environments, whereas it was detected in only about 10% of the analysed irrigated/lowland varieties (Chin *et al.*, 2010). Notably, the *Pup1* locus was detected in more than

80% of the drought-tolerant rice breeding lines analysed, suggesting that breeders have unknowingly selected for *Pup1*. Contrasting *Pup1* NILs were subsequently grown in two different P-deficient soils and environments. Under the applied aerobic growth conditions, NILs with the *Pup1* locus significantly out-yielded their counterparts without *Pup1*. Overall, these results provide strong evidence that *Pup1* has the potential to improve yield in P-deficient and/or drought-prone environments and in diverse genetic backgrounds (Chin *et al.*, 2010).

8.4 Improving Abiotic Stress Tolerance via Marker-assisted Selection

Several factors limit the possibility of obtaining reliable QTL data and, most importantly, their deployment in breeding programmes through marker-assisted selection (MAS). Among such factors, the environmental dependence of QTL expression is of utmost importance in order to obtain reproducible data and effectively assess the value of a particular QTL. This aspect is particularly relevant for stress-tolerance traits since, as compared with other traits, their phenotypic variance and the direction of the additive effect are greatly influenced by environmental factors and the intensity of the stress. In simpler words, the effect of the same QTL can markedly differ according to the prevailing environmental conditions (Collins *et al.*, 2008).

More in general terms, the heritability (i.e. the predictability of a trait phenotypic value between subsequent generations) of tolerance to abiotic stress is usually low to medium, and thus it should come aa no surprise that the effects of single QTLs contributing to the overall variability of stress tolerance can change drastically as a function of the prevailing environmental conditions and their effects on the crop (e.g. water status under conditions of water deficit). Therefore, unless these environmental conditions are properly monitored and accounted for, conclusions drawn based on QTLs characterized by large QTL–

environment interaction can be misleading. As an example, ABA controls stomatal conductance and transpiration and, therefore, the rate of soil and plant dehydration (Tuberosa *et al.*, 1994; Iuchi *et al.*, 2001), two factors that are known strongly to influence plant ABA levels (Quarrie, 1991). Accordingly, if our interest is to investigate the role of ABA accumulation in response to drought, the inherent (i.e. per se) capacity of genotypes to accumulate ABA should be assessed under experimental conditions that allow us to compare plants characterized by similar water status, hence – barring the presence of strong differences in capacity of the tested genotypes to adjust osmotically – exposed to water deficit of similar intensity. Clearly, these conditions are unattainable in field experiments; conversely, they can to a greater extent be met in experiments carried out in controlled conditions (e.g. growth chamber and greenhouse), the main limitation being that the rhizosphere in these experiments is at best a poor surrogate for the real one experienced by plants in the field. Therefore, selection and modelling for yield based on the effects of QTLs for ecophysiological traits that are highly sensitive to environmental conditions should duly account for the effects that such conditions may have on the expression of the traits. Eventually, this will help in assessing the genetic and environmental components of the association between yield and ecophysiological traits that influence the adaptive response of crops to abiotic constraints. These considerations reinforce the importance of evaluating the effects of QTLs for abiotic stress tolerance across a broad range of environments (Maccaferri *et al.*, 2008a, 2011a). Importantly, the presence of multiple abiotic constraints will further blur our capacity to predict correctly genotype preformance based on the information available on the effects of single QTLs. Only a handful of studies have addressed this important issue (LeDeaux *et al.*, 2006; MacMillan *et al.*, 2006a, b; Mittler, 2006; Ribaut *et al.*, 2007; Welcker *et al.*, 2007; Pinto *et al.*, 2010).

In addition to environment-dependent effects, other factors are likely to limit the utilization of QTLs for genetic improvement via MAS. An agronomically desirable QTL allele (i.e. one that increases yield) discovered in non-elite genetic material might not provide any benefit when introgressed into an elite background, because the allele may already be prevalent and in some cases ubiquitous in the elite gene pool. Additionally, the effects of the agronomically positive allele may not be transferable to elite backgrounds due to unfavourable epistatic interactions (Podlich *et al.*, 2004). With only a few notable exceptions (e.g. *Sub1*, a major submergence-tolerance QTL in rice (Ismail *et al.*, 2007), *Vgt1*, a major flowering-time QTL in maize (Vladutu *et al.*, 1999; Salvi *et al.*, 2007; Ducrocq *et al.*, 2008), when a QTL from a particular genetic background is transferred to a different background it usually shows much smaller effects that may disappear altogether, even under similar experimental conditions. Consequently, translating the knowledge gained from QTL and other molecular studies (e.g. transcriptome profiling; Fowler and Thomashow, 2002; Talamè *et al.*, 2007; Zhuang *et al.*, 2007; Fernandes *et al.*, 2008; Ergen *et al.*, 2009; Pariasca-Tanaka *et al.*, 2009; Cohen *et al.*, 2010; Narsai *et al.*, 2010; Chauhan *et al.*, 2011) into an improved cultivar is one of the most difficult, albeit rewarding if met, challenges faced by the scientific community (Parry *et al.*, 2005; Passioura, 2007; Tuberosa *et al.*, 2007a; Xu and Crouch, 2008; Luo, 2010). Several reviews have addressed the key issues to be considered in the adoption of the QTL approach for improving crop performance (Beavis, 1998; Hospital *et al.*, 2000; Bernardo *et al.*, 2006; Bernardo and Yu, 2007; Dwivedi *et al.*, 2007; Ragot and Lee, 2007; Xu and Crouch, 2008; Luo, 2010). The following sections describe QTL case studies for each of the three above-mentioned factors and provide a glimpse of how combining accurate and relevant phenotyping with modelling can help us better to deal with the above-mentioned complexity.

A major cause of inconsistency in QTL effects and outcome of MAS in different genetic backgrounds is due to epistasis (Reynolds and Tuberosa, 2008). The size of mapping populations is mostly inadequate for an effective detection of epistatic QTL interactions, which can represent an important source of variation for quantitative traits in autogamous crops (Maccaferri *et al.*, 2008a). Given their potential for impact on response to selection, empirical investigations to quantify the importance of epistasis are an important component of the design and optimization of any MAS strategy (Podlich *et al.*, 2004; Isobe *et al.*, 2007; Jannink *et al.*, 2009). Most MAS strategies have assumed that QTLs act independently, i.e. no interaction with other genes, and that the effects of such QTLs do not change after a number of selection cycles. Partially to overcome these problems, Podlich *et al.* (2004) have proposed the 'Mapping As You Go' approach, a mapping–MAS strategy that partly accounts for the presence of epistasis and QTL–environment interaction. In this case, MAS is implemented so that the estimated values of QTL alleles evolve as the current germplasm evolves over cycles of selection (Podlich *et al.*, 2004).

Hundreds of studies and reviews have described QTLs that influence tolerance to abiotic stresses. That notwithstanding, MAS has so far contributed marginally to the release of improved cultivars with greater tolerance to abiotic stresses, particularly drought. Improving crop performance under water-limiting conditions via MAS may also require consideration of QTLs for tolerance to abiotic and biotic factors that impair root growth and functions, such as nematodes in soybean (Ha *et al.*, 2007) and bread wheat (Langridge, 2005; Zwart *et al.*, 2010), high boron in barley and bread wheat (Langridge, 2005; Schnurbusch *et al.*, 2007; Sutton *et al.*, 2007) and aluminium toxicity in rice (Nguyen *et al.*, 2003) and sorghum (Magalhaes *et al.*, 2007). A common feature of cereal responses to drought, heat and cold stress near flowering and during the early stages of seed growth is a reduction of reproductive fertility due to partial sterility and/or early abortion. This loss of fertility has been attributed to different factors (e.g. low supply of photosynthates, high ABA, changes in the cell cycle following diverse signallings, etc.) acting alone and more likely

on reproductive fertility (Saini and Westgate, 2000; Zinselmeier *et al.*, 2002; McLaughlin and Boyer, 2004a, b; Boyer and McLaughlin, 2007; Fresneau *et al.*, 2007; Setter *et al.*, 2011). The QTL approach attempts to dissect out the genetic and physiological components affecting source–sink relationships under abiotic stress and to what extent these may influence yield (Prioul *et al.*, 1997, 1999; Pelleschi *et al.*, 2006; Miralles and Slafer, 2007; Welcker *et al.*, 2007).

8.4.1 Marker-assisted selection for drought tolerance

Although remarkable levels of dehydration tolerance have been obtained under laboratory conditions through genetic engineering of a number of functionally relevant signalling, metabolic and/or structural targets (Capell *et al.*, 2004; Jenks *et al.*, 2007; Rivero *et al.*, 2007, 2009; Zhu *et al.*, 2007; Mittler and Blumwald, 2010; Pasapula *et al.*, 2011), there is little evidence to show that characteristics enabling survival under the drastic water stress treatments typically imposed on the tested transgenic lines will provide any yield advantage under the milder stress conditions usually experienced in commercially productive fields. In contrast, exploitation of natural variation for drought-related traits has resulted in slow but unequivocal progress in crop performance (Blum, 1988; Ribaut *et al.*, 2004; Duvick, 2005; Reynolds and Tuberosa, 2008). Biomass accumulation is intrinsically linked to transpiration, because stomatal aperture and leaf area determine the rate of both photosynthesis and transpiration. Therefore, there is an inherent contradiction between biomass accumulation and stress avoidance via a reduction of transpiration. The maintenance of transpiration rate under water deficit is most often achieved by improving the size, architecture or hydraulic conductance of the root system (Maurel and Chrispeels, 2001; Price *et al.*, 2002a; Steele *et al.*, 2006; Ehlert *et al.*, 2009; Reynolds *et al.*, 2009a). This is the case when a more vigorous and/or deeper root system allows plants to access a greater amount of soil moisture. In contrast, when roots explore a limited volume of soil because of physical or chemical constraints (e.g. soil compaction, soil pH, etc.), a larger and/or deeper root system can be detrimental, because depletion of soil moisture occurs more rapidly, thereby causing severe stress at the end of the season, and because the assimilates invested in roots would be better invested in other organs. Accordingly, conflicting results have been reported in those studies that have investigated – either with or without a QTL approach – the association between root features and crop yield (Richards and Passioura, 1989; Bolanos and Edmeades, 1993; Bruce et al., 2002; Tuberosa et al., 2002b, c, 2003; Yue et al., 2005, 2006; MacMillan et al., 2006a, b; Steele et al., 2006, 2007; Landi et al., 2007, 2010).

To date, the most compelling story for the outcome of MAS in improving drought tolerance has unfolded in rice, where it all started with the identification of four major QTLs influencing root traits (Courtois et al., 2000). Subsequently, marker-assisted backcrossing (MABC) allowed for the introgression of the alleles for greater root length at these loci from Azucena into Kalinga III, an upland variety characterized by a rather shallow root system (Steele et al., 2006, 2007). All these efforts have resulted in the release of the first MAS-derived drought-tolerant rice variety – Birsa Vikas Dhan 111 (PY 84) – in the Indian state of Jharkhand (Ashraf, 2010). A major QTL (qtl12.1) for grain yield under drought, located within a 3 cM interval on chromosome 12, was tested in multiple locations. The relative effect of this QTL on grain yield increased as the intensity of drought stress increased, reaching an additive effect of more than 40% of the trial mean under severe drought stress. Additional QTLs governing grain yield under drought in the lowland ecosystem have been identified in the population Apo/Swarna. Marker RM520 on chr. 3 showed a large effect on grain yield under severe lowland drought stress. This QTL was also effective in yield improvement under mild stress in upland conditions and under mild and severe stress in lowland conditions. MAS

has begun for introgression of this QTL in Swarna.

Another strategy to maintain transpiration is to avoid leaf senescence (stay-green trait), thereby increasing the accumulated photosynthesis over the crop life cycle (Borrell et al., 2000a, b). In sorghum, four major QTLs that control stay-green and grain yield (Stg1–Stg4) have been identified (Harris et al., 2007) and the favourable alleles introgressed via MAS in elite materials. Growth maintenance under drought conditions allows better light interception by leaves, hence maintenance of photosynthesis, but also higher transpiration rate and, consequently, faster soil moisture depletion. Therefore, this is an appropriate strategy in many cases except for severe terminal water deficits. A high degree of genetic variability in sensitivity of leaf growth has been reported in maize (Welcker et al., 2007). In three maize mapping populations, QTLs of leaf growth sensitivity to water deficit largely overlapped with QTLs for leaf response to evaporative demand, suggesting common hydraulic mechanisms (Reymond et al., 2003; Sadok et al., 2007; Welcker et al., 2007). Importantly, in one maize mapping population, half of the QTLs for sensitivity of leaf growth overlapped with those for silk growth (Welcker et al., 2007), an important trait related to anthesis–silking interval (ASI), a trait negatively correlated with yield (Bolanos and Edmeades, 1996; Duvick, 2005). Ribaut and colleagues introgressed five QTL alleles for short ASI through MABC from a drought-tolerant donor to an elite, drought-susceptible line (Ribaut et al., 2004); although under severe drought the selected lines tested out-yielded the unselected control, and this advantage decreased at lower stress intensity and disappeared when water stress decreased yield to less than 40% (Ribaut and Ragot, 2007).

In pearl millet, three major QTLs for grain yield with low QTL–environment interactions were identified across a range of post-flowering moisture environments, leading Bidinger et al. (2007) to advocate MAS for introgressing and pyramiding of QTL alleles for high yield. Notably, a major QTL for grain yield per se and under drought conditions accounted for up to 32% of the phenotypic variation of grain yield observed among RILs. The effects of this QTL have been validated in two independent MABC programmes in which the 30% improvement in grain yield general combining ability expected of this QTL under terminal drought stress conditions was recovered in introgression lines based on the information provided by the markers flanking the QTL (Serraj, *et al.*, 2005; Yadav *et al.*, 2011).

Major QTLs for seed weight and grain yield at different moisture conditions have also been identified in rice (Wang *et al.*, 2006; Bernier *et al.*, 2007) and durum wheat (Maccaferri *et al.*, 2008a) and are being introgressed in different genetic backgrounds. In bread wheat, Fleury *et al.* (2010) have implemented a strategy where a specific environment is targeted and appropriate germplasm adapted to the chosen environment is selected, based on extensive definition of the morpho-physiological and molecular mechanisms of tolerance of the parents. This information was then used to create structured populations and develop models for QTL analysis, MAS and positional cloning.

In cotton, extensive work has been carried out by Saranga and colleagues for the identification of drought-related QTLs and their selection to improve yield under water-limited conditions (Levi *et al.*, 2009a, b). In particular, MAS has been deployed to introgress QTL alleles for yield and drought-related physiological traits such as $\Delta^{13}C$, osmotic potential (OP) and leaf chlorophyll content (Chl) between elite cultivars of the two elite cotton species cv. F-177 (*Gossypium barbadense*) and cv. Siv'on (*Gossypium hirsutum*) (Levi *et al.*, 2009a, b). After deriving a set of NILs, those introgressed with QTL alleles for high yield rarely exhibited an advantage in yield relative to the recipient parent, whereas several NILs had the expected phenotype in terms of lower OP, higher $\Delta^{13}C$ or high Chl. In *G. barbadense* genotypes, yield was correlated negatively with OP and positively with $\Delta^{13}C$ (but expressed as carbon isotope composition, $\delta^{13}C$) and stomatal conductance, SLW and Chl, whereas in *G. hirsutum* yield was negatively correlated with

SLW and Chl and positively with $\Delta^{13}C$. Based on these results, Levi *et al.* (2009b) concluded that MAS can improve drought-adaptive traits in cotton NILs, but that complementary conventional breeding is required to combine the introduced QTLs with high yield potential.

8.4.2 Marker-assisted selection for salinity tolerance

Major efforts in MAS to improve salinity tolerance in rice have been undertaken at the International Rice Research Institute (IRRI) for introgressing *Saltol*, a major QTL located on chr. 1. The initial MABC lines for *Saltol* were developed using FL478 as the donor for a high level of tolerance. Several MABC lines have now been developed at IRRI, with the FL478 *Saltol* allele involved in several popular cultivars. Although the introgression lines showed an increased level of tolerance, their tolerance was not as high as for the original FL478 donor (Thomson *et al.*, 2010). Further testing is under way to determine the relative tolerance effect of different *Saltol* alleles, as well as the ultimate effect that this seedling-stage tolerance will have on crop establishment and grain yield under field conditions, particularly in areas where salinity is high at the beginning of the season, such as in coastal areas during the monsoon season. In addition, once other QTLs for salinity tolerance become better characterized, these can be combined with *Saltol* using MABC in a QTL pyramiding scheme to increase the levels of salt tolerance. Accordingly, *Saltol* and other QTLs for seedling-stage tolerance and one for the reproductive stage are being targeted to develop varieties tolerant at both stages, and for better understanding of the physiology and genetics of tolerance at both stages.

8.4.3 Marker-assisted selection for submergence and anoxia tolerance

Following the identification of *Sub1*, MABC was used to efficiently convert submergent-susceptible rice varieties into tolerant ones in only three backcross generations.

Accordingly, markers were developed for introgressing *Sub1* into six popular varieties to meet the needs of farmers in flood-prone regions (Neeraja *et al.*, 2007; Bailey-Serres *et al.*, 2010). Field testing of six NILs pairs (with and without *Sub1* introgression) showed that *Sub1* is effective in all target environments and is independent of the recurrent parent background, with no effect on agronomic or quality aspects of these varieties under normal conditions. Relative to the submergent-intolerant lines, *Sub1* NILs had several-fold greater yield when flooding occurred for 12–18 days during the vegetative stage (Sarkar *et al.*, 2009; Singh *et al.*, 2009). The cultivation of these new varieties has significantly increased yield and food security for local farmers. These results clearly demonstrate the effectiveness of MAS for introgressing agronomically valuable QTL alleles into elite material. The success of this work is largely due to the major effect of the *Sub1* QTL and the stability of its effect under submergence conditions.

8.5 Future Perspectives

During the past decade, increasing attention has been devoted to the use of crop modelling for elucidating the genetic basis of genotype–management–environment (G–M–E) interaction at the level of the entire genotype and, more recently, at the level of single loci (Chapman *et al.*, 2003; Tardieu *et al.*, 2003; Reymond *et al.*, 2004; Cooper *et al.*, 2005, 2007; Hammer *et al.*, 2005; Semenov *et al.*, 2009; Ludwig and Asseng, 2010; Tardieu and Tuberosa, 2010; van Eeuwijk *et al.*, 2010). The objective is to predict, via modelling, yield differences among genotypes grown under different environmental conditions (Hammer *et al.*, 2006; Cooper *et al.*, 2009; Tardieu and Tuberosa, 2010). The benefits accrued by modelling studies are expected to increase as the complexity of the genetic control of traits increases, providing that it is possible to account for the effects of genetic interactions for prediction of trait variation (Cooper *et al.*, 2009). In their study on sorghum, Chapman *et al.* (2003) critically described an iterative genotype-to-pheno-type model-building process for linking the trait quantitative genetic models that can be identified by QTL mapping to the growth and development processes that are influenced by the putative traits that influence drought tolerance. Combining the quantitative genetic model and the crop growth model provides a genotype-to-phenotype multi-trait model that can simulate the contributions and value of traits and their QTLs to genetically improve crop performance in a particular target environment. Ultimately, modelling aims to predict the best allelic combinations able to optimize yield. Analogously to that observed in QTL studies, the predictive ability of modelling is strongly limited when two or more environmental variables vary simultaneously, i.e. a condition that is typically encountered by crops in the field (Mittler, 2006), differently from the controlled conditions commonly present in the experiments used to derive the growth parameters that define the model.

The main underlying assumption of the modelling approach is that yield and other functionally complex traits can be analysed and improved by dissecting them into simpler processes, and then by reassembling such processes to reconstruct, via modelling, a higher order of plant functionality and, ultimately, of yield itself. This approach is now more feasible in view of: (i) the improved phenotyping capacities that create the possibility of measuring the response of adaptive traits to environmental factors with both high accuracy and high throughput (Casadesus *et al.*, 2007; Montes *et al.*, 2007; Salekdeh *et al.*, 2009; Tester and Langridge, 2010); (ii) the availability of methods (e.g. model-assisted phenotyping and meta-analyses of large data sets) to dissect complex phenotypes into heritable traits that are stable characteristics of genotypes (Tardieu and Tuberosa, 2010); and (iii) modelling to simulate the effects of the genetic variability of physiological traits (e.g. leaf expansion under water deficit) under a broad range of climatic scenarios (Chenu *et al.*, 2008, 2009), following the identification of major yield QTLs across a series of field experiments (Vargas *et al.*, 2006). Models have been used to generate an index of the climatic environment (e.g. of drought stress) for breeding programme trials. In wheat and sorghum grown in northern Australia, it was demonstrated that mid-season drought generates large genotype by environment interaction (Chapman, 2008).

Apart from the few exceptions listed above, the vast majority of loci that affect crop yield per se have a rather limited effect (Mackill, 2007), particularly under environmentally constrained conditions that markedly reduce yield. Therefore, combining favourable alleles by MAS to achieve a significant improvement quickly becomes impractical and would excessively constrain the potential for achieving yield gain due to the action of other loci. In this case, it is preferable simply to select for the phenotype or yield itself or adopt genome-wide selection, rather than attempting to manipulate the trait by MAS at multiple loci (Bernardo *et al.*, 2006; Bernardo, 2008, 2010). Nowadays, genome selection is facilitated by the availability of large numbers of markers, particularly single nucleotide polymorphysms (SNPs) that are amenable to high-throughput profiling at very low cost.

8.6 Conclusions

Within the atmospheric and metereological conditions expected for the next few decades, growing evidence suggests that C_3 crops are likely to produce more harvestable products and that both C_3 and C_4 crops are likely to use less water with rising atmospheric $[CO_2]$ in the absence of stressful conditions (DaMatta *et al.*, 2010). None the less, the expected direct beneficial impact of elevated $[CO_2]$ on crop yield will probably be offset by other effects of climate change, such as elevated temperatures and altered rainfall patterns. Existing modelling results show that an increase in precipitation will increase crop yield and, what is more, crop yield is more sensitive to precipitation than temperature (Kang *et al.*, 2009). Therefore, the release of cultivars better adapted to a broader range of

environmental conditions will become an increasingly important goal of breeding projects worldwide. Compared with conventional breeding practices, the contribution in this direction of both physiological breeding and molecular breeding has somehow fallen short of expectations, though a number of notable exceptions have been presented in this chapter.

Adaptation of agricultural systems to climate change and attaininment of a suitable level of food security will be possible only through a strong public–private partnership (Antle and Capalbo, 2010) based on the mutual engagement of academia and industry, particularly the seed industry, when it comes to the release of improved cultivars better able to withstand the negative effects of environmental constraints. Genomics approaches and sequence-based breeding will expedite the dissection of the genetic basis of abiotic stress tolerance while providing unparalleled opportunities to tap into wild relatives of crops, hence expanding the reservoir of genetic diversity available to breeders (Glaszmann *et al.*, 2010). To what extent this will actually impact the release of improved cultivars will largely depend on a more complete and comprehensive understanding of the adaptive response of crops to abiotic stress, and on our capacity to integrate this information into breeding programmes via modelling or other approaches such as genome selection. In view of the complexity of yield, particularly under drought and other abiotically constrained conditions, we foresee that genome selection will provide the most powerful way to raise the yield potential to the levels required to keep up with the fast-increasing demand in food, fibre and fuel worldwide. MAS will remain a valid option for major loci (genes and/or QTLs) as long as their effects remain sufficiently predictable and economically viable. Additionally, the cloning of QTLs will become a more routine activity thanks to a more widespread utilization of high-throughput, accurate phenotyping (Berger *et al.*, 2010; Tuberosa, 2011b), sequencing (Varshney *et al.*, 2009b) and the identification of suitable candidate genes through 'omics' profiling (Cohen *et al.*, 2010; Urano *et al.*, 2010; Zargar *et al.*, 2010). Cloned QTLs will provide novel opportunites for genetic engineering of abiotic stress tolerance and for a more targeted search for novel alleles in wild germplasm (Salvi and Tuberosa, 2007).

Ultimately, reducing crop vulnerability to climate change and improving the sustainability of agricultural production will require a multidisciplinary and integrated approach (Tuberosa *et al.*, 2007a; Keating *et al.*, 2010; Parry and Hawskesford, 2010; Parry and Lea, 2010; Passioura, 2010) based on a number of different disciplines (e.g. genetics, breeding, agronomy, crop physiology, etc.) that will eventually allow breeders to undertake better-informed and more effective decisions.

References

Acuna, T.L.B., Lafitte, H.R. and Wade, L.J. (2008) Genotype × environment interactions for grain yield of upland rice backcross lines in diverse hydrological environments. *Field Crops Research* 108, 117–125.

Ainsworth, E.A. and Ort, D.R. (2010) How do we improve crop production in a warming world? *Plant Physiology* 154, 526–530.

Alonso-Blanco, C., Aarts, M.G.M., Bentsink, L., Keurentjes, J.J.B., Reymond, M., Vreugdenhil, D. *et al.* (2009) What has natural variation taught us about plant development, physiology, and adaptation? *Plant Cell* 21, 1877–1896.

Alpert, P. (2006) Constraints of tolerance: why are desiccation-tolerant organisms so small or rare? *Journal of Experimental Biology* 209, 1575–1584.

Andaya, V.C. and Mackill, D.J. (2003) QTLs conferring cold tolerance at the booting stage of rice using recombinant inbred lines from a japonica x indica cross. *Theoretical and Applied Genetics* 106, 1084–1090.

Andaya, V.C. and Tai, T.H. (2006) Fine mapping of the *qCTS12* locus, a major QTL for seedling cold tolerance in rice. *Theoretical and Applied Genetics* 113, 467–475.

Andersen, J.R. and Lubberstedt, T. (2003) Functional markers in plants. *Trends in Plant Science* 8, 554–560.

Antle, J.M. and Capalbo, S.M. (2010) Adaptation of agricultural and food systems to climate change: An economic and policy perspective. *Applied Economic Perspectives and Policy* 32, 386–416.

Araus, J.L., Brown, H.R., Febrero, A., Bort, J. and Serret, M.D. (1993) Ear photosynthesis, carbon isotope discrimination and the contribution of respiratory CO_2 to differences in grain mass in durum. *Plant Cell and Environment* 16, 383–392.

Araus, J.L., Amaro, T., Zuhair, Y. and Nachit, M.M. (1997) Effect of leaf structure and water status on carbon isotope discrimination in field-grown durum wheat. *Plant, Cell and Environment* 20, 1484–1494.

Araus, J.L., Slafer, G.A., Reynolds, M.P. and Royo, C. (2002) Plant breeding and drought in C_3 cereals: What should we breed for? *Annals of Botany* 89, 925–940.

Araus, J.L., Bort, J., Steduto, P., Villegas, D. and Royo, C. (2003) Breeding cereals for Mediterranean conditions: ecophysiological clues for biotechnology application. *Annals of Applied Biology* 142, 129–141.

Araus, J.L., Slafer, G.A., Royo, C. and Serret, M.D. (2008) Breeding for yield potential and stress adaptation in cereals. *Critical Reviews in Plant Sciences* 27, 377–412.

Archer, E.R.M., Oettle, N.M., Louw, R. and Tadross, M.A. (2008) 'Farming on the edge' in arid western South Africa: climate change and agriculture in marginal environments. *Geography* 93, 98–107.

Asghari, A., Mohammadi, S.A., Moghaddam, M. and Mohammaddoost, H.R. (2007) QTL analysis for cold resistance-related traits in *Brassica napus* using RAPD markers. *Journal of Food Agriculture and Environment* 5, 188–192.

Ashraf, M. (2010) Inducing drought tolerance in plants: Recent advances. *Biotechnology Advances* 28, 169–183.

Babu, R.C., Nguyen, B.D., Chamarerk, V., Shanmugasundaram, P., Chezhian, P., Jeyaprakash, P. *et al.* (2003) Genetic analysis of drought resistance in rice by molecular markers: Association between secondary traits and field performance. *Crop Science* 43, 1457–1469.

Bagge, M., Xia, X.C. and Lubberstedt, T. (2007) Functional markers in wheat – Commentary. *Current Opinion in Plant Biology* 10, 211–216.

Bailey-Serres, J., Fukao, T., Ronald, P., Ismail, A., Heuer, S. and Mackill, D. (2010) Submergence tolerant rice: SUB1's journey from landrace to modern cultivar. *Rice* 3, 138–147.

Barloy, D., Lemoine, J., Abelard, P., Tanguy, A.M., Rivoal, R. and Jahier, J. (2007) Marker-assisted pyramiding of two cereal cyst nematode resistance genes from *Aegilops variabilis* in wheat. *Molecular Breeding* 20, 31–40.

Barriere, Y., Gibelin, C., Argillier, O. and Mechin, V. (2001) Genetic analysis in recombinant inbred lines of early dent forage maize. I. QTL mapping for yield, earliness, starch and crude protein contents from per se value and top cross experiments. *Maydica* 46, 253–266.

Bartels, D. and Sunkar, R. (2005) Drought and salt tolerance in plants. *Critical Reviews in Plant Sciences* 24, 23–58.

Beavis, W.D. (1998) QTL analysis: power, precision, and accuracy. In: Paterson, A.H. (ed.) *Molecular Dissection of Complex Traits*. CRC Press, Boca Raton, Florida, pp. 145–162.

Berger, B., Parent, B. and Tester, M. (2010) High-throughput shoot imaging to study drought responses. *Journal of Experimental Botany* 61, 3519–3528.

Bernardo, R. (2008) Molecular markers and selection for complex traits in plants: Learning from the last 20 years. *Crop Science* 48, 1649–1664.

Bernardo, R. (2010) Genomewide selection with minimal crossing in self-pollinated crops. *Crop Science* 50, 624–627.

Bernardo, R. and Yu, J.M. (2007) Prospects for genome-wide selection for quantitative traits in maize. *Crop Science* 47, 1082–1090.

Bernardo, R., Moreau, L. and Charcosset, A. (2006) Number and fitness of selected individuals in marker-assisted and phenotypic recurrent selection. *Crop Science* 46, 1972–1980.

Bernier, J., Kumar, A., Ramaiah, V., Spaner, D. and Atlin, G. (2007) A large-effect QTL for grain yield under reproductive-stage drought stress in upland rice. *Crop Science* 47, 507–518.

Bernier, J., Serraj, R., Kumar, A., Venuprasad, R., Impa, S., Gowda, R.P.V. *et al.* (2009) The large-effect drought-resistance QTL qtl12.1 increases water uptake in upland rice. *Field Crops Research* 110, 139–146.

Bianchi-Hall, C.M., Carter, T.E.J., Bailey, M.A., Mian, M.A.R., Rufty, T.W., Ashley, D.A. *et al.* (2000) Aluminum tolerance associated with quantitative trait loci derived from soybean PI 416937 in hydroponics. *Crop Science* 40, 538–545.

Bidinger, F.R., Nepolean, T., Hash, C.T., Yadav, R.S. and Howarth, C.J. (2007) Quantitative trait loci for grain yield in pearl millet under variable postflowering moisture conditions. *Crop Science* 47, 969–980.

Blum, A. (1988) *Plant Breeding for Stress Environments*. CRC Press, Boca Raton, Florida.

Blum, A. (1996) Crop responses to drought and the interpretation of adaptation. *Plant Growth Regulation* 20, 135–148.

Blum, A. (1998) Improving wheat grain filling under stress by stem reserve mobilisation. *Euphytica* 100, 77–83.

Blum, A. (2009) Effective use of water (EUW) and not water-use efficiency (WUE) is the target of crop yield improvement under drought stress. *Field Crops Research* 112, 119–123.

Blum, A., Shpiler, L., Golan, G. and Mayer, J. (1989) Yield stability and canopy temperature of wheat genotypes under drought-stress. *Field Crops Research* 22, 289–296.

Blum, A., Sinmena, B., Mayer, J., Golan, G. and Shpiler, L. (1994) Stem reserve mobilization supports wheat-grain filling under heat-stress. *Australian Journal of Plant Physiology* 21, 771–781.

Bolanos, J. and Edmeades, G.O. (1993) Eight cycles of selection for drought tolerance in lowland tropical maize. 1. Responses in grain yield, biomass, and radiation utilization. *Field Crops Research* 31, 233–252.

Bolanos, J. and Edmeades, G.O. (1996) The importance of the anthesis-silking interval in breeding for drought tolerance in tropical maize. *Field Crops Research* 48, 65–80.

Borlaug, N.E. (2007) Sixty-two years of fighting hunger: personal recollections. *Euphytica* 157, 287–297.

Borlaug, N.E. and Dowswell, C.R. (2005) Feeding a world of ten billion people: a 21st century challenge. In: Tuberosa, R., Phillips, R.L. and Gale, M. (eds) *In the Wake of the Double Helix: From the Green Revolution to the Gene Revolution*. Avenue Media, Bologna, Italy, pp. 3–23.

Borrell, A.K., Hammer, G.L. and Douglas, A.C.L. (2000a) Does maintaining green leaf area in sorghum improve yield under drought? I. Leaf growth and senescence. *Crop Science* 40, 1026–1037.

Borrell, A.K., Hammer, G.L., and Henzell, R.G. (2000b) Does maintaining green leaf area in sorghum improve yield under drought? II. Dry matter production and yield. *Crop Science* 40, 1037–1048.

Boyer, J.S. (1982) Plant productivity and the environment. *Science* 218, 443–448.

Boyer, J.S. (1996) Advances in drought tolerance in plants. *Advances in Agronomy* 56, 187–218.

Boyer, J.S. and McLaughlin, J.E. (2007) Functional reversion to identify controlling genes in multigenic responses: analysis of floral abortion. *Journal of Experimental Botany* 58, 267–277.

Boyer, J.S. and Westgate, M.E. (2004) Grain yields with limited water. *Journal of Experimental Botany* 55, 2385–2394.

Brennan, J.P., Condon, A.G., Van Ginkel, M. and

Reynolds, M.P. (2007) An economic assessment of the use of physiological selection for stomatal aperture-related traits in the CIMMYT wheat breeding programme. *Journal of Agricultural Science* 145, 187–194.

Breseghello, F. and Sorrells, M.E. (2006) Association analysis as a strategy for improvement of quantitative traits in plants. *Crop Science* 46, 1323–1330.

Bruce, W.B., Edmeades, G.O. and Barker, T.C. (2002) Molecular and physiological approaches to maize improvement for drought tolerance. *Journal of Experimental Botany* 53, 13–25.

Buckler, E.S., Gaut, B.S. and McMullen, M.D. (2006) Molecular and functional diversity of maize. *Current Opinion in Plant Biology* 9, 172–176.

Buckler, E.S., Holland, J.B., Bradbury, P.J., Acharya, C.B., Brown, P.J., Browne, C. *et al.* (2009) The genetic architecture of maize flowering time. *Science* 325, 714–718.

Burke, J.M., Burger, J.C. and Chapman, M.A. (2007) Crop evolution: from genetics to genomics. *Current Opinion in Genetics and Development* 17, 525–532.

Byrt, C.S., Platten, J.D., Spielmeyer, W., James, R.A., Lagudah, E.S., Dennis, E.S. *et al.* (2007) HKT1;5-like cation transporters linked to Na+ exclusion loci in wheat, Nax2 and Kna1. *Plant Physiology* 143, 1918–1928.

Cairns, J.E., Audebert, A., Mullins, C.E. and Price, A.H. (2009) Mapping quantitative trait loci associated with root growth in upland rice (*Oryza sativa* L.) exposed to soil water-deficit in fields with contrasting soil properties. *Field Crops Research* 114, 108–118.

Campos, H., Cooper, A., Habben, J.E., Edmeades, G.O. and Schussler, J.R. (2004) Improving drought tolerance in maize: a view from industry. *Field Crops Research* 90, 19–34.

Capell, T., Bassie, L. and Christou, P. (2004) Modulation of the polyamine biosynthetic pathway in transgenic rice confers tolerance to drought stress. *Proceedings of the National Academy of Sciences of the United States of America* 101, 9909–9914.

Casadesus, J., Kaya, Y., Bort, J., Nachit, M.M., Araus, J.L., Amor, S. *et al.* (2007) Using vegetation indices derived from conventional digital cameras as selection criteria for wheat breeding in water-limited environments. *Annals of Applied Biology* 150, 227–236.

Casanoves, F., Baldessari, J. and Balzarini, M. (2005) Evaluation of multienvironment trials of peanut cultivars. *Crop Science* 45, 18–26.

Challinor, A.J., Simelton, E.S., Fraser, E.D.G., Hemming, D. and Collins, M. (2010) Increased crop failure due to climate change: assessing adaptation options using models and socio-economic data for wheat in China. *Environmental Research Letters* 5, 8.

Chao, S., Lazo, G.R., You, F., Crossman, C.C., Hummel, D.D., Lui, N. *et al.* (2006) Use of a large-scale Triticeae expressed sequence tag resource to reveal gene expression profiles in hexaploid wheat (*Triticum aestivum* L.). *Genome* 49, 531–544.

Chapman, S.C. (2008) Use of crop models to understand genotype by environment interactions for drought in real-world and simulated plant breeding trials. *Euphytica* 161, 195–208.

Chapman, S., Cooper, M., Podlich, D. and Hammer, G. (2003) Evaluating plant breeding strategies by simulating gene action and dryland environment effects. *Agronomy Journal* 95, 99–113.

Chardon, F., Virlon, B., Moreau, L., Falque, M., Joets, J., Decousset, L., Murigneux, A. and Charcosset, A. (2004) Genetic architecture of flowering time in maize as inferred from quantitative trait loci meta-analysis and synteny conservation with the rice genome. *Genetics* 168, 2169–2185.

Chauhan, H., Khurana, N., Tyagi, A.K., Khurana, J.P. and Khurana, P. (2011) Identification and characterization of high temperature stress responsive genes in bread wheat (*Triticum aestivum* L.) and their regulation at various stages of development. *Plant Molecular Biology* 75, 35–51.

Chaves, M.M. and Oliveira, M.M. (2004) Mechanisms underlying plant resilience to water deficits: prospects for water-saving agriculture. *Journal of Experimental Botany* 55, 2365–2384.

Chen, G.X., Pourkheirandish, M., Sameri, M., Wang, N., Nair, S., Shi, Y.L. *et al.* (2009) Genetic targeting of candidate genes for drought sensitive gene *eibi1* of wild barley (*Hordeum spontaneum*). *Breeding Science* 59, 637–644.

Chen, G.X., Krugman, T., Fahima, T., Chen, K.G., Hu, Y.G., Roder, M. *et al.* (2010) Chromosomal regions controlling seedling drought resistance in Israeli wild barley, *Hordeum spontaneum* C. Koch. *Genetic Resources and Crop Evolution* 57, 85–99.

Chenu, K., Chapman, S.C., Hammer, G.L., McLean, G., Salah, H.B.H. and Tardieu, F. (2008) Short-term responses of leaf growth rate to water deficit scale up to whole-plant and crop levels: an integrated modelling approach in maize. *Plant Cell and Environment* 31, 378–391.

Chenu, K., Chapman, S.C., Tardieu, F., McLean, G., Welcker, C. and Hammer, G.L. (2009) Simulating the yield impacts of organ-level quantitative trait loci associated with drought response in maize: a 'Gene-to-Phenotype' modeling approach. *Genetics* 183, 1507–1523.

Chin, J.H., Lu, X.C., Haefele, S.M., Gamuyao, R., Ismail, A., Wissuwa, M. *et al.* (2010) Development and application of gene-based markers for the major rice QTL Phosphorus uptake 1. *Theoretical and Applied Genetics* 120, 1073–1086.

Chung, J., Babka, H.L., Graef, G.L., Staswick, P.E., Lee, D.J., Cregan, P.B. *et al.* (2003) The seed protein, oil, and yield QTL on soybean linkage group I. *Crop Science* 43, 1053–1067.

Cohen, D., Bogeat-Triboulot, M.B., Tisserant, E., Balzergue, S., Martin-Magniette, M.L., Lelandais, G. *et al.* (2010) Comparative transcriptomics of drought responses in *Populus*: a meta-analysis of genome-wide expression profiling in mature leaves and root apices across two genotypes. *BMC Genomics* 11, 630.

Collins, N.C., Tardieu, F. and Tuberosa, R. (2008) Quantitative trait loci and crop performance under abiotic stress: Where do we stand? *Plant Physiology* 147, 469–486.

Colmer, T.D. and Voesenek, L. (2009) Flooding tolerance: suites of plant traits in variable environments. *Functional Plant Biology* 36, 665–681.

Comai, L., Young, K., Till, B.J., Reynolds, S.H., Greene, E.A., Codomo, C.A. *et al.* (2004) Efficient discovery of DNA polymorphisms in natural populations by ecotilling. *Plant Journal* 37, 778–786.

Condon, A.G., Richards, R.A. and Farquhar, G.D. (1993) Relationships between carbon isotope discrimination, water use efficiency and transpiration efficiency for dryland wheat. *Australian Journal of Agricultural Research* 44, 1693–1711.

Condon, A.G., Richards, R.A., Rebetzke, G.J. and Farquhar, G.D. (2002) Improving intrinsic water-use efficiency and crop yield. *Crop Science* 42, 122–131.

Condon, A.G., Richards, R.A., Rebetzke, G.J. and Farquhar, G.D. (2004) Breeding for high water-use efficiency. *Journal of Experimental Botany* 55, 2447–2460.

Cooper, M., Podlich, D.W. and Smith, O.S. (2005) Gene-to-phenotype models and complex trait genetics. *Australian Journal of Agricultural Research* 56, 895–918.

Cooper, M., Podlich, D.W. and Luo, L. (2007) Modeling QTL effects and MAS in plant breeding. In: Varshney, R.K. and Tuberosa, R. (eds) *Genomics-assisted Crop Improvement, Vol. 1: Genomics Approaches and Platforms.* Springer, New York, pp. 57–96.

Cooper, M., van Eeuwijk, F.A., Hammer, G.L., Podlich, D.W. and Messina, C. (2009) Modeling QTL for complex traits: detection and context for plant breeding. *Current Opinion in Plant Biology* 12, 231–240.

Coque, M. and Gallais, A. (2006) Genomic regions involved in response to grain yield selection at high and low nitrogen fertilization in maize. *Theoretical and Applied Genetics* 112, 1205–1220.

Coque, M., Martin, A., Veyrieras, J.B., Hirel, B. and Gallais, A. (2008) Genetic variation for N-remobilization and postsilking N-uptake in a set of maize recombinant inbred lines. 3. QTL detection and coincidences. *Theoretical and Applied Genetics* 117, 729–747.

Coudert, Y., Perin, C., Courtois, B., Khong, N.G. and Gantet, P. (2010) Genetic control of root development in rice, the model cereal. *Trends in Plant Science* 15, 219–226.

Courtois, B., McLaren, G., Sinha, P.K., Prasad, K., Yadav, R. and Shen, L. (2000) Mapping QTLs associated with drought avoidance in upland rice. *Molecular Breeding* 6, 55–66.

Courtois, B., Shen, L., Petalcorin, W., Carandang, S., Mauleon, R. and Li, Z. (2003) Locating QTLs controlling constitutive root traits in the rice population IAC 165 × Co39. *Euphytica* 134, 335–345.

Courtois, B., Ahmadi, N., Khowaja, F., Price, A., Rami, J.F., Frouin, J., Hamelia, C. and Ruiz, M. (2009) Rice root genetic architecture: Meta-analysis from a drought QTL database. *Rice* 2, 115–128.

Crossa, J., Burgueno, J., Dreisigacker, S., Vargas, M., Herrera-Foessel, S.A., Lillemo, M. *et al.* (2007) Association analysis of historical bread wheat germplasm using additive genetic covariance of relatives and population structure. *Genetics* 177, 1889–1913.

Cuellar-Ortiz, S.M., Arrieta-Montiel, M.D., Acosta-Gallegos, J. and Covarrubias, A.A. (2008) Relationship between carbohydrate partitioning and drought resistance in common bean. *Plant Cell and Environment* 31, 1399–1409.

Cui, K.H., Peng, S.B., Xing, Y.Z., Yu, S.B., Xu, C.G. and Zhang, Q. (2003) Molecular dissection of the genetic relationships of source, sink and transport tissue with yield traits in rice. *Theoretical and Applied Genetics* 106, 649–658.

DaMatta, F.M., Grandis, A., Arenque, B.C. and Buckeridge, M.S. (2010) Impacts of climate changes on crop physiology and food quality. *Food Research International* 43, 1814–1823.

Davies, W.J. (2007) Root growth response and functioning as an adaptation in water limiting soils. In: Jenks, M.A., Hasegawa, P.M. and

Mohan Jain, S. (eds) *Advances in Molecular Breeding Toward Drought and Salt Tolerant Crops*. Springer, Dordrecht, Netherlands, pp. 55–72.

de Dorlodot, S., Forster, B., Pages, L., Price, A., Tuberosa, R. and Draye, X. (2007) Root system architecture: opportunities and constraints for genetic improvement of crops. *Trends in Plant Science* 12, 474–481.

Delgado, J.A. and Berry, J.K. (2008) Advances in precision conservation. *Advances in Agronomy* 98, 1–44.

Den Herder, G., Van Isterdael, G., Beeckman, T. and De Smet, I. (2010) The roots of a new green revolution. *Trends in Plant Science* 15, 600–607.

Dhanda, S.S. and Munjal, R. (2006) Inheritance of cellular thermotolerance in bread wheat. *Plant Breeding* 125, 557–564.

Diab, A.A., Teulat-Merah, B., This, D., Ozturk, N.Z., Benscher, D. and Sorrells, M.E. (2004) Identification of drought-inducible genes and differentially expressed sequence tags in barley. *Theoretical and Applied Genetics* 109, 1417–1425.

Distelfeld, A., Li, C. and Dubcovsky, J. (2009) Regulation of flowering in temperate cereals. *Current Opinion in Plant Biology* 12, 178–184.

Doi, K., Izawa, T., Fuse, T., Yamanouchi, U., Kubo, T., Shimatani, Z. *et al.* (2004) *Ehd1*, a B-type response regulator in rice, confers short-day promotion of flowering and controls *FT-like* gene expression independently of *Hd1*. *Genes Development* 18, 926–936.

Du, W.J., Wang, M., Fu, S.X. and Yu, D.Y. (2009) Mapping QTLs for seed yield and drought susceptibility index in soybean (*Glycine max* L.) across different environments. *Journal of Genetics and Genomics* 36, 721–731.

Dubey, L., Prasanna, B.M. and Ramesh, B. (2009) Analysis of drought tolerant and susceptible maize genotypes using SSR markers tagging candidate genes and consensus QTLs for drought tolerance. *Indian Journal of Genetics and Plant Breeding* 69, 344–351.

Ducrocq, S., Madur, D., Veyrieras, J.B., Camus-Kulandaivelu, L., Kloiber-Maitz, M., Presterl, T. *et al.* (2008) Key impact of *Vgt1* on flowering time adaptation in maize: Evidence from association mapping and ecogeographical information. *Genetics* 178, 2433–2437.

Duvick, D.N. (2005) The contribution of breeding to yield advances in maize (*Zea mays* L.). *Advances in Agronomy* 86, 83–145.

Duvick, D.N. and Cassman, K.G. (1999) Post-green revolution trends in yield potential of temperate maize in the north-central United States. *Crop Science* 39, 1622–1630.

Dwivedi, S.L., Crouch, J.H., Mackill, D.J., Xu, Y.,

Blair, M.W., Ragot, M. *et al.* (2007) The molecularization of public sector crop breeding: Progress, problems, and prospects. *Advances in Agronomy* 95, 163–318.

Ebrahimi, A., Maury, P., Berger, M., Kiani, S.P., Nabipour, A., Shariati, F. *et al.* (2008) QTL mapping of seed-quality traits in sunflower recombinant inbred lines under different water regimes. *Genome* 51, 599–615.

Ebrahimi, A., Maury, P., Berger, M., Calmon, A., Grieu, P. and Sarrafi, A. (2009) QTL mapping of protein content and seed characteristics under water-stress conditions in sunflower. *Genome* 52, 419–430.

Ehdaie, B., Whitkus, R.W. and Waines, J.G. (2003) Root biomass, water-use efficiency, and performance of wheat–rye translocations of chromosomes 1 and 2 in spring bread wheat 'Pavon'. *Crop Science* 43, 710–717.

Ehdaie, B., Alloush, G.A. and Waines, J.G. (2008) Genotypic variation in linear rate of grain growth and contribution of stem reserves to grain yield in wheat. *Field Crops Research* 106, 34–43.

Ehlert, C., Maurel, C., Tardieu, F. and Simonneau, T. (2009) Aquaporin-mediated reduction in maize root hydraulic conductivity impacts cell turgor and leaf elongation even without changing transpiration. *Plant Physiology* 150, 1093–1104.

Ejeta, G. and Knoll, J.E. (2007) Marker-assisted selection in sorghum. In: Varshney, R.K. and Tuberosa, R. (eds) *Genomics-assisted Crop Improvemen, Vol 2: Genomics Applications in Crops*. Springer, Dordrecht, Netherlands, pp. 187–206.

Ella, E.S., Dionisio-Sese, M.L. and Ismail, A.M. (2010) Proper management improves seedling survival and growth during early flooding in contrasting rice genotypes. *Crop Science* 50, 1997–2008.

Ellis, R.P., Forster, B.P., Gordon, D.C., Handley, L.L., Keith, R.P., Lawrence, P. *et al.* (2002) Phenotype/genotype associations for yield and salt tolerance in a mapping population segregating for two dwarfing genes. *Journal of Experimental Botany* 53, 1–14.

Emrich, K., Price, A. and Piepho, H.P. (2008) Assessing the importance of genotype × environment interaction for root traits in rice using a mapping population III: QTL analysis by mixed models. *Euphytica* 161, 229–240.

Ergen, N.Z., Thimmapuram, J., Bohnert, H.J. and Budak, H. (2009) Transcriptome pathways unique to dehydration tolerant relatives of modern wheat. *Functional and Integrative Genomics* 9, 377–396.

Ersoz, E.S., Yu, J. and Buckler, E.S. (2007) Applications of linkage disequilibrium and association mapping. In: Varshney, R.K. and

Tuberosa, R. (eds) *Genomics-assisted Crop Improvemen, Vol. 1: Genomics Approaches and Platforms.* Springer, Dordrecht, Netherlands, pp. 97–120.

Eshed, Y. and Zamir, D. (1995) An introgression line population of *Lycopersicon pennellii* in the cultivated tomato enables the identification and fine mapping of yield-associated QTL. *Genetics* 141, 1147–1162.

Eshed, Y., Abu-Abied, M., Saranga, Y. and Zamir, D. (1992) *Lycopersicon esculentum* lines containing small overlapping introgressions from *L. pennellii. Theoretical and Applied Genetics* 83, 1027–1034.

Farooq, M., Kobayashi, N., Wahid, A., Ito, O. and Basra, S.M.A. (2009) Strategies for producing more rice with less water. *Advances in Agronomy* 101, 351–388.

Farquhar, G.D., Ehleringer, J.R. and Hubick, K.T. (1989) Carbon isotope discrimination and photosynthesis. *Annual Review of Plant Physiology and Plant Molecular Biology* 40, 503–537.

Fernandes, J., Morrow, D.J., Casati, P. and Walbot, V. (2008) Distinctive transcriptome responses to adverse environmental conditions in *Zea mays* L. *Plant Biotechnology Journal* 6, 782–798.

Fischer, R.A., Rees, D., Sayre, K.D., Lu, Z.M., Condon, A.G. and Larque Saavedra, A. (1998) Wheat yield progress associated with higher stomatal conductance and photosynthetic rate, and cooler canopies. *Crop Science* 38, 1467–1475.

Fleury, D., Jefferies, S., Kuchel, H. and Langridge, P. (2010) Genetic and genomic tools to improve drought tolerance in wheat. *Journal of Experimental Botany* 61, 3211–3222.

Flint-Garcia, S.A., Thuillet, A.C., Yu, J.M., Pressoir, G., Romero, S.M., Mitchell, S.E. *et al.* (2005) Maize association population: a high-resolution platform for quantitative trait locus dissection. *Plant Journal* 44, 1054–1064.

Flowers, T.J. and Flowers, S.A. (2005) Why does salinity pose such a difficult problem for plant breeders? *Agricultural Water Management* 78, 15–24.

Flowers, T.J., Koyama, M.L., Flowers, S.A., Sudhakar, C., Singh, K.P. and Yeo, A.R. (2000) QTL: their place in engineering tolerance of rice to salinity. *Journal of Experimental Botany* 51, 99–106.

Forster, B.P., Thomas, W.T.B. and Chloupek, O. (2005) Genetic controls of barley root systems and their associations with plant performance. *Aspects of Applied Biology* 73, 199–204.

Foulkes, M.J., Slafer, G.A., Davies, W.J., Berry, P.M., Sylvester-Bradley, R., Partre, P. *et al.* (2011) Raising yield potential of wheat. III. Optimizing partitioning to grain while maintaining lodging resistance. *Journal of Experimental Botany* 62, 469–486.

Fowler, S. and Thomashow, M.F. (2002) Arabidopsis transcriptome profiling indicates that multiple regulatory pathways are activated during cold acclimation in addition to the CBF cold response pathway. *Plant Cell* 14, 1675–1690.

Francia, E., Barabaschi, D., Tondelli, A., Laido, G., Rizza, F., Stanca, A.M., Busconi, M. *et al.* (2007) Fine mapping of a *HvCBF* gene cluster at the frost resistance locus Fr-H2 in barley. *Theoretical and Applied Genetics* 115, 1083–1091.

Fresneau, C., Ghashghaie, J. and Cornic, G. (2007) Drought effect on nitrate reductase and sucrose–phosphate synthase activities in wheat (*Triticum durum* L.): role of leaf internal CO_2. *Journal of Experimental Botany* 58, 2983–2992.

Frova, C. and Sari-Gorla, M. (1994) Quantitative trait loci (QTLs) for pollen thermotolerance detected in maize. *Molecular and General Genetics* 245, 424–430.

Frova, C., Krajewski, P., di Fonzo, N., Villa, M. and Sari-Gorla, M. (1999) Genetic analysis of drought tolerance in maize by molecular markers I. Yield components. *Theoretical and Applied Genetics* 99, 280–288.

Fujii, H. and Zhu, J.K. (2009) Arabidopsis mutant deficient in 3 abscisic acid-activated protein kinases reveals critical roles in growth, reproduction, and stress. *Proceedings of the National Academy of Sciences of the United States of America* 106, 8380–8385.

Fuller, M.P., Fuller, A.M., Kaniouras, S., Christophers, J. and Fredericks, T. (2007) The freezing characteristics of wheat at ear emergence. *European Journal of Agronomy* 26, 435–441.

Furukawa, J., Yamaji, N., Wang, H., Mitani, N., Murata, Y., Sato, K. *et al.* (2007) An aluminum-activated citrate transporter in barley. *Plant and Cell Physiology* 48, 1081–1091.

Gallais, A. and Hirel, B. (2004) An approach to the genetics of nitrogen use efficiency in maize. *Journal of Experimental Botany* 55, 295–306.

Gardiner, J., Schroeder, S., Polacco, M.L., Sanchez-Villeda, H., Fang, Z.W., Morgante, M. *et al.* (2004) Anchoring 9,371 maize expressed sequence tagged unigenes to the bacterial artificial chromosome contig map by two-dimensional overgo hybridization. *Plant Physiology* 134, 1317–1326.

Genc, Y., Oldach, K., Verbyla, A.P., Lott, G., Hassan, M., Tester, M. *et al.* (2010) Sodium exclusion

QTL associated with improved seedling growth in bread wheat under salinity stress. *Theoretical and Applied Genetics* 121, 877–894.

Giuliani, S., Sanguineti, M.C., Tuberosa, R., Bellotti, M., Salvi, S. and Landi, P. (2005) *Root-ABA1*, a major constitutive QTL, affects maize root architecture and leaf ABA concentration at different water regimes. *Journal of Experimental Botany* 56, 3061–3070.

Glaszmann, J.C., Kilian, B., Upadhyaya, H.D. and Varshney, R.K. (2010) Accessing genetic diversity for crop improvement. *Current Opinion in Plant Biology* 13, 167–173.

Goffinet, B. and Gerber, S. (2000) Quantitative trait loci: a meta–analysis. *Genetics* 155, 463–473.

Gonzalez-Martinez, S.C., Huber, D., Ersoz, E., Davis, J.M. and Neale, D.B. (2008) Association genetics in *Pinus taeda* L. II. Carbon isotope discrimination. *Heredity* 101, 19–26.

Graziani, M., Maccaferri, M., Sanguineti, M.C., Corneti, S., Stefanelli, S., Demontis, A. *et al.* (2011) Validation of *Qyld.Idw-3B*, a major QTL for grain yield and related morphophysiological traits in durum wheat. *Abstracts of the Plant and Animal Genomes XIX Conference*, 15–19 January 2011, San Diego, California, p. 294.

Gregory, P.J., Bengough, A.G., Grinev, D., Schmidt, S., Thomas, W.T.B., Wojciechowski, T. *et al.* (2009) Root phenomics of crops: opportunities and challenges. *Functional Plant Biology* 36, 922–929.

Groos, C., Robert, N., Bervas, E. and Charmet, G. (2003) Genetic analysis of grain protein-content, grain yield and thousand-kernel weight in bread wheat. *Theoretical and Applied Genetics* 106, 1032–1040.

Guan, Y.S., Serraj, R., Liu, S.H., Xu, J.L., Ali, J., Wang, W.S. *et al.* (2010) Simultaneously improving yield under drought stress and non-stress conditions: a case study of rice (*Oryza sativa* L.). *Journal of Experimental Botany* 61, 4145–4156.

Guo, J.F., Su, G.Q., Zhang, J.P. and Wang, G.Y. (2008) Genetic analysis and QTL mapping of maize yield and associate agronomic traits under semi-arid land condition. *African Journal of Biotechnology* 7, 1829–1838.

Gupta, P.K., Rustgi, S. and Kulwal, P. (2005) Linkage disequilibrium and association studies in higher plants: Present status and future prospects. *Plant Molecular Biology* 57, 461–485.

Gur, A., Semel, Y., Cahaner, A. and Zamir, D. (2004) Real time QTL of complex phenotypes in tomato interspecific introgression lines. *Trends in Plant Science* 9, 107–109.

Ha, B.K., Hussey, R.S. and Boerma, H.R. (2007)

Development of SNP assays for marker-assisted selection of two southern root-knot nematode resistance QTL in soybean. *Crop Science* 47, S73–S82.

Habash, D.Z., Bernard, S., Schondelmaier, J., Weyen, J. and Quarrie, S.A. (2007) The genetics of nitrogen use in hexaploid wheat: N utilisation, development and yield. *Theoretical and Applied Genetics* 114, 403–419.

Habash, D.Z., Kehel, Z. and Nachit, M. (2009) Genomic approaches for designing durum wheat ready for climate change with a focus on drought. *Journal of Experimental Botany* 60, 2805–2815.

Hammer, G.L., Chapman, S., van Oosterom, E. and Podlich, D.W. (2005) Trait physiology and crop modelling as a framework to link phenotypic complexity to underlying genetic systems. *Australian Journal of Agricultural Research* 56, 947–960.

Hammer, G., Cooper, M., Tardieu, F., Welch, S., Walsh, B., van Eeuwijk, F. *et al.* (2006) Models for navigating biological complexity in breeding improved crop plants. *Trends in Plant Science* 11, 587–593.

Hammer, G.., Dong, Z.S., McLean, G., Doherty, A., Messina, C., Schusler, J. *et al.* (2009) Can changes in canopy and/or root system architecture explain historical maize yield trends in the US corn belt? *Crop Science* 49, 299–312.

Hao, Z., Liu, X., Li, X., Xie, C., Li, M., Zhang, D. *et al.* (2009) Identification of quantitative trait loci for drought tolerance at seedling stage by screening a large number of introgression lines in maize. *Plant Breeding* 128, 337–341.

Hao, Z.F., Li, X.H., Liu, X.L., Xie, C.X., Li, M.S., Zhang, D.G. *et al.* (2010) Meta-analysis of constitutive and adaptive QTL for drought tolerance in maize. *Euphytica* 174, 165–177.

Harris, K., Subudhi, P.K., Borrell, A., Jordan, D., Rosenow, D., Nguyen, H. *et al.* (2007) Sorghum stay-green QTL individually reduce post-flowering drought-induced leaf senescence. *Journal of Experimental Botany* 58, 327–338.

Herve, D., Fabre, F., Berrios, E.F., Leroux, N., Al Chaarani, G., Planchon, C. *et al.* (2001) QTL analysis of photosynthesis and water status traits in sunflower (*Helianthus annuus* L.) under greenhouse conditions. *Journal of Experimental Botany* 52, 1857–1864.

Herve, P. and Serraj, R. (2009) Gene technology and drought: A simple solution for a complex trait? *African Journal of Biotechnology* 8, 1740–1749.

Hetz, W., Hochholdinger, F., Schwall, M. and Feix, G. (1996) Isolation and characterization of *rtcs*, a maize mutant deficient in the formation of nodal roots. *Plant Journal* 10, 845–857.

Heuer, S., Lu, X.C., Chin, J.H., Tanaka, J.P., Kanamori, H., Matsumoto, T. *et al.* (2009) Comparative sequence analyses of the major quantitative trait locus *phosphorus uptake 1* (*Pup1*) reveal a complex genetic structure. *Plant Biotechnology Journal* 7, 456–471.

Hirel, B., Bertin, P., Quiller, I., Bourdoncle, W., Attagnant, C., Dellay, C. *et al.* (2001) Towards a better understanding of the genetic and physiological basis for nitrogen use efficiency in maize. *Plant Physiology* 125, 1258–1270.

Hirel, B., Le Gouis, J., Ney, B. and Gallais, A. (2007) The challenge of improving nitrogen use efficiency in crop plants: towards a more central role for genetic variability and quantitative genetics within integrated approaches. *Journal of Experimental Botany* 58, 2369–2387.

Hochholdinger, F. and Tuberosa, R. (2009) Genetic and genomic dissection of maize root development and architecture. *Current Opinion in Plant Biology* 12, 172–177.

Hoekenga, O.A., Maron, L.G., Pineros, M.A., Cancado, G.M.A., Shaff, J., Kobayashi, Y. *et al.* (2006) *AtALMT1*, which encodes a malate transporter, is identified as one of several genes critical for aluminum tolerance in Arabidopsis. *Proceedings of the National Academy of Sciences of the United States of America* 103, 9738–9743.

Hospital, F. and Charcosset, A. (1997) Marker-assisted introgression of quantitative trait loci. *Genetics* 147, 1469–1485.

Hospital, F., Goldringer, I. and Openshaw, S. (2000) Efficient marker-based recurrent selection for multiple quantitative trait loci. *Genetical Research* 75, 357–368.

Huang, S.B., Spielmeyer, W., Lagudah, E.S., James, R.A., Platten, J.D., Dennis, E.S. *et al.* (2006) A sodium transporter (HKT7) is a candidate for *Nax1*, a gene for salt tolerance in durum wheat. *Plant Physiology* 142, 1718–1727.

Hund, A., Fracheboud, Y., Soldati, A., Frascaroli, E., Salvi, S. and Stamp, P. (2004) QTL controlling root and shoot traits of maize seedlings under cold stress. *Theoretical and Applied Genetics* 109, 618–629.

Hund, A., Frascaroli, E., Leipner, J., Jompuk, C., Stamp, P. and Fracheboud, Y. (2005) Cold tolerance of the photosynthetic apparatus: pleiotropic relationship between photosynthetic performance and specific leaf area of maize seedlings. *Molecular Breeding* 16, 321–331.

Hund, A., Ruta, N. and Liedgens, M. (2009a) Rooting depth and water use efficiency of

tropical maize inbred lines, differing in drought tolerance. *Plant and Soil* 318, 311–325.

Hund, A., Trachsel, S. and Stamp, P. (2009b) Growth of axile and lateral roots of maize: I development of a phenotying platform. *Plant and Soil* 325, 335–349.

Ikeda, H., Kamoshita, A. and Manabe, T. (2007) Genetic analysis of rooting ability of transplanted rice (*Oryza sativa* L.) under different water conditions. *Journal of Experimental Botany* 58, 309–318.

Ismail, A.M., Heuer, S., Thomson, M.J. and Wissuwa, M. (2007) Genetic and genomic approaches to develop rice germplasm for problem soils. *Plant Molecular Biology* 65, 547–570.

Isobe, S., Nakaya, A. and Tabata, S. (2007) Genotype matrix mapping: Searching for quantitative trait loci interactions in genetic variation in complex traits. *DNA Research* 14, 217–225.

Iuchi, S., Kobayashi, M., Taji, T., Naramoto, M., Seki, M., Kato, T. *et al.* (2001) Regulation of drought tolerance by gene manipulation of 9-cis-epoxycarotenoid dioxygenase, a key enzyme in abscisic acid biosynthesis in Arabidopsis. *Plant Journal* 27, 325–333.

Jackson, R.B., Sperry, J.S. and Dawson, T.E. (2000) Root water uptake and transport: using physiological processes in global predictions. *Trends in Plant Science* 5, 482–488.

Jagadish, S.V.K., Cairns, J., Lafitte, R., Wheeler, T.R., Price, A.H. and Craufurd, P.Q. (2010) Genetic analysis of heat tolerance at anthesis in rice. *Crop Science* 50, 1633–1641.

James, R.A., Davenport, R.J. and Munns, R. (2006) Physiological characterization of two genes for Na^+ exclusion in durum wheat, *Nax1* and *Nax2*. *Plant Physiology* 142, 1537–1547.

James, V.A., Neibaur, I. and Altpeter, F. (2008) Stress inducible expression of the DREB1A transcription factor from xeric, *Hordeum spontaneum* L. in turf and forage grass (*Paspalum notatum* Flugge) enhances abiotic stress tolerance. *Transgenic Research* 17, 93–104.

Jannink, J.L., Moreau, L., Charmet, G. and Charcosset, A. (2009) Overview of QTL detection in plants and tests for synergistic epistatic interactions. *Genetica* 136, 225–236.

Jenks, M.A, Hasegawa, P.M., and Mohan Jain, S. (2007) *Advances in Molecular Breeding Toward Drought and Salt Tolerant Crops.* Springer, Dordrecht, Netherlands.

Jiang, G.H., He, Y.Q., Xu, C.G., Li, X.H. and Zhang, Q. (2004) The genetic basis of stay-green in rice analyzed in a population of doubled haploid lines derived from an *indica* by *japonica* cross. *Theoretical and Applied Genetics* 108, 688–698.

Johnson, W.C., Jackson, L.E., Ochoa, O., van Wijk, R., Peleman, J., St. Clair, D.A. *et al.* (2000) Lettuce, a shallow-rooted crop, and *Lactuca serriola*, its wild progenitor, differ at QTL determining root architecture and deep soil water exploitation. *Theoretical and Applied Genetics* 101, 1066–1073.

Jones, H.G., Archer, N., Rotenberg, E. and Casa, R. (2003) Radiation measurement for plant ecophysiology. *Journal of Experimental Botany* 54, 879–889.

Jones, H.G., Serraj, R., Loveys, B.R., Xiong, L.Z., Wheaton, A. and Price, A.H. (2009) Thermal infrared imaging of crop canopies for the remote diagnosis and quantification of plant responses to water stress in the field. *Functional Plant Biology* 36, 978–989.

Jung, K.H., Seo, Y.S., Walia, H., Cao, P.J., Fukao, T., Canlas, P.E. *et al.* (2010) The submergence tolerance regulator *Sub1A* mediates stress-responsive expression of *AP2/ERF* transcription factors. *Plant Physiology* 152, 1674–1692.

Kaeppler, S.M., Parke, J.L., Mueller, S.M., Senior, L., Stuber, C. and Tracy, W.F. (2000) Variation among maize inbred lines and detection of quantitative trait loci for growth at low phosphorus and responsiveness to Arbuscular mycorrhizal fungi. *Crop Science* 40, 358–364.

Kahraman, A., Kusmenoglu, I., Aydin, N., Aydogan, A., Erskine, W. and Muehlbauer, F.J. (2004) QTL mapping of winter hardiness genes in lentil. *Crop Science* 44, 13–22.

Kamoshita, A., Babu, R.C., Boopathi, N.M. and Fukai, S. (2008) Phenotypic and genotypic analysis of drought-resistance traits for development of rice cultivars adapted to rainfed environments. *Field Crops Research* 109, 1–23.

Kang, Y.H., Khan, S. and Ma, X.Y. (2009) Climate change impacts on crop yield, crop water productivity and food security – a review. *Progress in Natural Science* 19, 1665–1674.

Kao, C.H., Zeng, Z.B. and Teasdale, R.D. (1999) Multiple interval mapping for quantitative trait loci. *Genetics* 152, 1203–1216.

Keating, B.A., Carberry, P.S., Bindraban, P.S., Asseng, S., Meinke, H. and Dixon, J. (2010) Eco-efficient agriculture: concepts, challenges, and opportunities. *Crop Science* 50, S109–S119.

Khavkin, E. and Coe, E. (1997) Mapped genomic locations for developmental functions and QTLs reflect concerted groups in maize (*Zea mays* L.) *Theoretical and Applied Genetics* 95, 343–352.

Kholova, J., Hash, C.T., Kakkera, A., Kocova, M.

and Vadez, V. (2010a) Constitutive water-conserving mechanisms are correlated with the terminal drought tolerance of pearl millet (*Pennisetum glaucum* L.) R. Br. *Journal of Experimental Botany* 61, 369–377.

Kholova, J., Hash, C.T., Kumar, P.L., Yadav, R.S., Kocova, M. and Vadez, V. (2010b) Terminal drought-tolerant pearl millet *Pennisetum glaucum* (L.) R. Br. have high leaf ABA and limit transpiration at high vapour pressure deficit. *Journal of Experimental Botany* 61, 1431–1440.

Khowaja, F.S. and Price, A.H. (2008) QTL mapping rolling, stomatal conductance and dimension traits of excised leaves in the Bala × Azucena recombinant inbred population of rice. *Field Crops Research* 106, 248–257.

Khowaja, F.S., Norton, G.J., Courtois, B. and Price, A.H. (2009) Improved resolution in the position of drought-related QTLs in a single mapping population of rice by meta-analysis. *BMC Genomics* 10, 276.

King, C.A., Purcell, L.C. and Brye, K.R. (2009) Differential wilting among soybean genotypes in response to water deficit. *Crop Science* 49, 290–298.

Kirigwi, F.M., Van Ginkel, M., Brown-Guedira, G., Gill, B.S., Paulsen, G.M. and Fritz, A.K. (2007) Markers associated with a QTL for grain yield in wheat under drought. *Molecular Breeding* 20, 401–413.

Kochian, L.V., Pineros, M.A. and Hoekenga, O.A. (2005) The physiology, genetics and molecular biology of plant aluminum resistance and toxicity. *Plant and Soil* 274, 175–195.

Kochian, L.V., Magalhaes, J., Liu, J., Guimaraes, C., Maron, L., Pineros, M. *et al.* (2011) The Comparative Genomics Challenge Initiative – translational research for improving aluminum tolerance and phosphorous efficiency in cereals. *Abstracts of the Plant and Animal Genomes XIX Conference*, 15–19 January 2011, San Diego, California, W141.

Kojima, S., Takahashi, Y., Kobayashi, Y., Monna, L., Sasaki, T., Araki, T. *et al.* (2002) *Hd3a*, a rice orthologue of the *Arabidopsis FT* gene, promotes transition to flowering downstream of *Hd1* under short-day conditions. *Plant and Cell Physiology* 43, 1096–1105.

Koyama, M.L., Levesley, A., Koebner, R.M.D., Flowers, T.J. and Yeo, A.R. (2001) Quantitative trait loci for component physiological traits determining salt tolerance in rice. *Plant Physiology* 125, 406–422.

Krill, A.M., Kirst, M., Kochian, L.V., Buckler, E.S. and Hoekenga, O.A. (2010) Association and linkage analysis of aluminum tolerance genes in maize. *PLoS One* 5, e9958.

Kuang, R.B., Liao, H., Yan, X.L. and Dong, Y.S. (2005) Phosphorus and nitrogen interactions in field-grown soybean as related to genetic attributes of root morphological and nodular traits. *Journal of Integrative Plant Biology* 47, 549–559.

Kuchel, H., Williams, K., Langridge, P., Eagles, H.A. and Jefferies, S.P. (2007a) Genetic dissection of grain yield in bread wheat. II. QTL-by-environment interaction. *Theoretical and Applied Genetics* 115, 1015–1027.

Kuchel, H., Williams, K.J., Langridge, P., Eagles, H.A. and Jefferies, S.P. (2007b) Genetic dissection of grain yield in bread wheat. I. QTL analysis. *Theoretical and Applied Genetics* 115, 1029–1041.

Kumar, A., Bernier, J., Verulkar, S., Lafitte, H.R. and Atlin, G.N. (2008) Breeding for drought tolerance: Direct selection for yield, response to selection and use of drought-tolerant donors in upland- and lowland-adapted populations. *Field Crops Research* 107, 221–231.

Kuroki, M., Saito, K., Matsuba, S., Yokogami, N., Shimizu, H., Ando, I. *et al.* (2007) A quantitative trait locus for cold tolerance at the booting stage on rice chromosome 8. *Theoretical and Applied Genetics* 115, 593–600.

Kuroki, M., Saito, K., Matsuba, S., Yokogami, N., Shimizu, H., Ando, I. *et al.* (2009) Quantitative trait locus analysis for cold tolerance at the booting stage in a rice cultivar, Hatsushizuku. *Jarq-Japan Agricultural Research Quarterly* 43, 115–121.

Kuromori, T., Miyaji, T., Yabuuchi, H., Shimizu, H., Sugimoto, E., Kamiya, A. *et al.* (2010) ABC transporter AtABCG25 is involved in abscisic acid transport and responses. *Proceedings of the National Academy of Sciences of the United States of America* 107, 2361–2366.

Lafitte, H.R., Courtois, B. and Arraudeau, M. (2002) Genetic improvement of rice in aerobic systems: progress from yield to genes. *Field Crops Research* 75, 171–190.

Lafitte, H.R., Guan, Y.S., Yan, S. and Li, Z.K. (2007) Whole plant responses, key processes, and adaptation to drought stress: the case of rice. *Journal of Experimental Botany* 58, 169–175.

Lanceras, J.C., Pantuwan, G., Jongdee, B. and Toojinda, T. (2004) Quantitative trait loci associated with drought tolerance at reproductive stage in rice. *Plant Physiology* 135, 384–399.

Landi, P., Sanguineti, M.C., Darrah, L.L., Giuliani, M.M., Salvi, S., Conti, S. *et al.* (2002) Detection of QTLs for vertical root pulling resistance in maize and overlap with QTLs for root traits in hydroponics and for grain yield under different water regimes. *Maydica* 47, 233–243.

Landi, P., Sanguineti, M.C., Salvi, S., Giuliani, S., Bellotti, M., Maccaferri, M. *et al.* (2005) Validation and characterization of a major QTL affecting leaf ABA concentration in maize. *Molecular Breeding* 15, 291–303.

Landi, P., Sanguineti, M.C., Liu, C., Li, Y., Wang, T.Y., Giuliani, S. *et al.* (2007) Root-ABA1 QTL affects root lodging, grain yield, and other agronomic traits in maize grown under well-watered and water-stressed conditions. *Journal of Experimental Botany* 58, 319–326.

Landi, P., Giuliani, S., Salvi, S., Ferri, M., Tuberosa, R. and Sanguineti, M.C. (2010) Characterization of *root-yield-1.06*, a major constitutive QTL for root and agronomic traits in maize across water regimes. *Journal of Experimental Botany* 61, 3553–3562.

Langridge, P. (2005) Molecular breeding of wheat and barley. In: Tuberosa, R., Phillips, R.L. and Gale, M. (eds) *In the Wake of the Double Helix: From the Green Revolution to the Gene Revolution.* Avenue Media, Bologna, Italy, pp. 279–286.

Langridge, P. and Fleury, D. (2011) Making the most of 'omics' for crop breeding. *Trends in Biotechnology* 29, 33–40.

Laperche, A., Brancourt-Hulmel, M., Heumez, E., Gardet, O., Hanocq, E., Devienne-Barret, F. *et al.* (2007) Using genotype × nitrogen interaction variables to evaluate the QTL involved in wheat tolerance to nitrogen constraints. *Theoretical and Applied Genetics* 115, 399–415.

Lebreton, C., Lazic-Jancic, V., Steed, A., Pekic, S. and Quarrie, S.A. (1995) Identification of QTL for drought responses in maize and their use in testing causal relationships between traits. *Journal of Experimental Botany* 46, 853–865.

LeDeaux, J.R., Graham, G.I. and Stuber, C.W. (2006) Stability of QTLs involved in heterosis in maize when mapped under several stress conditions. *Maydica* 51, 151–167.

Leung, H. (2008) Stressed genomics – bringing relief to rice fields. *Current Opinion in Plant Biology* 11, 201–208.

Levi, A., Ovnat, L., Paterson, A.H. and Saranga, Y. (2009a) Photosynthesis of cotton near-isogenic lines introgressed with QTLs for productivity and drought related traits. *Plant Science* 177, 88–96.

Levi, A., Paterson, A., Barak, V., Yakir, D., Wang, B., Chee, P. and Saranga, Y. (2009b) Field evaluation of cotton near-isogenic lines introgressed with QTLs for productivity and drought related traits. *Molecular Breeding* 23, 179–195.

Levitt, J. (1972) *Responses of Plants to Environmental Stresses.* Academic Press, New York.

Li, S.X., Wang, Z.H., Hu, T.T., Gao, Y.J. and Stewart, B.A. (2009) Nitrogen in dryland soils of China and its management. *Advances in Agronomy* 101, 123–181.

Li, X.H., Li, X.H., Hao, Z.F., Tian, Q.Z. and Zhang, S.H. (2005) Consensus map of the QTL relevant to drought tolerance of maize under drought conditions. *Scientia Agricultura Sinica* 38, 882–890.

Li, Y.D., Wang, Y.J., Tong, Y.P., Gao, J.G., Zhang, J.S. and Chen, S.Y. (2005) QTL mapping of phosphorus deficiency tolerance in soybean (*Glycine max* L. Merr.). *Euphytica* 142, 137–142.

Li, Z.K., Fu, B.Y., Gao, Y.M., Xu, J.L., Ali, J., Lafitte, H.R. *et al.* (2005) Genome-wide introgression lines and their use in genetic and molecular dissection of complex phenotypes in rice (*Oryza sativa* L.). *Plant Molecular Biology* 59, 33–52.

Liu, G.L., Mei, H.W., Liu, H.Y., Yu, X.Q., Zou, G.H. and Luo, L.J. (2010) Sensitivities of rice grain yield and other panicle characters to late-stage drought stress revealed by phenotypic correlation and QTL analysis. *Molecular Breeding* 25, 603–613.

Liu, H.Y., Zou, G.H., Liu, G.L., Hu, S.P., Li, M.S., Yu, X.Q. *et al.* (2005) Correlation analysis and QTL identification for canopy temperature, leaf water potential and spikelet fertility in rice under contrasting moisture regimes. *Chinese Science Bulletin* 50, 317–326.

Liu, L., Lafitte, R. and Guan, D. (2004) Wild *Oryza* species as potential sources of drought-adaptive traits. *Euphytica* 138, 149–161.

Liu, L.F., Mu, P., Li, X.Q., Qu, Y.Y., Wang, Y. and Li, Z.C. (2008) Localization of QTL for basal root thickness in japonica rice and effect of marker-assisted selection for a major QTL. *Euphytica* 164, 729–737.

Lopes, M.S. and Reynolds, M.P. (2010) Partitioning of assimilates to deeper roots is associated with cooler canopies and increased yield under drought in wheat. *Functional Plant Biology* 37, 147–156.

Lorens, G.F., Bennett, J.M. and Loggale, L.B. (1987) Differences in drought resistance between two corn hybrids. 1. Water relations and root length density. *Agronomy Journal* 79, 802–807.

Lu, Z., Percy, R.G., Qualset, C.O. and Zeiger, E. (1998) Stomatal conductance predicts yields in irrigated Pima cotton and bread wheat grown at high temperatures. *Journal of Experimental Botany* 49, 543–560.

Ludlow, M.M. and Muchow, R.C. (1990) A critical-evaluation of traits for improving crop yields in water-limited environments. *Advances in Agronomy* 43, 107–153.

Ludwig, F. and Asseng, S. (2010) Potential benefits of early vigor and changes in phenology in wheat to adapt to warmer and drier climates. *Agricultural Systems* 103, 127–136.

Luo, L.J. (2010) Breeding for water-saving and drought-resistance rice (WDR) in China. *Journal of Experimental Botany* 61, 3509–3517.

Luquet, D., Clement-Vidal, A., Fabre, D., This, D., Sonderegger, N. and Dingkuhn, M. (2008) Orchestration of transpiration, growth and carbohydrate dynamics in rice during a dry-down cycle. *Functional Plant Biology* 35, 689–704.

Maccaferri, M., Sanguineti, M.C., Noli, E. and Tuberosa, R. (2005) Population structure and long-range linkage disequilibrium in a durum wheat elite collection. *Molecular Breeding* 15, 271–289.

Maccaferri, M., Sanguineti, M.C., Corneti, S., Araus, J.L., Ben Salem, M., Bort, J. *et al.* (2008a) Quantitative trait loci for grain yield and adaptation of durum wheat (*Triticum durum* Desf.) across a wide range of water availability. *Genetics* 178, 489–511.

Maccaferri, M., Sanguineti, M.C., Giuliani, S. and Tuberosa, R. (2008b) Genomics of tolerance to abiotic stress in the Triticeae. In: Feuillet, C. and Muehlbauer, G. (eds) *Triticeae Genomics*. Springer, Dordrecht, Netherlands, pp. 259–318.

Maccaferri, M., Graziani, M., Sanguineti, M.C., Demontis, A., Paux, E., Salse, J. *et al.* (2011a) Characterization and progress in the fine mapping of *Qyld.Idw-3B*, a major QTL for grain yield in durum wheat. *Abstracts of the Plant and Animal Genomes XIX Conference*, 15–19 January 2011, San Diego, California, W347.

Maccaferri, M., Sanguineti, M.C., Demontis, A., El-Ahmed, A., Garcia del Moral, L., Maalouf, F. *et al.* (2011b) Association mapping in durum wheat grown across a broad range of water regimes. *Journal of Experimental Botany* 62, 409–438.

Mackill, D.J. (2007) Molecular markers and marker-assisted selection in rice. In: Varshney, R.K. and Tuberosa, R. (eds) *Genomics-assisted Crop Improvement, Vol 2: Genomics Applications in Crops*. Springer, Dordrecht, Netherlands, pp. 147–168.

Mackill, D.J., Nguyen, H.T. and Zhang, J.X. (1999) Use of molecular markers in plant improvement programs for rainfed lowland rice. *Field Crops Research* 64, 177–185.

MacMillan, K., Emrich, K., Piepho, H.P., Mullins, C.E. and Price, A.H. (2006a) Assessing the importance of genotype × environment interaction for root traits in rice using a mapping population II: conventional QTL analysis. *Theoretical and Applied Genetics* 113, 953–964.

MacMillan, K., Emrich, K., Piepho, H.P., Mullins, C.E. and Price, A.H. (2006b) Assessing the importance of genotype × environment interaction for root traits in rice using a mapping population. I: a soil-filled box screen. *Theoretical and Applied Genetics* 113, 977–986.

Magalhaes, J.V. (2010) How a microbial drug transporter became essential for crop cultivation on acid soils: aluminium tolerance conferred by the multidrug and toxic compound extrusion (MATE) family. *Annals of Botany* 106, 199–203.

Magalhaes, J.V., Liu, J., Guimaraes, C.T., Lana, U.G.P., Alves, V.M.C., Wang, Y.H. *et al.* (2007) A gene in the multidrug and toxic compound extrusion (MATE) family confers aluminum tolerance in sorghum. *Nature Genetics* 39, 1156–1161.

Maggio, A., Hasegawa, P.M., Bressan, R.A., Consiglio, M.F. and Joly, R.J. (2001) Unravelling the functional relationship between root anatomy and stress tolerance. *Australian Journal of Plant Physiology* 28, 999–1004.

Mahalakshmi, V. and Bidinger, F.R. (2002) Evaluation of stay-green sorghum germplasm lines at ICRISAT. *Crop Science* 42, 965–974.

Manavalan, L.P., Guttikonda, S.K., Tran, L.S.P. and Nguyen, H.T. (2009) Physiological and molecular approaches to improve drought resistance in soybean. *Plant and Cell Physiology* 50, 1260–1276.

Mano, Y., Muraki, M., Fujimori, M., Takamizo, T. and Kindiger, B. (2005a) Identification of QTL controlling adventitious root formation during flooding conditions in teosinte (*Zea mays* ssp. *huehuetenangensis*) seedlings. *Euphytica* 142, 33–42.

Mano, Y., Omori, F., Muraki, M. and Takamizo, T. (2005b) QTL mapping of adventitious root formation under flooding conditions in tropical maize (*Zea mays* L.) seedlings. *Breeding Science* 55, 343–347.

Mano, Y., Omori, F., Takamizo, T., Kindiger, B., Bird, R.M., Loaisiga, C.H. *et al.* (2007) QTL mapping of root aerenchyma formation in seedlings of a maize × rare teosinte "Zea *nicaraguensis*" cross. *Plant and Soil* 295, 103–113.

Marino, R., Ponnaiah, M., Krajewski, P., Frova, C., Gianfranceschi, L., Pe, M.E. *et al.* (2009) Addressing drought tolerance in maize by transcriptional profiling and mapping. *Molecular Genetics and Genomics* 281, 163–179.

Maron, L.G., Pineros, M.A., Guimaraes, C.T., Magalhaes, J.V., Pleiman, J.K., Mao, C.Z. *et al.* (2010) Two functionally distinct members of the MATE (multi-drug and toxic compound extrusion) family of transporters potentially underlie two major aluminum tolerance QTLs in maize. *Plant Journal* 61, 728–740.

Martin, B., Nienhuis, J., King, G. and Schaefer, A. (1989) Restriction fragment length polymorphisms associated with water use efficiency in tomato. *Science* 243, 1725–1728.

Masle, J., Gilmore, S.R. and Farquhar, G.D. (2005) The *ERECTA* gene regulates plant transpiration efficiency in *Arabidopsis*. *Nature* 436, 866–870.

Matos, M., Camacho, M.V., Perez-Flores, V., Pernaute, B., Pinto-Carnide, O. and Benito, C. (2005) A new aluminum tolerance gene located on rye chromosome arm 7RS. *Theoretical and Applied Genetics* 111, 360–369.

Maurel, C. and Chrispeels, M.J. (2001) Aquaporins: a molecular entry into plant water relations. *Plant Physiology* 125, 135–138.

McDonald, G.K., Eglinton, J.K. and Barr, A.R. (2010) Assessment of the agronomic value of QTL on chromosomes 2H and 4H linked to tolerance to boron toxicity in barley (*Hordeum vulgare* L.). *Plant and Soil* 326, 275–290.

McLaughlin, J.E. and Boyer, J.S. (2004a) Glucose localization in maize ovaries when kernel number decreases at low water potential and sucrose is fed to the stems. *Annals of Botany* 94, 75–86.

McLaughlin, J.E. and Boyer, J.S. (2004b) Sugar-responsive gene expression, invertase activity, and senescence in aborting maize ovaries at low water potentials. *Annals of Botany* 94, 675–689.

McMullen, M.D., Kresovich, S., Villeda, H.S., Bradbury, P., Li, H.H., Sun, Q. *et al.* (2009) Genetic properties of the maize nested association mapping population. *Science* 325, 737–740.

Mei, H.W., Luo, L.J., Ying, C.S., Wang, Y.P., Yu, X.Q., Guo, L.B. *et al.* (2003) Gene actions of QTLs affecting several agronomic traits resolved in a recombinant inbred rice population and two testcross populations. *Theoretical and Applied Genetics* 107, 89–101.

Melchinger, A.E., Utz, H.F. and Schon, C.C. (1998) Quantitative trait locus (QTL) mapping using different testers and independent population samples in maize reveals low power of QTL detection and large bias in estimates of QTL effects. *Genetics* 149, 383–403.

Merah, O., Deleens, E., Al Hakimi, A. and Monneveux, P. (2001) Carbon isotope discrimination and grain yield variations among tetraploid wheat species cultivated under contrasting precipitation regimes. *Journal of Agronomy and Crop Science–Zeitschrift für Acker und Pflanzenbau* 186, 129–134.

Miralles, D.J. and Slafer, G.A. (2007) Sink limitations to yield in wheat: how could it be reduced? *Journal of Agricultural Science* 145, 139–149.

Mittler, R. (2006) Abiotic stress, the field environment and stress combination. *Trends in Plant Science* 11, 15–19.

Mittler, R. and Blumwald, E. (2010) Genetic engineering for modern agriculture: Challenges and perspectives. *Annual Review of Plant Biology* 61, 443–462.

Mohamed, M.F., Keutgen, N., Tawfik, A.A. and Noga, G. (2002) Dehydration-avoidance responses of tepary bean lines differing in drought resistance. *Journal of Plant Physiology* 159, 31–38.

Moncada, P., Martinez, C.P., Borrero, J., Chatel, M., Gauch, H., Guimaraes, E. *et al.* (2001) Quantitative trait loci for yield and yield components in an *Oryza sativa* × *Oryza rufipogon* BC$_2$F$_2$ population evaluated in an upland environment. *Theoretical and Applied Genetics* 102, 41–52.

Montes, J., Melchinger, A. and Reif, J. (2007) Novel throughput phenotyping platforms in plant genetic studies. *Trends in Plant Science* 12, 433–436.

Moore, J.P., Le, N.T., Brandt, W.F., Driouich, A. and Farrant, J.M. (2009) Towards a systems-based understanding of plant desiccation tolerance. *Trends in Plant Science* 14, 110–117.

Morgan, J.W. (1984) Osmoregulation and water stress in higher plants. *Annual Review of Plant Physiology* 35, 299–319.

Morison, J.I.L., Baker, N.R., Mullineaux, P.M. and Davies, W.J. (2008) Improving water use in crop production. *Philosophical Transactions of the Royal Society B – Biological Sciences* 363, 639–658.

Munns, R. and Tester, M. (2008) Mechanisms of salinity tolerance. *Annual Review of Plant Biology* 59, 651–681.

Munns, R., Rebetzke, G.J., Husain, S., James, R.A. and Hare, R.A. (2003) Genetic control of sodium exclusion in durum wheat. *Australian Journal of Agricultural Research* 54, 627–635.

Narasimhamoorthy, B., Gill, B.S., Fritz, A.K., Nelson, J.C. and Brown-Guedira, G.L. (2006) Advanced backcross QTL analysis of a hard winter wheat x synthetic wheat population. *Theoretical and Applied Genetics* 112, 787–796.

Narsai, R., Castleden, I. and Whelan, J. (2010) Common and distinct organ and stress responsive transcriptomic patterns in *Oryza sativa* and *Arabidopsis thaliana*. *BMC Plant Biology* 10, 262.

Navakode, S., Weidner, A., Lohwasser, U., Roder, M.S. and Borner, A. (2008) Molecular mapping of quantitative trait loci (QTLs) controlling aluminium tolerance in bread wheat. *Euphytica* 166, 283–290.

Neeraja, C.N., Maghirang-Rodriguez, R., Pamplona, A., Heuer, S., Collard, B.C.Y., Septiningsih, E.M. *et al.* (2007) A marker-assisted backcross approach for developing submergence-tolerant rice cultivars. *Theoretical and Applied Genetics* 115, 767–776.

Nguyen, B.D., Brar, D.S., Bui, B.C., Nguyen, T.V., Pham, L.N. and Nguyen, H.T. (2003) Identification and mapping of the QTL for aluminum tolerance introgressed from the new source, *Oryza rufipogon* Griff., into indica rice (*Oryza sativa* L.). *Theoretical and Applied Genetics* 106, 583–593.

Nguyen, H.T. and Blum, A. (2004) *Physiology and Biotechnology Integration for Plant Breeding*. Marcel Dekker, Inc., New York.

Nguyen, H.T., Babu, R.C. and Blum, A. (1997) Breeding for drought resistance in rice: Physiology and molecular genetics considerations. *Crop Science* 37, 1426–1434.

Nguyen, T.T.T., Klueva, N., Chamareck, V., Aarti, A., Magpantay, G., Millena, A.C.M. *et al.* (2004) Saturation mapping of QTL regions and identification of putative candidate genes for drought tolerance in rice. *Molecular Genetics and Genomics* 272, 35–46.

Ninamango-Cárdenas, F.E., Guimaraes, C.T., Martins, P.R., Parentoni, S.N., Carneiro, N.P., Lopes, M.A. *et al.* (2003) Mapping QTLs for aluminum tolerance in maize. *Euphytica* 130, 223–232.

Norton, G.J. and Price, A.H. (2009) Mapping of quantitative trait loci for seminal root morphology and gravitropic response in rice. *Euphytica* 166, 229–237.

Ochoa, I.E., Blair, M.W. and Lynch, J.P. (2006) QTL analysis of adventitious root formation in common bean under contrasting phosphorus availability. *Crop Science* 46, 1609–1621.

O'Toole, J.C. and Bland, W.L. (1987) Genotypic variation in crop plant root systems. *Advances in Agronomy* 41, 91–145.

Ozturk, Z.N., Talame, V., Deyholos, M., Michalowski, C.B., Galbraith, D.W., Gozukirmizi, N. *et al.* (2002) Monitoring large-scale changes in transcript abundance in drought- and salt-stressed barley. *Plant Molecular Biology* 48, 551–573.

Pakniyat, H., Powell, W., Baird, E., Handley, L.L., Robinson, D., Scrimgeour, C.M. *et al.* (1997) AFLP variation in wild barley (*Hordeum spontaneum* C. Koch) with reference to salt tolerance and associated ecogeography. *Genome* 40, 332–341.

Pantuwan, G., Fukai, S., Cooper, M., Rajatasereekul, S. and O'Toole, J.C. (2002) Yield response of rice (*Oryza sativa* L.) genotypes to different types of drought under rainfed lowlands – Part 3. Plant factors contributing to drought resistance. *Field Crops Research* 73, 181–200.

Paran, I. and Zamir, D. (2003) Quantitative traits in plants: beyond the QTL. *Trends in Genetics* 19, 303–306.

Pariasca-Tanaka, J., Satoh, K., Rose, T., Mauleon, R. and Wissuwa, M. (2009) Stress response versus stress tolerance: A transcriptome analysis of two rice lines contrasting in tolerance to phosphorus deficiency. *Rice* 2, 167–185.

Park, S.Y., Fung, P., Nishimura, N., Jensen, D.R., Fujii, H., Zhao, Y. *et al.* (2009) Abscisic acid inhibits type 2C protein phosphatases via the PYR/PYL family of START proteins. *Science* 324, 1068–1071.

Parry, M.A.J. and Hawkesford, M.J. (2010) Food security: increasing yield and improving resource use efficiency. *Proceedings of the Nutrition Society* 69, 592–600.

Parry, M.A.J. and Lea, P.J. (2009) Food security and drought. *Annals of Applied Biology* 155, 299–300.

Parry, M.A.J., Flexas, J. and Medrano, H. (2005) Prospects for crop production under drought: research priorities and future directions. *Annals of Applied Biology* 147, 211–226.

Parry, M.A., Reynolds, M., Salvucci, M.E., Raines, C., Andralojc, P.J., Zhu, X.-G. *et al.* (2011) Raising yield potential of wheat. II. Increasing photosynthetic capacity and efficiency. *Journal of Experimental Botany* 62, 453–467.

Pasapula, V., Shen, G.X., Kuppu, S., Paez-Valencia, J., Mendoza, M., Hou, P. *et al.* (2011) Expression of an Arabidopsis vacuolar H+-pyrophosphatase gene (*AVP1*) in cotton improves drought- and salt tolerance and increases fibre yield in the field conditions. *Plant Biotechnology Journal* 9, 88–99.

Passioura, J. (1996) Drought and drought tolerance. *Plant Growth Regulation* 20, 79–83.

Passioura, J.B. (2002) Environmental biology and crop improvement. *Functional Plant Biology* 29, 537–546.

Passioura, J. (2004) Increasing crop productivity when water is scarce: From breeding to field management. In: *Proceedings of the 4th International Crop Science Congress: New Directions for a Diverse Planet*, Brisbane, Australia, pp. 1–17.

Passioura, J. (2006) Increasing crop productivity when water is scarce – from breeding to field

management. *Agricultural Water Management* 80, 176–196.

Passioura, J. (2007) The drought environment: physical, biological and agricultural perspectives. *Journal of Experimental Botany* 58, 113–117.

Passioura, J.B. (2010) Scaling up: the essence of effective agricultural research. *Functional Plant Biology* 37, 585–591.

Passioura, J.B. and Angus, J.F. (2010) Improving productivity of crops in water-limited environments. *Advances in Agronomy* 106, 37–75.

Passioura, J.B., Spielmeyer, W. and Bonnett, D.G. (2007) Requirements for success in marker-assisted breeding for drought-prone environments. In: Jenks, M.A., Hasegawa, P.M. and Mohan Jain, S. (eds) *Advances in Molecular Breeding Toward Drought and Salt Tolerant Crops*. Springer, Dordrecht, Netherlands, pp. 479–500.

Pelleschi, S., Guy, S., Kim, J.Y., Pointe, C., Mahe, A., Barthes, L. *et al.* (1999) *Ivr2*, a candidate gene for a QTL of vacuolar invertase activity in maize leaves. Gene-specific expression under water stress. *Plant Molecular Biology* 39, 373–380.

Pelleschi, S., Leonardi, A., Rocher, J.P., Cornic, G., de Vienne, D., Thévenot, C. *et al.* (2006) Analysis of the relationships between growth, photosynthesis and carbohydrate metabolism using quantitative trait loci (QTLs) in young maize plants subjected to water deprivation. *Molecular Breeding* 17, 21–39.

Pennisi, E. (2008) Plant sciences – Corn genomics pops wide open. *Science* 319, 1333.

Pflieger, S., Lefebvre, V. and Causse, M. (2001) The candidate gene approach in plant genetics: A review. *Molecular Breeding* 7, 275–291.

Pinheiro, H.A., DaMatta, F.M., Chaves, A.R.M., Loureiro, M.E. and Ducatti, C. (2005) Drought tolerance is associated with rooting depth and stomatal control of water use in clones of *Coffea canephora*. *Annals of Botany* 96, 101–108.

Pinto, R.S., Reynolds, M.P., Mathews, K.L., McIntyre, C.L., Olivares-Villegas, J.J. and Chapman, S.C. (2010) Heat and drought adaptive QTL in a wheat population designed to minimize confounding agronomic effects. *Theoretical and Applied Genetics* 121, 1001–1021.

Platten, J.D., Cotsaftis, O., Berthomieu, P., Bohnert, H., Davenport, R.J., Fairbairn, D.J. *et al.* (2006) Nomenclature for HKT transporters, key determinants of plant salinity tolerance. *Trends in Plant Science* 11, 372–374.

Podlich, D.W., Winkler, C.R. and Cooper, M. (2004) Mapping as you go: an effective approach for marker-assisted selection of complex traits. *Crop Science* 44, 1560–1571.

Prasanna, B.M., Beiki, A.H., Sekhar, J.C., Srinivas, A. and Ribaut, J.M. (2009) Mapping QTLs for component traits influencing drought stress tolerance of maize (*Zea mays* L) in India. *Journal of Plant Biochemistry and Biotechnology* 18, 151–160.

Presterl, T., Ouzunova, M., Schmidt, W., Moller, E.M., Rober, F.K., Knaak, C. *et al.* (2007) Quantitative trait loci for early plant vigour of maize grown in chilly environments. *Theoretical and Applied Genetics* 114, 1059–1070.

Price, A.H. and Tomos, A.D. (1997) Genetic dissection of root growth in rice (*Oryza sativa* L.) II: Mapping quantitative trait loci using molecular markers. *Theoretical and Applied Genetics* 95, 143–152.

Price, A., Young, E. and Tomos, A. (1997) Quantitative trait loci associated with stomatal conductance, leaf rolling and heading date mapped in upland rice (*Oryza sativa*). *New Phytologist* 137, 83–91.

Price, A.H., Steele, K.A., Moore, B.J., Barraclough, P.B. and Clark, L.J. (2000) A combined RFLP and AFLP linkage map of upland rice (*Oryza sativa* L.) used to identify QTLs for root-penetration ability. *Theoretical and Applied Genetics* 100, 49–56.

Price, A.H., Cairns, J.E., Horton, P., Jones, H.G. and Griffiths, H. (2002a) Linking drought-resistance mechanisms to drought avoidance in upland rice using a QTL approach: progress and new opportunities to integrate stomatal and mesophyll responses. *Journal of Experimental Botany* 53, 989–1004.

Price, A.H., Steele, K.A., Gorham, J., Bridges, J.M., Moore, B.J., Evans, J.L. *et al.* (2002b) Upland rice grown in soil-filled chambers and exposed to contrasting water-deficit regimes I. Root distribution, water use and plant water status. *Field Crops Research* 76, 11–24.

Prioul, J.L., Quarrie, S., Causse, M. and deVienne, D. (1997) Dissecting complex physiological functions through the use of molecular quantitative genetics. *Journal of Experimental Botany* 48, 1151–1163.

Prioul, J.L., Pelleschi, S., Sene, M., Thevenot, C., Causse, M., de Vienne, D. *et al.* (1999) From QTLs for enzyme activity to candidate genes in maize. *Journal of Experimental Botany* 50, 1281–1288.

Qi, B., Korir, P., Zhao, T.J., Yu, D.Y., Chen, S.Y. and Gai, J.Y. (2008) Mapping quantitative trait loci associated with aluminum toxin tolerance in NJRIKY recombinant inbred line population of soybean (*Glycine max*). *Journal of Integrative Plant Biology* 50, 1089–1095.

Quarrie, S.A. (1991) Implications of genetic differences in ABA accumulation for crop production. In: Davies, W.J. and Jones, H.G. (eds) *Abscisic Acid: Physiology and Biochemistry*. Bios Scientific Publishers, Oxford, UK, pp. 227–243.

Quarrie, S.A., Laurie, D.A., Zhu, J.H., Lebreton, C., Semikhodskii, A., Steed, A. *et al.* (1997) QTL analysis to study the association between leaf size and abscisic acid accumulation in droughted rice leaves and comparisons across cereals. *Plant Molecular Biology* 35, 155–165.

Quarrie, S.A., Stojanovic, J. and Pekic, S. (1999) Improving drought resistance in small grained cereals: A case study, progress and prospects. *Plant Growth Regulation* 29, 1–21.

Quarrie, S.A., Steed, A., Calestani, C., Semikhodskii, A., Lebreton, C., Chinoy, C. *et al.* (2005) A high-density genetic map of hexaploid wheat (*Triticum aestivum* L.) from the cross Chinese Spring × SQ1 and its use to compare QTLs for grain yield across a range of environments. *Theoretical and Applied Genetics* 110, 865–880.

Quarrie, S.A., Quarrie, S.P., Radosevic, R., Rancic, D., Kaminska, A., Barnes, J.D. *et al.* (2006) Dissecting a wheat QTL for yield present in a range of environments: from the QTL to candidate genes. *Journal of Experimental Botany* 57, 2627–2637.

Rafalski, J.A. (2010) Association genetics in crop improvement. *Current Opinion in Plant Biology* 13, 174–180.

Ragot, M. and Lee, M. (2007) Marker-assisted selection in maize: current status, potential, limitations and perspectives from the private and public sectors. In: Guimarães, E., Ruane, J., Scherf, B., Sonnino, A. and Dargie, J. (eds) *Marker-assisted Selection – Current Status and Future Perspectives in Crops, Livestock, Forestry and Fish*. FAO, Rome, pp. 117–150.

Raju, N.L., Gnanesh, B.N., Lekha, P., Jayashree, B., Pande, S., Hiremath, P.J. *et al.* (2010) The first set of EST resources for gene discovery and marker development in pigeonpea (*Cajanus cajan* L.). *BMC Plant Biology* 10, 45.

Raman, H., Zhang, K.R., Cakir, M., Appels, R., Garvin, D.F., Maron, L.G. *et al.* (2005) Molecular characterization and mapping of ALMT1, the aluminium-tolerance gene of bread wheat (*Triticum aestivum* L.). *Genome* 48, 781–791.

Rebetzke, G., Condon, A., Farquhar, G., Appels, R. and Richards, R. (2008a) Quantitative trait loci for carbon isotope discrimination are repeatable across environments and wheat mapping populations. *Theoretical and Applied Genetics* 118, 123–137.

Rebetzke, G.J., van Herwaarden, A.F., Jenkins, C., Weiss, M., Lewis, D., Ruuska, S. *et al.* (2008b) Quantitative trait loci for water-soluble carbohydrates and associations with agronomic traits in wheat. *Australian Journal of Agricultural Research* 59, 891–905.

Reid, R. (2010) Can we really increase yields by making crop plants tolerant to boron toxicity? *Plant Science* 178, 9–11.

Reinheimer, J.L., Barr, A.R. and Eglinton, J.K. (2004) QTL mapping of chromosomal regions conferring reproductive frost tolerance in barley (*Hordeum vulgare* L.). *Theoretical and Applied Genetics* 109, 1267–1274.

Ren, Z.H., Gao, J.P., Li, L.G., Cai, X.L., Huang, W., Chao, D.Y. *et al.* (2005) A rice quantitative trait locus for salt tolerance encodes a sodium transporter. *Nature Genetics* 37, 1141–1146.

Rengasamy, P. (2006) World salinization with emphasis on Australia. *Journal of Experimental Botany* 57, 1017–1023.

Reymond, M., Muller, B., Leonardi, A., Charcosset, A. and Tardieu, F. (2003) Combining quantitative trait loci analysis and an ecophysiological model to analyze the genetic variability of the responses of maize leaf growth to temperature and water deficit. *Plant Physiology* 131, 664–675.

Reymond, M., Muller, B. and Tardieu, F. (2004) Dealing with the genotype x environment interaction via a modelling approach: a comparison of QTLs of maize leaf length or width with QTLs of model parameters. *Journal of Experimental Botany* 55, 2461–2472.

Reynolds, M.P. and Pfeiffer, W.H. (2000) Applying physiological strategies to improve yield potential. In: Royo, C., Nachit, M.M., di Fonzo, N. and Araus, J.L. (eds) *Durum wheat improvement in the Mediterranean region: New challenges. Options Méditerranéennes* 40, 95–103.

Reynolds, M. and Tuberosa, R. (2008) Translational research impacting on crop productivity in drought-prone environments. *Current Opinion in Plant Biology* 11, 171–179.

Reynolds, M.P., Mujeeb-Kazi, A. and Sawkins, M. (2005) Prospects for utilising plant-adaptive mechanisms to improve wheat and other crops in drought- and salinity-prone environments. *Annals of Applied Biology* 146, 239–259.

Reynolds, M., Foulkes, M.J., Slafer, G.A., Berry, P., Parry, M.A.J., Snape, J.W. *et al.* (2009a) Raising yield potential in wheat. *Journal of Experimental Botany* 60, 1899–1918.

Reynolds, M., Manes, Y., Izanloo, A. and Langridge, P. (2009b) Phenotyping approaches for physiological breeding and gene discovery in wheat. *Annals of Applied Biology* 155, 309–320.

Reynolds, M., Bonnett, D., Chapman, S.C., Furbank, R.T., Manès, Y., Mather, D.E. *et al.* (2011) Raising yield potential of wheat. I. Overview of a consortium approach and breeding strategies. *Journal of Experimental Botany* 62, 439–452.

Ribaut, J.M. (2006) *Drought Adaptation in Cereals.* The Harworth Press, Inc., Binghamton, New York.

Ribaut, J.M. and Ragot, M. (2007) Marker-assisted selection to improve drought adaptation in maize: the backcross approach, perspectives, limitations, and alternatives. *Journal of Experimental Botany* 58, 351–360.

Ribaut, J.M., Hoisington, D., Banziger, M., Setter, T. and Edmeades, G. (2004) Genetic dissection of drought tolerance in maize: a case study. In: Nguyen, H.T. and Blum, A. (eds) *Physiology and Biotechnology Integration for Plant Breeding.* Marcel Dekker, Inc., New York, pp. 571–609.

Ribaut, J.M., Fracheboud, Y., Monneveux, P., Banziger, M., Vargas, M. and Jiang, C.J. (2007) Quantitative trait loci for yield and correlated traits under high and low soil nitrogen conditions in tropical maize. *Molecular Breeding* 20, 15–29.

Richards, R.A. (1996) Defining selection criteria to improve yield under drought. *Plant Growth Regulation* 20, 157–166.

Richards, R.A. (2000) Selectable traits to increase crop photosynthesis and yield of grain crops. *Journal of Experimental Botany* 51, 447–458.

Richards, R. (2004) Drought tolerant wheat. *Food Australia* 56, 65.

Richards, R.A. (2006) Physiological traits used in the breeding of new cultivars for water-scarce environments. *Agricultural Water Management* 80, 197–211.

Richards, R.A. (2008) Genetic opportunities to improve cereal root systems for dryland agriculture. *Plant Production Science* 11, 12–16.

Richards, R.A. and Passioura, J.B. (1989) A breeding program to reduce the diameter of the major xylem vessel in the seminal roots of wheat and its effect on grain yield in rain-fed environments. *Australian Journal of Agricultural Research* 40, 943–950.

Richards, R., Rebetzke, G., Condon, A. and van Herwaarden, A. (2002) Breeding opportunities for increasing the efficiency of water use and crop yield in temperate cereals. *Crop Science* 42, 111–121.

Richards, R.A., Rebetzke, G.J., Watt, M., Condon, A.G., Spielmeyer, W. and Dolferus, R. (2010)

Breeding for improved water productivity in temperate cereals: phenotyping, quantitative trait loci, markers and the selection environment. *Functional Plant Biology* 37, 85–97.

Rivero, R.M., Kojima, M., Gepstein, A., Sakakibara, H., Mittler, R., Gepstein, S. *et al.* (2007) Delayed leaf senescence induces extreme drought tolerance in a flowering plant. *Proceedings of the National Academy of Sciences of the United States of America* 104, 19631–19636.

Rivero, R.M., Shulaev, V. and Blumwald, E. (2009) Cytokinin-dependent photorespiration and the protection of photosynthesis during water deficit. *Plant Physiology* 150, 1530–1540.

Roitsch, T. (1999) Source-sink regulation by sugar and stress. *Current Opinion in Plant Biology* 2, 198–206.

Rostoks, N., Ramsay, L., MacKenzie, K., Cardle, L., Bhat, P.R., Roose, M.L. *et al.* (2006) Recent history of artificial outcrossing facilitates whole-genome association mapping in elite inbred crop varieties. *Proceedings of the National Academy of Sciences of the United States of America* 103, 18656–18661.

Royo, A., Gil, L. and Pardos, J.A. (2001) Effect of water stress conditioning on morphology, physiology and field performance of *Pinus halepensis* Mill. seedlings. *New Forests* 21, 127–140.

Royo, C., Maccaferri, M., Alvaro, F., Moragues, M., Sanguineti, M.C., Tuberosa, R. *et al.* (2010) Understanding the relationships between genetic and phenotypic structures of a collection of elite durum wheat accessions. *Field Crops Research* 119, 91–105.

Ruan, C.J., Xu, X.X., Shao, H.B. and Jaleel, C.A. (2010) Germplasm-regression-combined (GRC) marker-trait association identification in plant breeding: a challenge for plant biotechnological breeding under soil water deficit conditions. *Critical Reviews in Biotechnology* 30, 192–199.

Rus, A., Baxter, I., Muthukumar, B., Gustin, J., Lahner, B., Yakubova, E. *et al.* (2006) Natural variants of *AtHKT1* enhance Na+ accumulation in two wild populations of Arabidopsis. *Plos Genetics* 2, 1964–1973.

Ruta, N., Liedgens, M., Fracheboud, Y., Stamp, P. and Hund, A. (2010a) QTLs for the elongation of axile and lateral roots of maize in response to low water potential. *Theoretical and Applied Genetics* 120, 621–631.

Ruta, N., Stamp, P., Liedgens, M., Fracheboud, Y. and Hund, A. (2010b) Collocations of QTLs for seedling traits and yield components of tropical maize under water stress conditions. *Crop Science* 50, 1385–1392.

Sadok, W., Naudin, P., Boussuge, B., Muller, B., Welcker, C. and Tardieu, F. (2007) Leaf growth rate per unit thermal time follows QTL-dependent daily patterns in hundreds of maize lines under naturally fluctuating conditions. *Plant Cell and Environment* 30, 135–146.

Sadras, V.O. and Angus, J.F. (2006) Benchmarking water-use efficiency of rainfed wheat in dry environments. *Australian Journal of Agricultural Research* 57, 847–856.

Saini, H.S. and Westgate, M.E. (2000) Reproductive development in grain crops during drought. *Advances in Agronomy* 68, 59–96.

Saint Pierre, C., Crossa, J., Manes, Y. and Reynolds, M.P. (2010) Gene action of canopy temperature in bread wheat under diverse environments. *Theoretical and Applied Genetics* 120, 1107–1117.

Saito, K., Hayano-Saito, Y., Kuroki, M. and Sato, Y. (2010) Map-based cloning of the rice cold tolerance gene Ctb1. *Plant Science* 179, 97–102.

Salekdeh, G.H., Reynolds, M., Bennett, J. and Boyer, J. (2009) Conceptual framework for drought phenotyping during molecular breeding. *Trends in Plant Science* 14, 488–496.

Salem, K.F.M., Röder, M.S. and Börner, A. (2007) Identification and mapping quantitative trait loci for stem reserve mobilisation in wheat (*Triticum aestivum* L.). *Cereal Research Communications* 35, 1367–1374.

Salvi, S. and Tuberosa, R. (2007) Cloning QTLs in plants. In: Varshney, R.K. and Tuberosa, R. (eds) *Genomics-assisted Crop Improvement, Vol 1: Genomics Approaches and Platforms*. Springer, Dordrecht, Netherlands, pp. 207–226.

Salvi, S., Sponza, G., Morgante, M., Tomes, D., Niu, X., Fengler, K.A. *et al.* (2007) Conserved noncoding genomic sequences associated with a flowering-time quantitative trait locus in maize. *Proceedings of the National Academy of Sciences of the United States of America* 104, 11376–11381.

Salvi, S., Castelletti, S. and Tuberosa, R. (2010) An updated consensus MAP for flowering time QTLs in maize. *Maydica* 54, 501–512.

Salvi, S., Corneti, S., Bellotti, M., Carraro, N., Sanguineti, M.C., Castelletti, S. *et al.* (2011a) Genetic dissection of maize phenology using an intraspecific introgression library. *BMC Plant Biology* 11, 4.

Salvi, S., Ricciolini, C., Carraro, N., Presterl, T., Ouzunova, M. and Tuberosa, R. (2011b) Genetic control of seminal root architecture in maize. *Abstracts of the Plant and Animal Genomes XIX Conference*, 15–19 January 2011, San Diego, California, W503.

Sanguineti, M.C., Tuberosa, R., Landi, P., Salvi, S., Maccaferri, M., Casarini, E. *et al.* (1999) QTL analysis of drought related traits and grain yield in relation to genetic variation for leaf abscisic acid concentration in field-grown maize. *Journal of Experimental Botany* 50, 1289–1297.

Sanguineti, M.C., Li, S., Maccaferri, M., Corneti, S., Rotondo, F., Chiari, T. *et al.* (2007) Genetic dissection of seminal root architecture in elite durum wheat germplasm. *Annals of Applied Biology* 151, 291–305.

Saranga, Y., Jiang, C., Wright, R., Yakir, D. and Paterson, A.H. (2004) Genetic dissection of cotton physiological responses to arid conditions and their inter-relationships with productivity. *Plant Cell Environment* 27, 263–277.

Sarkar, R.K., Panda, D., Reddy, J.N., Patnaik, S.S.C., Mackill, D.J. and Ismail, A.M. (2009) Performance of submergence tolerant rice (*Oryza sativa*) genotypes carrying the Sub1 quantitative trait locus under stressed and non-stressed natural field conditions. *Indian Journal of Agricultural Sciences* 79, 876–883.

Sasaki, T., Ryan, P.R., Delhaize, E., Hebb, D.M., Ogihara, Y., Kawaura, K. *et al.* (2006) Sequence upstream of the wheat (*Triticum aestivum* L.) ALMT1 gene and its relationship to aluminum resistance. *Plant and Cell Physiology* 47, 1343–1354.

Sawkins, M.C., Farmer, A.D., Hoisington, D., Sullivan, J., Tolopko, A., Jiang, Z. *et al.* (2004) Comparative Map and Trait Viewer (CMTV): an integrated bioinformatic tool to construct consensus maps and compare QTL and functional genomics data across genomes and experiments. *Plant Molecular Biology* 56, 465–480.

Schmidhuber, J. and Tubiello, F.N. (2007) Global food security under climate change. *Proceedings of the National Academy of Sciences of the United States of America* 104, 19703–19708.

Schnurbusch, T., Collins, N.C., Eastwood, R.F., Sutton, T., Jefferies, S.P. and Langridge, P. (2007) Fine mapping and targeted SNP survey using rice–wheat gene colinearity in the region of the *Bo1* boron toxicity tolerance locus of bread wheat. *Theoretical and Applied Genetics* 115, 451–461.

Semenov, M.A., Martre, P. and Jamieson, P.D. (2009) Quantifying effects of simple wheat traits on yield in water-limited environments using a modelling approach. *Agricultural and Forest Meteorology* 149, 1095–1104.

Serraj, R., Hash, C.T., Rizvi, S.M.H., Sharma, A., Yadav, R.S. and Bidinger, F.R. (2005) Recent advances in marker-assisted selection for drought tolerance in pearl millet. *Plant Production Science* 8, 334–337.

Serraj, R., Kumar, A., McNally, K.L., Slamet-Loedin, I., Bruskiewich, R., Mauleon, R. *et al.* (2009) Improvement of drought resistance in rice. *Advances in Agronomy* 103, 41–99.

Setter, T.L., Flannigan, B.A. and Melkonian, J. (2001) Loss of kernel set due to water deficit and shade in maize: Carbohydrate supplies, abscisic acid, and cytokinins. *Crop Science* 41, 1530–1540.

Setter, T.L., Yan, J., Warburton, M., Ribaut, J.-M., Xu, Y., Sawkins, M. *et al.* (2011) Genetic association mapping identifies single nucleotide polymorphisms in genes that affect abscisic acid levels in maize floral tissues during drought. *Journal of Experimental Botany* 62, 701–716.

Sharp, R.E. and Davies, W.J. (1985) Root growth and water uptake by maize plants in drying soil. *Journal of Experimental Botany* 36, 1441–1456.

Shavrukov, Y., Gupta, N.K., Miyazaki, J., Baho, M.N., Chalmers, K.J., Tester, M. *et al.* (2010) HvNax3 – a locus controlling shoot sodium exclusion derived from wild barley (*Hordeum vulgare* ssp. *spontaneum*). *Functional and Integrative Genomics* 10, 277–291.

She, K.C., Kusano, H., Koizumi, K., Yamakawa, H., Hakata, M., Imamura, T. *et al.* (2010) A novel factor FLOURY ENDOSPERM2 is involved in regulation of rice grain size and starch quality. *Plant Cell* 22, 3280–3294.

Shinozaki, K. and Yamaguchi-Shinozaki, K. (2000) Molecular responses to dehydration and low temperature: differences and cross-talk between two stress signaling pathways. *Current Opinion in Plant Biology* 3, 217–223.

Shinozaki, K. and Yamaguchi-Shinozaki, K. (2007) Gene networks involved in drought stress response and tolerance. *Journal of Experimental Botany* 58, 221–227.

Sinclair, T.R. and Muchow, R.C. (2001) System analysis of plant traits to increase grain yield on limited water supplies. *Agronomy Journal* 93, 263–270.

Sinclair, T.R., Purcell, L.C. and Sneller, C.H. (2004) Crop transformation and the challenge to increase yield potential. *Trends in Plant Science* 9, 70–75.

Singh, N., Dang, T.T.M., Vergara, G.V., Pandey, D.M., Sanchez, D., Neeraja, C.N. *et al.* (2010) Molecular marker survey and expression analyses of the rice submergence-tolerance gene *SUB1A*. *Theoretical and Applied Genetics* 121, 1441–1453.

Singh, S., Mackill, D.J. and Ismail, A.M. (2009) Responses of *SUB1* rice introgression lines to submergence in the field: Yield and grain quality. *Field Crops Research* 113, 12–23.

Slafer, G.A. (2003) Genetic basis of yield as viewed from a crop physiologist's perspective. *Annals of Applied Biology* 142, 117–128.

Snape, J.W., Foulkes, M.J., Simmonds, J., Leverington, M., Fish, L.J., Wang, Y. *et al.* (2007) Dissecting gene x environmental effects on wheat yields via QTL and physiological analysis. *Euphytica* 154, 401–408.

Somers, D.J., Banks, T., DePauw, R., Fox, S., Clarke, J., Pozniak, C. *et al.* (2007) Genome-wide linkage disequilibrium analysis in bread wheat and durum wheat. *Genome* 50, 557–567.

Specht, J.E., Chase, K., Macrander, M., Graef, G.L., Chung, J., Markwell, J.P. *et al.* (2001) Soybean response to water: A QTL analysis of drought tolerance. *Crop Science* 41, 493–509.

Srivastava, A., Kumar, S.N. and Aggarwal, P.K. (2010) Assessment on vulnerability of sorghum to climate change in India. *Agriculture Ecosystems and Environment* 138, 160–169.

Steele, K.A., Price, A.H., Shashidhar, H.E. and Witcombe, J.R. (2006) Marker-assisted selection to introgress rice QTLs controlling root traits into an Indian upland rice variety. *Theoretical and Applied Genetics* 112, 208–221.

Steele, K.A., Virk, D.S., Kumar, R., Prasad, S.C. and Witcombe, J.R. (2007) Field evaluation of upland rice lines selected for QTLs controlling root traits. *Field Crops Research* 101, 180–186.

Su, J.Y., Xiao, Y.M., Li, M., Liu, Q.Y., Li, B., Tong, Y.P. *et al.* (2006) Mapping QTLs for phosphorus-deficiency tolerance at wheat seedling stage. *Plant and Soil* 281, 25–36.

Subashri, M., Robin, S., Vinod, K.K., Rajeswari, S., Mohanasundaram, K. and Raveendran, T.S. (2009) Trait identification and QTL validation for reproductive stage drought resistance in rice using selective genotyping of near flowering RILs. *Euphytica* 166, 291–305.

Subudhi, P.K., Rosenow, D.T. and Nguyen, H.T. (2000) Quantitative trait loci for the stay green trait in sorghum (*Sorghum bicolor* L. Moench): consistency across genetic backgrounds and environments. *Theoretical and Applied Genetics* 101, 733–741.

Sunarpi, J., Horie, T., Motoda, J., Kubo, M., Yang, H., Yoda, K. *et al.* (2005) Enhanced salt tolerance mediated by AtHKT1 transporter-induced Na$^+$ unloading from xylem vessels to xylem parenchyma cells. *Plant Journal* 44, 928–938.

Sutton, T., Baumann, U., Hayes, J., Collins, N.C., Shi, B.J., Schnurbusch, T. *et al.* (2007) Boron-toxicity tolerance in barley arising from efflux

transporter amplification. *Science* 318, 1446–1449.

Szalma, S.J., Hostert, B.M., LeDeaux, J.R., Stuber, C.W. and Holland, J.B. (2007) QTL mapping with near-isogenic lines in maize. *Theoretical and Applied Genetics* 114, 1211–1228.

Takahashi, Y., Shomura, A., Sasaki, T. and Yano, M. (2001) *Hd6*, a rice quantitative trait locus involved in photoperiod sensitivity, encodes the a subunit of protein kinase CK2. *Proceedings of the National Academy of Sciences USA* 98, 7922–7927.

Takai, T., Ohsumi, A., San-oh, Y., Laza, M.R.C., Kondo, M., Yamamoto, T. *et al.* (2009) Detection of a quantitative trait locus controlling carbon isotope discrimination and its contribution to stomatal conductance in japonica rice. *Theoretical and Applied Genetics* 118, 1401–1410.

Talamè, V., Ozturk, N.Z., Bohnert, H.J. and Tuberosa, R. (2007) Barley transcript profiles under dehydration shock and drought stress treatments: a comparative analysis. *Journal of Experimental Botany* 58, 229–240.

Tambussi, E.A., Bort, J. and Araus, J.L. (2007) Water use efficiency in C$_3$ cereals under Mediterranean conditions: a review of physiological aspects. *Annals of Applied Biology* 150, 307–321.

Tan, L.B., Liu, F.X., Xue, W., Wang, G.J., Ye, S., Zhu, Z.F. *et al.* (2007) Development of *Oryza rufipogon* and *O. sativa* introgression lines and assessment for yield-related quantitative trait loci. *Journal of Integrative Plant Biology* 49, 871–884.

Tanksley, S.D. (1993) Mapping polygenes. *Annual Review of Genetics* 27, 205–233.

Taramino, G. and Tingey, S. (1996) Simple sequence repeats for germplasm analysis and mapping in maize. *Genome* 39, 277–287.

Taramino, G., Sauer, M., Stauffer, J.L., Multani, D., Niu, X.M., Sakai, H. *et al.* (2007) The maize (*Zea mays* L.) *RTCS* gene encodes a LOB domain protein that is a key regulator of embryonic seminal and post-embryonic shoot-borne root initiation. *Plant Journal* 50, 649–659.

Tardieu, F. and Tuberosa, R. (2010) Dissection and modelling of abiotic stress tolerance in plants. *Current Opinion in Plant Biology* 13, 206–212.

Tardieu, F., Muller, B. and Reymond, M. (2003) Leaf growth regulation under drought: combining ecophysiological modelling, QTL analysis and search for mechanisms. *Journal of Experimental Botany* 54, 18–19.

Tester, M. and Langridge, P. (2010) Breeding technologies to increase crop production in a changing world. *Science* 327, 818–822.

Teulat, B., Merah, O., Sirault, X., Borries, C., Waugh, R. and This, D. (2002) QTLs for grain carbon isotope discrimination in field-grown barley. *Theoretical and Applied Genetics* 106, 118–126.

Thevenot, C., Simond-Cote, E., Reyss, A., Manicacci, D., Trouverie, J., Le Guilloux, M. *et al.* (2005) QTLs for enzyme activities and soluble carbohydrates involved in starch accumulation during grain filling in maize. *Journal of Experimental Botany* 56, 945–958.

Thomas, H. and Howarth, C.J. (2000) Five ways to stay green. *Journal of Experimental Botany* 51, 329–337.

Thomson, M.J., de Ocampo, M., Egdane, J., Rahman, M.A., Sajise, A.G., Adorada, D.L. *et al.* (2010) Characterizing the *Saltol* quantitative trait locus for salinity tolerance in rice. *Rice* 3, 148–160.

Till, B.J., Comai, L. and Henikoff, S. (2007) TILLING and EcoTILLING for crop improvement. In: Varshney, R.K. and Tuberosa, R. (eds) *Genomics-assisted Crop Improvement, Vol 1: Genomics Approaches and Platforms.* Springer, Dordrecht, Netherlands, pp. 333–350.

Tollenaar, M. and Lee, E.A. (2006) Dissection of physiological processes underlying grain yield in maize by examining genetic improvement and heterosis. *Maydica* 51, 399–408.

Tollenaar, M. and Wu, J. (1999) Yield improvement in temperate maize is attributable to greater stress tolerance. *Crop Science* 39, 1597–1604.

Tondelli, A., Francia, E., Barabaschi, D., Aprile, A., Skinner, J.S., Stockinger, E.J. *et al.* (2006) Mapping regulatory genes as candidates for cold and drought stress tolerance in barley. *Theoretical and Applied Genetics* 112, 445–454.

Trachsel, S., Messmer, R., Stamp, P., Ruta, N. and Hund, A. (2010) QTLs for early vigor of tropical maize. *Molecular Breeding* 25, 91–103.

Tripathy, J.N., Zhang, J., Robin, S., Nguyen, T.T. and Nguyen, H.T. (2000) QTLs for cell-membrane stability mapped in rice (*Oryza sativa* L.) under drought stress. *Theoretical and Applied Genetics* 100, 1197–1202.

Trouverie, J. and Prioul, J.L. (2006) Increasing leaf export and grain import capacities in maize plants under water stress. *Functional Plant Biology* 33, 209–218.

Tuberosa, R. (2011) Phenotyping drought-stressed crops: key concepts, issues and approaches. In: Monneveux, P. and Ribaut, J.M. (eds) *Drought Phenotyping in Crops: From Theory to Practice.* CIMMYT/Generation Challenge Programme, Mexico, pp. 3–35.

Tuberosa, R. and Salvi, S. (2006) Genomics-based approaches to improve drought tolerance of crops. *Trends in Plant Science* 11, 405–412.

Tuberosa, R., Sanguineti, M.C. and Landi, P. (1994) Abscisic acid concentration in the leaf and xylem sap, leaf water potential, and stomatal conductance in drought-stressed maize. *Crop Science* 34, 1557–1563.

Tuberosa, R., Parentoni, S., Kim, T.S., Sanguineti, M.C. and Phillips, R.L. (1998) Mapping QTLs for ABA concentration in leaves of a maize cross segregating for anthesis date. *Maize Genetics Cooperation Newsletter* 72, 72–73.

Tuberosa, R., Gill, B.S. and Quarrie, S.A. (2002a) Cereal genomics: ushering in a brave new world. *Plant Molecular Biology* 48, 445–449.

Tuberosa, R., Salvi, S., Sanguineti, M.C., Landi, P., Maccaferri, M. and Conti, S. (2002b) Mapping QTLs regulating morpho-physiological traits and yield: Case studies, shortcomings and perspectives in drought-stressed maize. *Annals of Botany* 89, 941–963.

Tuberosa, R., Sanguineti, M.C., Landi, P., Michela Giuliani, M., Salvi, S. and Conti, S. (2002c) Identification of QTLs for root characteristics in maize grown in hydroponics and analysis of their overlap with QTLs for grain yield in the field at two water regimes. *Plant Molecular Biology* 48, 697–712.

Tuberosa, R., Grillo, S. and Ellis, R.P. (2003) Unravelling the genetic basis of drought tolerance in crops. In: Sanità di Toppi, L. and Pawlik-Skowronska, B. (eds) *Abiotic Stresses in Plants.* Kluwer Academic Publishers, Dordrecht, Netherlands, pp. 71–122.

Tuberosa, R., Giuliani, S., Parry, M.A.J. and Araus, J.L. (2007a) Improving water use efficiency in Mediterranean agriculture: what limits the adoption of new technologies? *Annals of Applied Biology* 150, 157–162.

Tuberosa, R., Salvi, S., Giuliani, S., Sanguineti, M.C., Bellotti, M., Conti, S. *et al.* (2007b) Genome-wide approaches to investigate and improve maize response to drought. *Crop Science* 47, S120–S141.

Tuberosa, R., Graner, A. and Varshney, R.K. (2011a) Genomics of plant genetic resources: an introduction. *Plant Genetic Resources: Characterization and Utilization* 9, 151–154.

Tuberosa, R., Salvi, S., Giuliani S., Sanguineti, M.C., Frascaroli, E., Conti, S. *et al.* (2011b) Genomics of root architecture and functions in maize. In: Costol, A. and Varshney, K. (eds) *Root Genomics.* Springer, Dordrecht, Netherlands, pp. 179–204.

Tuinstra, M.R., Grote, E.M., Goldsbrough, P.B. and Ejeta, G. (1997) Genetic analysis of post-flowering drought tolerance and components of grain development in *Sorghum bicolor* (L.) Moench. *Molecular Breeding* 3, 439–448.

Turner, L.B., Cairns, A.J., Armstead, I.P., Ashton, J., Skot, K., Whittaker, D. *et al.* (2006) Dissecting the regulation of fructan metabolism in perennial ryegrass (*Lolium perenne*) with quantitative trait locus mapping. *New Phytologist* 169, 45–57.

Turner, L.B., Farrell, M., Humphreys, M.O. and Dolstra, O. (2010) Testing water-soluble carbohydrate QTL effects in perennial ryegrass (*Lolium perenne* L.) by marker selection. *Theoretical and Applied Genetics* 121, 1405–1417.

Turner, N.C. (1997) Further progress in crop water relations. *Advances in Agronomy* 528, 293–338.

Ulloa, M., Cantrell, R.G., Percy, R.G., Zeiger, E. and Lu, Z.M. (2000) QTL analysis of stomatal conductance and relationship to lint yield in an interspecific cotton. *Journal of Cotton Science* 4, 10–18.

Urano, K., Kurihara, Y., Seki, M. and Shinozaki, K. (2010) 'Omics' analyses of regulatory networks in plant abiotic stress responses. *Current Opinion in Plant Biology* 13, 132–138.

Valliyodan, B. and Nguyen, H.T. (2006) Understanding regulatory networks and engineering for enhanced drought tolerance in plants. *Current Opinion in Plant Biology* 9, 189–195.

van Buuren, M.L., Salvi, S., Morgante, M., Serhani, B. and Tuberosa, R. (2002) Comparative genomic mapping between a 754 kb region flanking *DREB1A* in *Arabidopsis thaliana* and maize. *Plant Molecular Biology* 48, 741–750.

van Eeuwijk, F.A., Bink, M., Chenu, K. and Chapman, S.C. (2010) Detection and use of QTL for complex traits in multiple environments. *Current Opinion in Plant Biology* 13, 193–205.

Vargas, M., van Eeuwijk, F.A., Crossa, J. and Ribaut, J.M. (2006) Mapping QTLs and QTL x environment interaction for CIMMYT maize drought stress program using factorial regression and partial least squares methods. *Theoretical and Applied Genetics* 112, 1009–1023.

Varshney, R.K. and Dubey, A. (2009) Novel genomic tools and modern genetic and breeding approaches for crop improvement. *Journal of Plant Biochemistry and Biotechnology* 18, 127–138.

Varshney, R.K. and Tuberosa, R. (2007) Genomics-assisted crop improvement: an overview. In: Varshney, R.K. and Tuberosa, R. (eds) *Genomics-assisted Crop Improvement, Vol 1: Genomics Approaches and Platforms*. Springer, Dordrecht, Netherlands, pp. 1–12.

Varshney, R.K., Hiremath, P.J., Lekha, P.,

Kashiwagi, J., Balaji, J., Deokar, A.A. *et al.* (2009a) A comprehensive resource of drought- and salinity-responsive ESTs for gene discovery and marker development in chickpea (*Cicer arietinum* L.). *BMC Genomics* 10, 523.

Varshney, R.K., Nayak, S.N., May, G.D. and Jackson, S.A. (2009b) Next-generation sequencing technologies and their implications for crop genetics and breeding. *Trends in Biotechnology* 27, 522–530.

Venuprasad, R., Bool, M.E., Dalid, C.O., Bernier, J., Kumar, A. and Atlin, G.N. (2009) Genetic loci responding to two cycles of divergent selection for grain yield under drought stress in a rice breeding population. *Euphytica* 167, 261–269.

Verulkar, S.B., Mandal, N.P., Dwivedi, J.L., Singh, B.N., Sinha, P.K., Mahato, R.N. *et al.* (2010) Breeding resilient and productive genotypes adapted to drought-prone rainfed ecosystem of India. *Field Crops Research* 117, 197–208.

Vladutu, C., McLaughlin, J. and Phillips, R.L. (1999) Fine mapping and characterization of linked quantitative trait loci involved in the transition of the maize apical meristem from vegetative to generative structures. *Genetics* 153, 993–1007.

Vuylsteke, M., Mank, R., Antonise, R., Bastiaans, E., Senior, M.L., Stuber, C.W. *et al.* (1999) Two high-density AFLP® linkage maps of *Zea mays* L.: analysis of distribution of AFLP markers. *Theoretical and Applied Genetics* 99, 921–935.

Wang, J.K., Wan, X.Y., Crossa, J., Crouch, J., Weng, J.F., Zhai, H.Q. *et al.* (2006) QTL mapping of grain length in rice (*Oryza sativa* L.) using chromosome segment substitution lines. *Genetical Research* 88, 93–104.

Wang, J.P., Raman, H., Zhou, M.X., Ryan, P.R., Delhaize, E., Hebb, D.M. *et al.* (2007) High-resolution mapping of the Alp locus and identification of a candidate gene HvMATE controlling aluminium tolerance in barley (*Hordeum vulgare* L.). *Theoretical and Applied Genetics* 115, 265–276.

Wassmann, R., Jagadish, S.V.K., Heuer, S., Ismail, A., Redona, E., Serraj, R. *et al.* (2009) Climate change affecting rice production: the physiological and agronomic basis for possible adaptation strategies. In: *Advances in Agronomy, Vol. 101*. Elsevier Academic Press, Inc., San Diego, California, pp. 59–122.

Welch, J.R., Vincent, J.R., Auffhammer, M., Moya, P.F., Dobermann, A. and Dawe, D. (2010) Rice yields in tropical/subtropical Asia exhibit large but opposing sensitivities to minimum and maximum temperatures. *Proceedings of the National Academy of Sciences of the United States of America* 107, 14562–14567.

Welcker, C., Boussuge, B., Bencivenni, C., Ribaut,

J.M. and Tardieu, F. (2007) Are source and sink strengths genetically linked in maize plants subjected to water deficit? A QTL study of the responses of leaf growth and of anthesis–silking interval to water deficit. *Journal of Experimental Botany* 58, 339–349.

Wight, C.P., Kibite, S., Tinker, N.A. and Molnar, S.J. (2006) Identification of molecular markers for aluminium tolerance in diploid oat through comparative mapping and QTL analysis. *Theoretical and Applied Genetics* 112, 222–231.

Williams, K.J., Willsmore, K.L., Olson, S., Matic, M. and Kuchel, H. (2006) Mapping of a novel QTL for resistance to cereal cyst nematode in wheat. *Theoretical and Applied Genetics* 112, 1480–1486.

Wissuwa, M., Wegner, J., Ae, N. and Yano, M. (2002) Substitution mapping of *Pup1*: a major QTL increasing phosphorus uptake of rice from a phosphorus-deficient soil. *Theoretical and Applied Genetics* 105, 890–897.

Wissuwa, M., Gamat, G. and Ismail, A.M. (2005) Is root growth under phosphorus deficiency affected by source or sink limitations? *Journal of Experimental Botany* 56, 1943–1950.

Witcombe, J.R., Hollington, P.A., Howarth, C.J., Reader, S. and Steele, K.A. (2008) Breeding for abiotic stresses for sustainable agriculture. *Philosophical Transactions of the Royal Society B – Biological Sciences* 363, 703–716.

Xiao, Y.N., Li, X.H., George, M.L., Li, M.S., Zhang, S.H. and Zheng, Y.L. (2005) Quantitative trait locus analysis of drought tolerance and yield in maize in China. *Plant Molecular Biology Reporter* 23, 155–165.

Xing, Y.Z. and Zhang, Q.F. (2010) Genetic and molecular bases of rice yield. In: *Annual Review of Plant Biology, Vol. 61*. Annual Reviews, Palo Alto, California, pp. 421–442.

Xiong, W., Declan, C., Erda, L., Xu, Y.L., Ju, H., Jiang, J.H. *et al.* (2009) Future cereal production in China: The interaction of climate change, water availability and socio-economic scenarios. *Global Environmental Change – Human and Policy Dimensions* 19, 34–44.

Xiong, Y.W., Fei, S.Z., Arora, R., Brummer, E.C., Barker, R.E., Jung, G.W. *et al.* (2007) Identification of quantitative trait loci controlling winter hardiness in an annual x perennial ryegrass interspecific hybrid population. *Molecular Breeding* 19, 125–136.

Xu, K.N. and Mackill, D.J. (1996) A major locus for submergence tolerance mapped on rice chromosome 9. *Molecular Breeding* 2, 219–224.

Xu, K., Xu, X., Fukao, T., Canlas, P., Maghirang-Rodriguez, R., Heuer, S. *et al.* (2006) *Sub1A* is an ethylene-response-factor-like gene that confers submergence tolerance to rice. *Nature* 442, 705–708.

Xu, W.W., Subudhi, P.K., Crasta, O.R., Rosenow, D.T., Mullet, J.E. and Nguyen, H.T. (2000) Molecular mapping of QTLs conferring stay-green in grain sorghum (*Sorghum bicolor* L. Moench). *Genome* 43, 461–469.

Xu, X., Martin, B., Comstock, J.P., Vision, T.J., Tauer, C.G., Zhao, B. *et al.* (2008) Fine mapping a QTL for carbon isotope composition in tomato. *Theoretical and Applied Genetics* 117, 221–233.

Xu, Y. (2010) *Molecular Plant Breeding*. CAB International, Wallingford, UK.

Xu, Y.B. and Crouch, J.H. (2008) Marker-assisted selection in plant breeding: From publications to practice. *Crop Science* 48, 391–407.

Xue, W.Y., Xing, Y.Z., Weng, X.Y., Zhao, Y., Tang, W.J., Wang, L. *et al.* (2008) Natural variation in *Ghd7* is an important regulator of heading date and yield potential in rice. *Nature Genetics* 40, 761–767.

Xue, Y., Jiang, L., Su, N., Wang, J.K., Deng, P., Ma, J.F. *et al.* (2007) The genetic basic and fine-mapping of a stable quantitative-trait loci for aluminium tolerance in rice. *Planta* 227, 255–262.

Xue-Xuan, X., Hong-Bo, S., Yuan-Yuan, M., Gang, X., Jun-Na, S., Dong-Gang, G. *et al.* (2010) Biotechnological implications from abscisic acid (ABA) roles in cold stress and leaf senescence as an important signal for improving plant sustainable survival under abiotic-stressed conditions. *Critical Reviews in Biotechnology* 30, 222–230.

Yadav, R.S., Sehgal, D. and Vadez, V. (2011) Using genetic mapping and genomics approaches in understanding and improving drought tolerance in pearl millet. *Journal of Experimental Botany* 62, 397–408.

Yan, X.L., Liao, H., Beebe, S.E., Blair, M.W. and Lynch, J.P. (2004) QTL mapping of root hair and acid exudation traits and their relationship to phosphorus uptake in common bean. *Plant and Soil* 265, 17–29.

Yang, D.L., Jing, R.L., Chang, X.P. and Li, W. (2007a) Identification of quantitative trait loci and environmental interactions for accumulation and remobilization of water-soluble carbohydrates in wheat (*Triticum aestivum* L.) stems. *Genetics* 176, 571–584.

Yang, D.L., Jing, R.L., Chang, X.P. and Li, W. (2007b) Quantitative trait loci mapping for chlorophyll fluorescence and associated traits in wheat (*Triticum aestivum*). *Journal of Integrative Plant Biology* 49, 646–654.

Yang, J., Sears, R.G., Gill, B.S. and Paulsen, G.M. (2002) Quantitative and molecular characterization of heat tolerance in hexaploid wheat. *Euphytica* 126, 275–282.

Yano, M., Katayose, Y., Ashikari, M., Yamanouchi, U., Monna, L., Fuse, T. *et al.* (2000) *Hd1*, a major photoperiod sensitivity quantitative trait locus in rice, is closely related to the Arabidopsis flowering time gene *CONSTANS. Plant Cell* 12, 2473–2484.

Yoo, S.C., Cho, S.H., Zhang, H., Paik, H.C., Lee, C.H., Li, J. *et al.* (2007) Quantitative trait loci associated with functional stay-green SNU-SG1 in rice. *Molecules and Cells* 24, 83–94.

Yoshida, T., Fujita, Y., Sayama, H., Kidokoro, S., Maruyama, K., Mizoi, J. *et al.* (2010) AREB1, AREB2, and ABF3 are master transcription factors that cooperatively regulate ABRE-dependent ABA signaling involved in drought stress tolerance and require ABA for full activation. *Plant Journal* 61, 672–685.

Yu, J.M., Holland, J.B., McMullen, M.D. and Buckler, E.S. (2008) Genetic design and statistical power of nested association mapping in maize. *Genetics* 178, 539–551.

Yue, B., Xiong, L.Z., Xue, W.Y., Xing, Y.Z., Luo, L.J. and Xu, C.G. (2005) Genetic analysis for drought resistance of rice at reproductive stage in field with different types of soil. *Theoretical and Applied Genetics* 111, 1127–1136.

Yue, B., Xue, W.Y., Xiong, L.Z., Yu, X.Q., Luo, L.J., Cui, K.H. *et al.* (2006) Genetic basis of drought resistance at reproductive stage in rice: Separation of drought tolerance from drought avoidance. *Genetics* 172, 1213–1228.

Zamir, D. (2001) Improving plant breeding with exotic genetic libraries. *Nature Reviews Genetics* 2, 983–989.

Zargar, S.M., Nazir, M., Agrawal, G.K., Kim, D.W. and Rakwal, R. (2010) Silicon in plant tolerance against environmental stressors: Towards crop improvement using omics approaches. *Current Proteomics* 7, 135–143.

Zhang, F.S., Shen, J.B., Zhang, J.L., Zuo, Y.M., Li, L. and Chen, X.P. (2010) Rhizosphere processes and management for improving nutrient use efficiency and crop productivity: Implications for China. *Advances in Agronomy* 107, 1–32.

Zhang, J., Zheng, H.G., Aarti, A., Pantuwan, G., Nguyen, T.T., Tripathy, J.N. *et al.* (2001) Locating genomic regions associated with components of drought resistance in rice: comparative mapping within and across species. *Theoretical and Applied Genetics* 103, 19–29.

Zhang, W.P., Shen, X.Y., Wu, P., Hu, B. and Liao, C.Y. (2001) QTLs and epistasis for seminal root

length under a different water supply in rice (*Oryza sativa* L.). *Theoretical and Applied Genetics* 103, 118–123.

Zhang, X., Zhou, S.X., Fu, Y.C., Su, Z., Wang, X.K. and Sun, C.Q. (2006) Identification of a drought tolerant introgression line derived from dongxiang common wild rice (*O. rufipogon* Griff.). *Plant Molecular Biology* 62, 247–259.

Zheng, B.S., Yang, L., Zhang, W.P., Mao, C.Z., Wu, Y.R., Yi, K.K. *et al.* (2003) Mapping QTLs and candidate genes for rice root traits under different water-supply conditions and comparative analysis across three populations. *Theoretical and Applied Genetics* 107, 1505–1515.

Zhu, C., Gore, M., Buckler, E.S. and Yu, J. (2008) Status and prospects of association mapping in plants. *The Plant Genome* 1, 5–20.

Zhu, J.M., Kaeppler, S.M. and Lynch, J.P. (2005) Mapping of QTL controlling root hair length in maize (*Zea mays* L.) under phosphorus deficiency. *Plant and Soil* 270, 299–310.

Zhu, J.M., Mickelson, S.M., Kaeppler, S.M. and Lynch, J.P. (2006) Detection of quantitative trait loci for seminal root traits in maize (*Zea mays* L.) seedlings grown under differential phosphorus levels. *Theoretical and Applied Genetics* 113, 1–10.

Zhu, J.M., Alvarez, S., Marsh, E.L., LeNoble, M.E., Cho, I.J., Sivaguru, M. *et al.* (2007) Cell wall proteome in the maize primary root elongation zone. II. Region-specific changes in water soluble and lightly ionically bound proteins under water deficit. *Plant Physiology* 145, 1533–1548.

Zhuang, Y., Ren, G.J., Yue, G.D., Li, Z.X., Qu, X., Hou, G.H. *et al.* (2007) Effects of water-deficit stress on the transcriptomes of developing immature ear and tassel in maize. *Plant Cell Reports* 26, 2137–2147.

Zinselmeier, C., Sun, Y.J., Helentjaris, T., Beatty, M., Yang, S., Smith, H. *et al.* (2002) The use of gene expression profiling to dissect the stress sensitivity of reproductive development in maize. *Field Crops Research* 75, 111–121.

Zou, G.H., Mei, H.W., Liu, H.Y., Liu, G.L., Hu, S.P., Yu, X.Q. *et al.* (2005) Grain yield responses to moisture regimes in a rice population: association among traits and genetic markers. *Theoretical and Applied Genetics* 112, 106–113.

Zwart, R.S., Thompson, J.P., Milgate, A.W., Bansal, U.K., Williamson, P.M., Raman, H. *et al.* (2010) QTL mapping of multiple foliar disease and root-lesion nematode resistances in wheat. *Molecular Breeding* 26, 107–124.

9 Crop Management to Cope with Global Change: a Systems Perspective Aided by Information Technologies

G.S. McMaster and J.C. Ascough II

9.1 Introduction

Managing complex agricultural systems presents many challenges. On a broad scale, opportunities and risks in crop management in times of global change must consider not only biological factors, but also economic, environmental, sociological and policy concerns. For effective adoption of management guidelines, developing comprehensive guiding principles and adjusting them for the specific system in the local environment (e.g. climate, soils) is required. Precision agriculture further reduces the spatial scale of management to the sub-field level. Not surprisingly, addressing the multitude of factors comprehensively and across multiple agricultural systems is difficult at best. A frequent approach is to reduce the problem into smaller scales and areas of concern, as is being done in this book by focusing on the crop–eco-physiological perspectives of risks and opportunities for agriculture in times of global change. Despite narrowing the scale and focus of interest, numerous risks and opportunities addressed in this book, such as breeding and molecular biology advances in cultivar selection, physiology and agronomic practices, exist and pose challenges for selecting strategies.

The rapid expansion of information technologies shows great promise in aiding our abilities to capture diverse knowledge from all spatial and temporal scales and areas of concern and to integrate this knowledge into tools for further developing knowledge in a comprehensive systems perspective. While this is the promise of information technologies, it is nevertheless a somewhat daunting and ambitious endeavour. It requires understanding of multiple processes and linking these together in a quantitative manner. The ideology dating to at least Aristotle that 'the sum is greater than the parts' reflects the perspective that building systems from the bottom up has difficult obstacles to overcome. Fortunately, methods are being developed to address these obstacles, making the existing knowledge on many processes derived from extensive research more available when developing information technologies. The chapters in this book provide information ranging from molecular biology, to breeding, to physiology, to general agronomy related to various aspects of crop stress management and climate change of different systems. Whether, and how, this and other information is captured in our information technologies is essential in using these technologies for help in managing risk and opportunities of climate change. This is important for several reasons, including the fact that experimentation alone cannot adequately address many climate change concerns, and that mitigation and policy necessarily required predictive tools for assessing possible future scenarios. This chapter provides a general overview of the diversity of information technologies and how these technologies have captured knowledge related to assessing crop–eco-physiological responses to global change.

Following this, the adoption and application of these technologies to explore opportunities and risks in crop management are discussed.

9.2 Information Technologies

Information technologies comprise a broad base of technologies designed to provide information and assist in understanding and managing agricultural systems for a variety of users (e.g. producers, consultants, scientists, agribusiness, action agencies and policy makers). Information technologies include computer simulation models, regression/statistical models, decision support technology, integrated assessment technology, Internet-based information systems (e.g. university, government, industry and private Internet pages), information databases and database archival systems. As might be expected, clear distinction among information technologies is often difficult.

To varying degrees, all information technologies have some means of (i) interacting with the user, and (ii) accessing or 'creating' the underlying knowledge or science base. The importance and emphasis among these two components are largely dependent on the intended target users of the technologies, which determine the objectives of the technology. Advances in computer science, technology and sociology/psychology of user behaviour continue to improve both of these components. Fortunately, the underlying knowledge used in these technologies is able to draw upon decades of agricultural research. Challenges remain in quantifying the research, how system components interact and incorporating the latest research findings.

For the purposes of this chapter, we focus mainly on information technologies related to crop simulation models, decision support technology, integrated assessment technology and some critical information databases. As this is still very broad, often we draw upon our experiences in developing and applying these technologies.

9.2.1 Crop simulation models (CSM)

Crop simulation models (CSM) have received extensive attention for the past four decades, and a vast diversity of models and approaches exist for simulating crop yield and production as a function of various environmental variables. Most modelling work has focused on major international crops such as wheat, maize and rice, with few or no CSM available for minor crops. Today, CSM cover scales from the gene to the agroecosystem, and only a broad overview will be presented here.

At the most fundamental level, CSM typically simulate yield (Y) as the function of daily growth rate (GR) that is partitioned to the yield component (P) and integrated over a daily time step from emergence (*emerge*) through physiological maturity (*maturity*):

$$Y = \int_{emerge}^{maturity} GR \cdot P \qquad (9.1)$$

Equation 9.1 can be implemented in numerous ways, depending on the model objectives and interests of the model developers.

A generic representation showing major processes and inputs often considered in a CSM for implementing Equation 9.1 is shown in Fig. 9.1. If CSM are to be used in assessing crop eco-physiological responses to climate change, then interactions among the environmental variables of temperature, water, light, CO_2 and nutrients are essential, and the role of management practices in altering these environmental variables is desired to assess management opportunities. Therefore, critical environmental inputs to most CSM include temperature, precipitation, solar radiation and nitrogen (other nutrients are usually not considered). Water and nitrogen balance sub-models that consider the influence of soil properties and weather interact with the plant growth component, whereby a variety of plant processes (e.g. growth, partitioning, phenology) are simulated to implement Equation 9.1.

As a general rule, the earliest CSM tended to focus on the scale of whole-plant growth, with little detail on processes at lower scales.

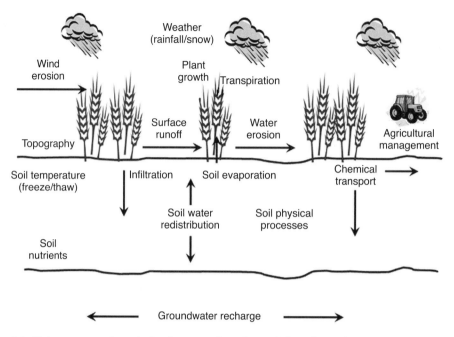

Fig. 9.1. Major processes in agricultural systems (from Ascough II *et al.*, 2008).

These models tended to be energy- or light-driven models in determining the growth rate, and this approach remains popular today. The fundamental approach simulated leaf area index on a daily time step, which was used to capture energy/sunlight and produce biomass that was then distributed to basic plant components of leaves (providing the feedback to the cycle), stems, roots and seeds. Partitioning coefficients were often used to allocate the biomass produced, and phenology sub-models were essential in improving the accuracy predicting timing of sources and sinks, and changing partitioning coefficients based on developmental stage. Creation of biomass was done in various ways, ranging from an energy–biomass conversion factor (e.g. Monteith, 1977; Williams *et al.*, 1989; Stockle *et al.*, 2003) to transpiration-based (e.g. Tanner and Sinclair, 1983) to more detailed photosynthetic sub-models (e.g. Farquhar *et al.*, 1980; Porter, 1984, 1993; Acock, 1991; Grant, 2001; Yin and Struik, 2009). A summary of these various approaches is provided by Boote and Loomis (1991).

Phenology modelling has been one of the most successful areas in crop simulation modelling, and the ability to simulate genotype phenology across a broad range of environments is quite reliable for major crops such as wheat. Many alternative approaches exist for predicting phenology, and approaches differ in input requirements and number of developmental stages simulated. Essentially, all models are based on the thermal time approach, with some emphasizing the role of vernalization and photoperiod more than others. One area of divergence in phenology sub-models, particularly for small-grain cereals, is whether leaf number or strictly thermal time is used to estimate the time interval between developmental stages. In phenology sub-models, temperature effects are well considered, but rarely are the effects of water and nutrient availability or CO_2 concentration considered.

As crop simulation modelling progressed, greater attention was focused on representing plant processes below the whole-plant level. In general, energy- or light-driven modelling emphasized functional physiology, particularly for assessing energy balance and leaf functioning at the individual organ level (e.g. Norman, 1979; Grant, 2001). Considerable research on crop development during the 1970s and 1980s undoubtedly spurred interest in alternative modelling approaches based on more developmentally driven approaches, beginning in the 1980s. Developmentally based approaches recognized the long-held view that plant development is orderly and predictable based on basic units that dynamically appear, grow and senesce over time (Gray, 1879; Prusinkiewicz, 1998; Forster *et al.*, 2007). The basic unit for building a plant is the phytomer, most commonly defined as the leaf, node, internode and axillary bud (Wilhelm and McMaster, 1995). To illustrate this perspective, the grass plant begins growth with the appearance of the main shoot from the seed. Successive leaves appear, grow and senesce on the main shoot, and each axillary bud associated with each leaf may initiate a new shoot if conditions are favourable to do so, with leaf dynamics following the main shoot pattern. Similarly, development of the inflorescence on a shoot is orderly and predictable. Summaries of these general concepts have been provided by Rickman and Klepper (1995) and McMaster (1997, 2005, 2009).

Largely dependent on modelling objectives, CSM tend to be either crop-specific or more generic to be able to simulate many crops. Examples of crop-specific models for simulating wheat are Sirius (Jamieson *et al.*, 1998b), ARFCWheat1/2 (Porter, 1984, 1993; Weir *et al.*, 1984), SUCROS (van Laar *et al.*, 1992), SWHEAT (van Keulen and Seligman, 1987), WINTER WHEAT (Baker *et al.*, 1985), MODWht (Rickman *et al.*, 1996) and SHOOTGRO (McMaster *et al.*, 1991, 1992a, b; Wilhelm *et al.*, 1993; Zalud *et al.*, 2003). Models such as EPIC (Williams *et al.*, 1989), WEPP (Arnold *et al.*, 1995), CropSyst (Stockle *et al.*, 2003), AquaCrop (Steduto *et al.*, 2009) and GPFARM (McMaster *et al.*, 2002, 2003; Andales *et al.*, 2003; Shaffer *et al.*, 2004; Ascough II *et al.*, 2007) simulate many crops. These 'generic' CSM tend to have fewer required parameters and inputs, and are typically easier to set up and use. Some

models such as APSIM (Keating *et al.*, 2003) and DSSAT/CSM/CERES-Wheat/CropSim (Ritchie and Otter, 1985; Ritchie, 1991; Hunt and Pararajasingham, 1995; Jones *et al.*, 2003; Hoogenboom *et al.*, 2004) are crop-specific models that have been developed to simulate numerous crops, although they may simulate processes differently depending on the crop.

Regardless of modelling approach and goals, determining plant parameters and how to address the genotype by environment interaction are common concerns for all models. With the explosion of genome mapping (e.g. Arabidopsis Genome Initiative, 2000) and molecular biology research, opportunities for understanding and resolving these issues are emerging (e.g. White and Hoogenboom, 2003; Edmeades *et al.*, 2004; White, 2006). For example, the presence or absence of known alleles influencing a trait can be used to determine the parameters used in the algorithm representing the process (White *et al.*, 2004a, b) or the response to environmental factors (Weiss *et al.*, 2009). Clarification of gene networks controlling processes such as time of flowering has considerably advanced our understanding and simulation of these processes (e.g. Welch *et al.*, 2003).

Crop simulation modelling is increasingly benefiting from the advent of object-oriented design and programming languages such as C++ and Java, in terms of developing and maintaining models as well as in providing greater flexibility in representing plant processes within models. Initial efforts tended to view the plant as a collection of objects that equate to leaf, stem, root and seed components (Sequeira *et al.*, 1991, 1997). Recent attempts have begun to incorporate the phytomer approach of building plant canopies into the object-oriented design that can also be scaled up, or aggregated, into lower levels of resolution, such as the seed component of earlier designs (Prusinkiewicz, 1998; Dingkuhn *et al.*, 2005; Vos *et al.*, 2007; McMaster and Hargreaves, 2009). Other attempts have focused on modular or component-oriented modelling and are discussed in more detail in the next section.

9.2.2 Modular components and modelling framework technology

Crop simulation model developers have long recognized the value of a modular style of model development. For instance, the early mechanistic models from the 'school of de Wit' were designed with a high degree of modularity in mind. Modular approaches to CSM development are advantageous in that they: (i) facilitate substitution and reusability of different model components; (ii) encourage documentation and sharing of code; (iii) permit linkage of components written in different programming languages; (iv) allow greater flexibility in model updates and maintenance; and (v) increase collaboration opportunities between different model development groups.

In general, however, modular or component-based programming techniques have been slow to filter down to the modelling community where different approaches to CSM sub-model development (e.g. crop growth, nutrient cycling, greenhouse gas emissions, etc.) have resulted in the proliferation of codes and models as described above. Most current CSM are still monolithic, difficult to update as new knowledge becomes available and are not easily extensible if new problems and modelling needs arise. Notable exceptions are the DSSAT 4 CSM and the APSIM modelling platforms. The basis for the DSSAT CSM design is a modular structure in which components are structured to allow for easy replacement or addition of modules, and there are common components for soil water, soil nitrogen, weather and competition for light and water among the soil, plants and atmosphere (Hoogenboom et al., 2004). The APSIM modelling framework has the ability to integrate models derived in disparate research efforts via the implementation of a 'plug-in-pull-out' approach, e.g. the user can configure a model by choosing a set of sub-models from a suite of crop, soil and utility modules (Keating et al., 2003). Thus, any logical combination of modules can be specified by the user 'plugging in' required modules and 'pulling out' any modules no longer required.

Recently, as computer scientists trained in modern software engineering methods have entered the natural resource, environmental science and ecology disciplines, the APSIM 'plug-and-play' approach has been taken one step further with the development of formalized modular modelling frameworks. These frameworks bring together suites or libraries of modules, have architectures designed to fit well with the basic and natural structures of environmental problem situations and can maintain reusability and compatibility of both science and auxiliary (e.g. parameter estimation, output visualization) components. A range of modelling frameworks with different capabilities currently exists, including the Open Modelling Interface (OpenMI, Gregersen et al., 2007), Earth System Modelling Framework (ESMF, Hill et al., 2004), Framework for Risk Analysis in Multimedia Environmental Systems (FRAMES, Castleton and Gelston, 2003) and the Object Modeling System (OMS, David et al., 2002).

The overall goal of the majority of framework development efforts is to allow the agricultural, environmental and ecological modelling communities to focus more on the science components and the system being modelled, thereby allowing core development, interpretation and management requirements to be addressed more fully. Code modularization is desirable and should be a key concept in simulation model development. The component-oriented and modular approach of the modelling frameworks (and the modules/models implemented in them) should provide a basis for more efficient and collaborative CSM development in the future.

9.2.3 Decision support technology (DST)

As the name implies, DST is focused on providing decision support for a user (e.g. producers, consultants, scientists, agribusiness, action agencies, policy makers). With such diversity of intended users, considerable diversity exists among DST and clear distinction between CSM and DST is often blurred. Much of the promise of CSM to transfer science and technology from the modellers to non-modellers has not been fully realized, for a number of reasons. Certainly the complexity of most CSM, difficulty and time required setting up and using CSM and 'non-transparency' in using CSM to address specific problems are part of the explanation. DST therefore emphasizes the interface between the underlying technology and science and the user to address these problems of CSM and provide the user with information to aid in decision making.

The complexity and array of problems that a DST addresses vary, and GPFARM can illustrate one type of a broad-based DST. The GPFARM project was initiated in the late 1980s with the objective of providing farmers and ranchers with a strategic planning tool for managing specific fields on their enterprise (Ascough II *et al.*, 2007). Working with the producers from the beginning included determining the information they needed for making management decisions (e.g. production,

economic costs and returns, environmental impacts), providing the information in a manner that they understood and developing a DST that was 'quick and easy' to set up and run. To achieve this, a Microsoft Windows™-based graphical user interface (GUI) provided the linkage between the science, economic and environmental components, managed the underlying databases, aided in setting up and running the DST and provided the output to the user (Fig. 9.2).

One approach that was used for simulating crop production in the GPFARM science component was both to directly incorporate codes from existing CSM and simplify/modify other codes as necessary. Particularly important was providing default parameters for soil and plant processes for the user. However, one problem with the GPFARM approach, and with other similar projects such as APSIM/Farmscape, was the underlying premise that providing the software for the user was sufficient to lead to adoption of the technology. For many reasons discussed below, this was rarely the case. One attempted solution in the GPFARM project was to provide training for

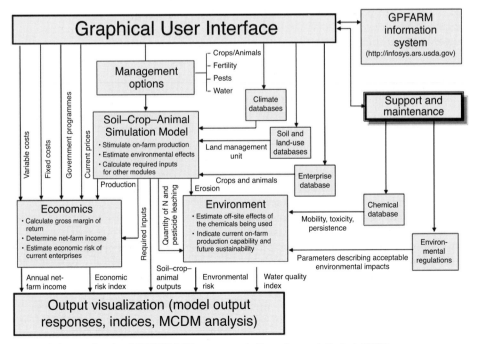

Fig. 9.2. Scheme of major GPFARM DST components (from Ascough II *et al.*, 2007).

potential users and then have them set up the DST for their farm enterprise. This approach worked with slightly greater success. The next step tried was to help the user set up their particular farm/ranch and then either have them run GPFARM and change various scenarios or, what soon became apparent, we would run various scenarios for them and then discuss the results. Clearly this model was not feasible as few, if any, research groups have the resources to provide this level of support and training.

Experiences such as ours lent encouragement to a parallel approach to DST emphasizing simpler technology that focused on a reduced problem-solving set since complex and highly detailed process-level CSM/DST are generally too difficult for consultants, producers or policy makers to use directly. When considering examples such as WISDEM for weed population dynamics and yield reduction (Canner *et al.*, 2002, 2009), PhenologyMMS for predicting crop developmental stages (McMaster *et al.*, 2011; http://arsagsoftware.ars.usda.gov), SCALES for converting between growth staging scales (Harrell *et al.*, 1993, 1998), the Potato Calculator/Wheat Calculator (Armour *et al.*, 2002, 2004) and AmaizeN for N fertilizer management (Li *et al.*, 2009), one common feature is that the science simulation component was greatly reduced and sometimes eliminated, i.e. an integrated research information database was created as a DST core in place of a simulation model.

Using an integrated research information database in a DST involves using a CSM evaluated against available experimental data to generate production and environmental impacts of different management practices for soil types, weather conditions and cropping systems outside experimental limits. The model-generated information is then combined with experimental data and long-term management experience of farmers and field professionals to create a database. These databases are often then combined with a socio-economic analysis package or other tools (e.g. multi-objective decision analysis) in order to conduct a trade-off analysis between conflicting objectives such as production, economic return and environmental quality. Overall, this type of approach is very flexible in generating site-specific management recommendations and helps solve the problem of having to interpret complicated model output.

9.2.4 Integrated assessment technology (IAT)

Meeting agricultural and environmental policy challenges for cropping system management requires approaches that assess socio-economic concerns and environmental impacts integratively (Bland, 1999). Integrated assessment technology (IAT) is being used increasingly to integrate the numerous dimensions surrounding agro-ecosystem management, including the consideration of multiple issues and stakeholders, the key disciplines within and between the human and natural sciences, multiple and cascading scales (both spatially and temporally) of agricultural system behaviour, models of the different agricultural system components, and multiple databases (Fig. 9.3). Although there appears to be no universally agreed-upon definition in the literature of what constitutes IAT, there seems to be a consensus that IAT:

- is a feedback-driven, interdisciplinary and participatory (i.e. stakeholder involvement) process;
- is an iterative process of investigation and recommendation that stresses the importance of communication from scientists to decision makers;
- explicitly accommodates linkages between the natural and human environment, and between research and policy; and
- uses the latest scientific tools including CSM, systems simulation, remotely sensed data and other forms of information technology to assemble, integrate and synthesize data from a wide range of sources and across a wide range of spatial and temporal scales.

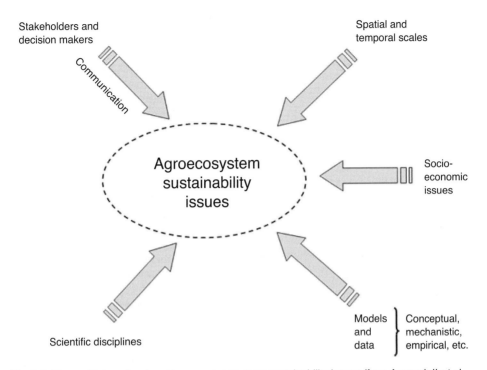

Stakeholders and
decision makers

Spatial and
temporal scales

Communication

Agroecosystem
sustainability
issues

Socio-
economic
issues

Scientific disciplines

Models
and
data

} Conceptual,
mechanistic,
empirical, etc.

Fig. 9.3. Types of integration to address agroecosystem sustainability issues (from Ascough II *et al.*, 2008).

Agricultural systems around the globe are continuously changing as a result of population demographics, climate fluctuations and the introduction of new agrotechnologies. There is consensus that CSM are needed to support sustainability within various agricultural sectors, and even more importantly to enhance the contribution of agricultural systems to sustainable development of societies at large. The integrated assessment and modelling (IAM) process attempts to assess the impacts of policies, technologies or societal trends on the environmental, economic and social sustainability of a system (Parker *et al.*, 2002). IAM is a methodology that combines quantitative models representing different aspects of sub-systems and scales into an IAT framework (Parker *et al.*, 2002). Quantitative models used in an IAM study originate from different disciplines, operate on different spatial and temporal scales and require diverse (and sometimes overlapping) data sources. Model integration within an IAM project requires that all input and

output data of each model should be integrated. Examples of IAM related to the assessment of climate change impacts are given in Weyant *et al.* (1996) and Cohen (1997).

More importantly, in current IAM approaches the earlier forms of systems modelling are being replaced with new integrated models that incorporate a multifaceted approach that considers ecological, social and economic values when addressing sustainable usage of agricultural resources. Examples of this are recent IAT analyses of climate change impacts on whole-farm systems (e.g. Rivington *et al.*, 2007). In these analyses, whole-system agricultural modelling frameworks were combined with a stakeholder-driven participatory process in order to assess potential effects of future climate change on agroecosystem land use and management patterns. The System for Environmental and Agricultural Modelling; Linking European Science and Society Integrated Framework (SEAMLESS-IF) is an example of IAT that uses sustainability indicators (economic, environmental and social) and agricultural systems evaluation (quantitative models, tools and databases) to assess and compare alternative agricultural and environmental policy options (Van Ittersum *et al.*, 2008). The goal of SEAMLESS-IF is to facilitate translation of policy questions into alternative scenarios that can be evaluated (through the above indicators) to capture the key economic, environmental, social and institutional issues of the questions at stake. The indicators in turn are assessed using an intelligent linkage of quantitative models that have been designed to simulate aspects of agricultural systems at specific scales (i.e. point or field scale, farm, region, etc.).

A review of the literature shows an increasing number of IAT exercises for solving crop management problems; however, data availability is a key issue requiring additional research in the future. Therefore, an important step towards improving IAT efficacy and usability is to further integrate CSM with field research. An agricultural system typically involves complex interactions among several different components and factors (Fig. 9.1). These interactions need interdisciplinary field research and quantification with the help of conceptual and process models. Integration of CSM with field research has the potential significantly to enhance efficiency of agricultural research and raise agricultural science and technology to the next level. The integration will benefit both field research and CSM in the following ways:

- Promote efficient and effective transfer of field research results to different soil conditions, climatic conditions and alternative cropping and management systems outside the experimental design.
- Encourage a whole-systems approach to field research that examines major component interactions and facilitates better understanding of cause-and-effect relationships and quantification of experimental results.
- Assist field researchers to focus on identified fundamental knowledge gaps to make field research more efficient, and

provide needed field evaluation and improvement of CSM before delivery to potential users.

A desirable vision for agricultural research and technology transfer is to have close integration between new field research, CSM and other components for IAT (Fig. 9.4). After a CSM has undergone thorough evaluation and both modellers and field scientists are satisfied with the results, it should be advanced to the application stage (with the goal of further model improvement through exposure to differing field conditions).

9.2.5 Information databases

In addition to requiring crop parameters, CSM, DST and IAT technologies are usually heavily dependent on input information such as weather and soil properties. Information databases often are used for providing the required data. For example, databases for soil properties are available from the USDA Natural Resources Conservation Service.

Numerous options exist for obtaining weather data. Historical weather data are readily available from local weather stations or regional networks, research stations, etc. (e.g. CoAgMet for Colorado; http://climate.colostate.edu/~coagmet). Stochastic weather generators are also available to generate weather data for a location (e.g. WGEN, Richardson and Wright, 1984; CLIGEN, Yu, 2000). Global circulation models (GCM) are frequently used for predicting future weather under different climate scenarios. Another alternative for testing possible future climate scenarios is to use historical weather data and: (i) alter the data, such as temperature, by certain systematic amounts to test the effects of increased or decreased values; or (ii) select certain years with certain characteristics (e.g. wetter or drier than normal, hotter or colder than normal).

As data sources on agricultural systems are distributed across institutions, scien-

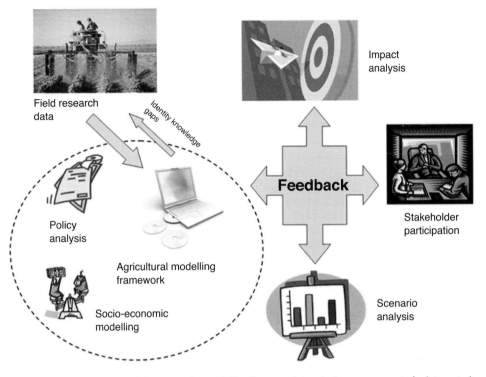

Fig. 9.4. Interactions among field research, modelling frameworks and other components for integrated agroecosystem assessment (from Ascough II *et al.*, 2008).

tists, who are required to integrate the data, typically extract data from the original sources in an impromptu manner. This practice is certainly prone to inaccuracy, and a paradigm shift is needed to overcome technical, conceptual and institutional problems. To support IAM/IAT applications, there is a need for an integrated information database on agricultural systems, which consistently combines data from different sources and ensures easy availability of data. Janssen *et al.* (2009) describe the SEAMLESS integrated database on European agricultural systems and demonstrate the use of the data in the database for calculating indicators and for model inputs. For the USA, a web-based data retrieval application was developed (Sustaining the Earth's Watersheds, Agricultural Research Data System, or STEWARDS) to increase the availability and accessibility of scientific data to the research community. The STEWARDS application is GIS-based and

couples temporal and spatial aspects of data collected from each site within a watershed (Steiner *et al.*, 2009). The STEWARDS database and software design accommodates research data with heterogeneous characteristics and format, and captures rich descriptive information that is important in understanding the data from complex, dynamic research programmes. The database includes soil, water, climate, land management and socio-economic data from multiple watersheds across the USA and can provide data commonly needed for hydrological modelling and assessments.

9.2.6 Adoption of information technologies

Intended users of information technologies range from producers and consultants to scientists, agribusiness, action agencies and policy makers. Despite the promise of information technologies, adoption of these technologies by such a diverse group of potential users has had variable success (e.g. Carberry *et al.*, 2009; Dalgliesh *et al.*, 2009; Hochman *et al.*, 2009; McCown *et al.*, 2009). Adoption partly depends on the abilities of the user, ease of use, purpose for using the technologies and whether he/she feels there is sufficient value-added for the effort in using them. For example, consultants and producers are more likely to use specific DST and information database technologies for weather, commodity prices and specific management issues such as weed, fertilizer and irrigation management. Scientists and action agencies are more likely than producers and consultants to use CSM and more detailed DST technologies.

As information technologies evolve, and means of delivering them to users improve, much has been learned about obstacles to adoption and possible solutions to reduce barriers to adoption. Summarizing some of the experiences of GPFARM (Ascough II *et al.*, 2005) and other DST such as APSIM/Farmscape (Carberry *et al.*, 2002; Keating *et al.*, 2003), several key points merit mention:

- The DST development process must carefully consider the scope of the DST in relation to the human and fiscal resources available (e.g. assessment of personnel available for developing, evaluating, implementing and maintaining DST that matches the scope, scale and complexity of the project). Formal project management and software engineering protocols and tools are increasingly essential as the scope, scale and complexity of the project increase.
- Careful attention to the intended target user group(s) is required by: (i) refining the scope of the project and matching the proposed technology appropriately with the user; and (ii) gathering input from a broad spectrum of potential users when performing a requirements analysis. A close working relationship with the target user group(s) should be maintained throughout the project development cycles.

- Mission creep is a common problem with information technology projects, as both the technology developers and target user group(s) are often overly ambitious. Constant vigilance regarding the project's scope, scale and complexity is key to maintaining focus and delivering the project in a timely manner.
- Capabilities for rapid updating of major components (e.g. simulation model or sub-models, databases) and addressing newly emerging questions or problems using the DST are an absolute necessity.
- An appropriate compromise between scientific rigour and simplicity is the eternal conundrum facing all information technology projects. Successful adoption of information technologies is greatly increased as simplicity increases. Therefore, simpler tools or integrated research information databases generated from CSM may be preferable rather than directly incorporating detailed CSM technology into the DST for some project purposes.

Many of these points apply to other information technologies as well. However, they do not consider the issues being debated when considering the science underlying these technologies, particularly for CSM (Baker, 1996):

- Careful distinction whether the CSM is primarily science oriented (i.e. increase understanding and test hypotheses) or engineering oriented (i.e. address specific agricultural problems and probably more of a DST or IAT technology) is important in developing the technology (Passioura, 1996).
- Balance in complexity between system components and the problem or question is important (Boote *et al.*, 1996; Monteith, 1996). As complexity increases, the likelihood increases that measurements needed for model calibration and validation become increasingly difficult, if not impossible, to obtain.
- Improving the linkage between experimentalists and modellers is important for both enhancing the model and

transferring information back to the experimentalist (Ascough II *et al.*, 2008).

- Overselling the promise of the CSM (or any information technology) is an unfortunate historical problem of CSM projects (Sinclair and Seligman, 1996). Frequently, insufficient understanding of the science, or inadequate implementation of existing knowledge, has resulted in unrealized promise of CSM.

9.3 Applying Information Technologies to Crop Stress Management and Climate Change

The overview of information technologies presented above illustrates how they vary in quantifying existing knowledge and their approach for exploring the complex and dynamic nature of agroecosystems. These technologies have been widely applied to evaluate the impacts of abiotic and biotic factors and management on crop production across a wide range of agroecosystems. Even a brief survey of the literature or Internet quickly overwhelms one with the magnitude of work that has been done. Major information technology projects such as GPFARM, DSSAT/CSM, APSIM/Farmscape, etc. have been applied extensively and can illustrate applications of information technologies.

Many assessments have used historical or current weather to evaluate crop responses to different soils and management practices. Examples include evaluating cropping systems (e.g. Jamieson *et al.*, 1991; Andales *et al.*, 2003; Zalud *et al.*, 2003; Anapalli *et al.*, 2005b; Ascough II *et al.*, 2007); nutrient management (e.g. Stockle and Debaeke, 1997; Shaffer *et al.*, 2004); planting date (Anapalli *et al.*, 2005a); evaluation of water deficits or drought (e.g. Semenov *et al.*, 2009); and impacts of weather variability on crop productivity (e.g. Tao *et al.*, 2009a). In most instances only a single information technology is used in the evaluation, but examples such as Jamieson *et al.* (1998a) have compared simulation results from five CSM for wheat grown under drought.

An important application of information technologies has been to use them for 'what if', or gaming, situations for evaluating possible future climatic scenarios or alternative management options, and can cover the spatial scales from the individual plant to a watershed. Some applications have come from large collaborative efforts of international scientists working on projects such as the GCTE project to evaluate the effects of global change on terrestrial ecosystems (GCTE Focus 3, 1996; Porter *et al.*, 2007). These efforts often use GCM or free-air carbon dioxide enrichment (FACE) experimental field data (e.g. Li *et al.*, 1997, 2000; Wall *et al.*, 2006) to test crop responses to CO_2, water, nitrogen and temperature (e.g. Grant *et al.*, 1995, 1999; Ewert *et al.*, 1999, 2002; Tubiello *et al.*, 1999; Grossman-Clarke *et al.*, 2001; Asseng *et al.*, 2004). Other individuals and projects have addressed more specific crop responses, or focused on specific regions, to future climate change scenarios (e.g. Stockle, 1992; Savabi and Stockle, 2001; Dhungana *et al.*, 2006; Challinor and Wheeler, 2008; Chavas *et al.*, 2009; Tao *et al.*, 2009b). Many GCM predict an increase in the frequency and degree of extreme climatic events (e.g. temperature and precipitation) for many crop production systems. Information technologies have been used to assess possible impacts of extreme weather events in various ways, such as running two GCM to evaluate the effect of extreme weather events on the hessian fly-free date and wheat production (Weiss *et al.*, 2003; Baenziger *et al.*, 2004).

Climate change is expected to provide challenges to crop breeding programmes (e.g. Araus *et al.*, 2002). Information technologies are emerging to assist breeding programmes such as the *in silico* work (e.g. Chapman and Barreto, 1996, Chapman *et al.*, 2000, 2002, 2003; Hammer *et al.*, 2004) and gene-based modelling work (e.g. White *et al.*, 2008). Climate change is likely to impact end use quality, and scientists such as Asseng and Milroy (2006), Martre *et al.* (2006), Weiss and Moreno-Sotomayer (2006) and White (2006) have worked on simulating weather effects on wheat quality. Phenological responses to climate change will vary by genotype, and work on improving phenological model response to varying water deficits, temperature and photoperiod should help in cultivar and trait selection (e.g. McMaster and Wilhelm, 2003; McMaster et al., 2008, 2009).

9.4 Future Opportunities and Obstacles in Using Information Technologies

The compelling potential of information technologies is increasingly being realized, and most certainly they have aided in our understanding and exploration of crop production management and climate change. This optimistic view is tempered by realism that much work is needed to further realize the immense potential of information technologies. All information technologies have strengths and weaknesses. When using these technologies, it is prudent that the user be aware of the limitations and maintain a healthy dose of scepticism when examining the output results. Considerable remaining challenges, if not conundrums, are how to (i) capture existing knowledge and (ii) deliver this knowledge to the user so that the tools are adopted and properly applied to exploring crop eco-physiological responses to an ever-changing climate. Successfully meeting these challenges has many obstacles that must be overcome.

Quantifying knowledge, particularly when significant uncertainty or knowledge gaps exist, speaks to the heart of the scientific process, but is critical in maintaining a solid science-based foundation within information technologies. Attention towards collecting appropriate data and making those available for the development, testing and release of information technologies are essential to enhancing the science base. Further improvements in the science base (and adoption) would be expected as greater linkage between experimentalists, users and IT developers occurs. Continued efforts to transfer scientific knowledge to integrated research information databases should increase confidence in information technologies that

rely heavily on this underlying science-based approach.

Capturing rapidly emerging new knowledge is further complicated by information technologies that are too often unwieldy and monolithic in structure. Opportunities for removal, or lowering, of obstacles to adoption of information technologies are needed. Furthermore, means of integrating possibly differing output results from different technology tools is a challenge. Of great concern is how to assist the user in: (i) making sense of possibly contradictory results from information technology tools; and (ii) summarizing what can be an overwhelming amount of 'noise' into something of value to the user (i.e. 'information'). In fairness to information technology tools, these dilemmas apply to experimental research as well. Strengthening the collaboration between technology developers, target users and experimentalists will not only aid in adoption of technologies but create greater understanding by all and improved technologies.

References

Acock, B. (1991) Modelling canopy photosynthetic response to carbon dioxide, light interception, temperature, and leaf traits. In: Boote, K.J. and Loomis, R.S. (eds) *Modeling Crop Photosynthesis – from Biochemistry to Canopy*. CSSA Special Publication Number 19, Madison, Wisconsin, pp. 41–55.

Anapalli, S.S., Ma, L., Nielsen, D.C., Vigil, M.F. and Ahuja, L.R. (2005a) Simulating planting date effects on corn production using RZWQM and CERES-Maize models. *Agronomy Journal* 97, 58–71.

Anapalli, S.S., Nielsen, D.C., Ma, L., Ahuja, L.R., Vigil, M.F. and Halvorson, A.D. (2005b) Effectiveness of RZWQM for simulating alternative Great Plains cropping systems. *Agronomy Journal* 97, 1183–1193.

Andales, A.A., Ahuja, L.R. and Peterson, G.A. (2003) Evaluation of GPFARM for dryland cropping systems in eastern Colorado. *Agronomy Journal* 95, 1510–1524.

Arabidopsis Genome Initiative (2000) Analysis of the genome sequence of the flowering plant *Arabidopsis thaliana*. *Nature* 408, 796–815.

Araus, J.L., Slafer, G.A., Reynolds, M.P. and Royo,

C. (2002) Plant breeding and drought in C_3 cereals: What should we breed for? *Annals of Botany* 89, 925–940.

Armour, T., Jamieson, P.D. and Zyskowski, R.F. (2002) Testing the Sirius Wheat Calculator. *Agronomy New Zealand* 32, 1–6.

Armour, T., Jamieson, P.D. and Zyskowski, R.F. (2004) Using the Sirius Wheat Calculator to manage wheat quality – the Canterbury experience. *Agronomy New Zealand* 34, 171–176.

Arnold, J.G., Weltz, M.A., Alberts, E.E. and Flanagan, D.C. (1995) Plant growth component. In: Flanagan, D.C., Nearing, M.A. and Laflen, J.M. (eds) *USDA-Water Erosion Prediction Project: Hillslope Profile and Watershed Model Documentation*, NSERL Report No. 10. USDA-ARS National Soil Erosion Research Laboratory, West Lafayette, Indiana, pp. 8.1–8.41.

Ascough II, J.C., Dunn, G.H., McMaster, G.S., Ahuja, L.R. and Andales, A.A. (2005) Producers, decision support systems and GPFARM: Lessons learned from a decade of development. In: Zerger, A. and Argent, R.M. (eds) *MODSIM 2005 International Congress on Modelling and Simulation.* Modelling and Simulation Society of Australia and New Zealand, December 2005, pp. 170–176 (http://www.mssanz.org.au/modsim05/papers/ascough_1.pdf).

Ascough II, J.C., McMaster, G.S., Andales, A.A., Hansen, N.C. and Sherod, L.A. (2007) Evaluating GPFARM crop growth, soil water, and soil nitrogen components for Colorado dryland locations. *Transactions of the American Society of Agricultural and Biological Engineers* 50, 1565–1578.

Ascough II, J.C., Ahuja, L.R., McMaster, G.S., Ma, L. and Andales, A.A. (2008) Ecological models: Agriculture models. In: Jorgensen, S.E. and Fath, B.D. (eds-in-chief) *Encyclopedia of Ecology* Vol. 1. Elsevier, Amsterdam, pp. 85–95.

Asseng, S. and Milroy, S.P. (2006) Simulation of environmental and genetic effects on grain protein concentration in wheat. *European Journal of Agronomy* 25, 119–128.

Asseng, S., Jamieson, P.D., Kimball, B., Pinter, P., Sayre, K., Bowden, J.W. *et al.* (2004) Simulated wheat growth affected by rising temperature, increased water deficit and elevated atmospheric CO_2. *Field Crops Research* 85, 85–102.

Baenziger, P.S., McMaster, G.S., Wilhelm, W.W., Weiss, A. and Hays, C.J. (2004) Putting genes into genetic coefficients. *Field Crops Research* 90, 133–143.

Baker, D.N., Whisler, F.D., Parton, W.J., Klepper, E.L., Cole, C.V., Willis, W.O. *et al.* (1985) The development of WINTER WHEAT: A physical physiological process model. In: Willis, W.O. (ed.) *Wheat Yield Improvement.* USDA-ARS Publ. 38, National Technical Information Service, Springfield, Virginia, pp. 176–187.

Baker, J.M. (1996) Use and abuse of crop simulation models. *Agronomy Journal* 88, 689.

Bland, W.L. (1999) Toward integrated assessment in agriculture. *Agricultural Systems* 60, 157–167.

Boote, K.J. and Loomis, R.S. (1991) *Modeling Crop Photosynthesis – from Biochemistry to Canopy.* CSSA Special Publication Number 19. Madison, Wisconsin.

Boote, K.J., Jones, J.W. and Pickering, N.B. (1996) Potential uses and limitation of crop models. *Agronomy Journal* 88, 704–716.

Canner, S.R., Wiles, L.J. and McMaster, G.S. (2002) Weed reproduction model parameters may be estimated from crop yield loss data. *Weed Science* 50, 763–772.

Canner, S.R., Wiles, L.J., Erskine, R.H., McMaster, G.S., Dunn, G.H. and Ascough II, J.A. (2009) Modelling with limited data: the influence of crop rotation and management on weed communities and crop yield loss. *Weed Science* 57, 175–186.

Carberry, P.S., Hochman, Z., McCown, R.L., Dalgliesh, N.P., Foale, M.A., Poulton, P.L. *et al.* (2002) The FARMSCAPE approach to decision support: farmers', advisers', researchers' monitoring, simulation, communication and performance evaluation. *Agricultural Systems* 74, 141–177.

Carberry, P.S., Hochman, Z., Hunt, J.R., Dalgiesh, N.P., McCown, R.L., Whish, J.P.M. *et al.* (2009) Re-inventing model-based decision support with Australian dryland farmers. 3. Relevance of APSIM to commercial crops. *Crop and Pasture Science* 60, 1044–1056.

Castleton, K.J. and Gelston, G.M. (2003) FRAMES 2.x – Science to Solutions Multimodel Operating System. PNNL Working Paper SA-39446, Pacific Northwest National Laboratory, Richland, Washington.

Challinor, A.J. and Wheeler, T.W. (2008) Use of a crop model ensemble to quantify CO_2 stimulation of water-stressed and well-watered crops. *Agricultural and Forest Meteorology* 148, 1062–1077.

Chapman, S.C. and Barreto, H.J. (1996) Using simulation models and spatial databases to improve the efficiency of plant breeding programs. In: Cooper, M. and Hammer, G.L. (eds) *Plant Adaptation and Crop Improvement.* CAB International, Wallingford, UK, pp. 563–587.

Chapman, S.C., Cooper, M., Hammer, G.L. and

Butler, D.G. (2000) Genotype by environment interactions affecting grain sorghum. II. Frequencies of different seasonal patterns of drought stress are related to location effects on hybrid yields. *Australian Journal of Agricultural Research* 51, 209–221.

Chapman, S.C., Cooper, M. and Hammer, G.L. (2002) Using crop simulation to generate genotype by environment interaction effects for sorghum in water-limited environments. *Australian Journal of Agricultural Research* 53, 379–389.

Chapman, S.C., Cooper, M., Podlich, D. and Hammer, G.L. (2003) Evaluating plant breeding strategies by simulating gene action and dryland environment effects. *Agronomy Journal* 95, 99–113.

Chavas, D.R., Izaurralde, R.C., Thomson, A.M. and Gao, X. (2009) Long-term climate change impacts on agricultural productivity in eastern China. *Agricultural and Forest Meteorology* 149, 1118–1128.

Cohen, S.J. (1997) Scientist–stakeholder collaboration in integrated assessment of climate change: lessons from a case study of Northwest Canada. *Environmental Modelling and Assessment* 2, 281–293.

Dalgliesh, N.P., Foale, M.A. and McCown, R.L. (2009) Re-inventing model-based decision support with Australian dryland farmers. 2. Pragmatic provision of soil information for paddock-specific simulation and farmer decision making. *Crop and Pasture Science* 60, 1031–1043.

David, O., Markstrom, S.L., Rojas, K.W., Ahuja, L.R. and Schneider, I.W. (2002) The object modelling system. In: Ahuja, L.R., Ma L. and Howell, T.A. (eds) *Agricultural System Models in Field Research and Technology Transfer*. Lewis Publishers, Boca Raton, Florida, pp. 317–330.

Dhungana, P., Eskridge, K.M., Weiss, A. and Baenziger, P.S. (2006) Designing crop technology for a future climate: An example using response surface methodology and the CERES-Wheat model. *Agricultural Systems* 87, 63–79.

Dingkuhn, M., Luquer, D., Quilot, B. and de Reffye, P. (2005) Environmental and genetic control of morphogenesis in crops: Towards models simulating phenotypic plasticity. *Australian Journal of Agricultural Research* 56, 1289–1302.

Edmeades, G.O., McMaster, G.S., White, J.W. and Campos, H. (2004) Genomics and the physiologist: Bridging the gap between genes and crop response. *Field Crops Research* 90, 5–18.

Ewert, F., van Oijen, M. and Porter, J.R. (1999) Simulation of growth and development processes of spring wheat in response to CO_2 and ozone for different sites and years in Europe using mechanistic crop simulation models. *European Journal of Agronomy* 10, 231–247.

Ewert, F., Rodriguez, D., Jamieson, P.D., Semenov, M.A., Mitchell, R.A.C., Goudriaan, J. *et al.* (2002) Modelling the effects of CO_2 and drought on wheat for different climatic conditions and cultivars. *Agriculture, Ecosystems, and Environment* 93, 249–266.

Farquhar, G.D., von Caemmerer, S. and Berry, J.A. (1980) A biochemical model of photosynthetic CO_2 assimilation in leaves of C_3 species. *Planta* 149, 78–90.

Forster, B.P., Franckowiak, J.D., Lundqvist, U., Lyon, J., Pitkethly, I. and Thomas, W.T.B. (2007) The barley phytomer. *Annals of Botany* 100, 725–733.

GCTE Focus 3 (1996) *Global Change and Terrestrial Ecosystems*. Report No. 2. GCTE Focus 3 Wheat Network: 1996 Model and Experimental Metadata. GCTE Focus 3 Office, Wallingford, UK.

Grant, R.F. (2001) A review of the Canadian ecosystem model – *ecosys*. In: Shaffer, M.J., Ma, L. and Hansen, S. (eds) *Modeling Carbon and Nitrogen Dynamics for Soil Management*. Lewis Publications, Boca Raton, Florida, pp. 173–263.

Grant, R.F., Kimball, B.A., Pinter Jr, P.J., Wall, G.W., Garcia, R.L., LaMorte, R.L. *et al.* (1995) Carbon dioxide effects on crop energy balance: testing *ecosys* with a Free-Air CO_2 Enrichment (FACE) experiment. *Agronomy Journal* 87, 446–457.

Grant, R.F., Wall, G.W., Kimball, B.A., Frumau, K.F.A., Pinter, P.J., Hunsaker, D.J. *et al.* (1999) Crop water relations under different CO_2 and irrigation: testing of *ecosys* with the free air CO_2 enrichment (FACE) experiment. *Agricultural and Forest Meteorology* 95, 27–51.

Gray, A. (1879) *Structural Botany*. Ivsion, Blakeman, Taylor, and Company, New York.

Gregersen, J.B., Gijsbers, P.J.A. and Westen, S.J.P. (2007) OpenMI: Open modelling interface. *Journal of Hydroinformatics* 9, 175–191.

Grossman-Clarke, S., Pinter, P.J., Kartschall, T., Kimball, B.A., Hunsaker, D.J., Wall, G.W. *et al.* (2001) Modelling a spring wheat crop under elevated CO_2 and drought. *New Phytologist* 150, 315–335.

Hammer, G.L., Sinclair, T.R., Chapman, S.C. and van Oosterom, E. (2004) On systems thinking, systems biology, and the *in silico* plant. *Plant Physiology* 134, 909–911.

Harrell, D.M., Wilhelm, W.W. and McMaster, G.S. (1993) SCALES: A computer program to convert among three developmental stage scales for wheat. *Agronomy Journal* 85, 758–763.

Harrell, D.M., Wilhelm, W.W. and McMaster, G.S. (1998) SCALES 2: Computer program to convert among developmental stage scales for corn and small grains. *Agronomy Journal* 90, 235–238.

Hill, C., DeLuca, C., Balaji, V., Suarez, M. and da Silva, A. (2004) The architecture of the Earth System Modelling Framework. *Computers, Science and Engineering* 6, 18–28.

Hochman, Z., van Rees, H., Carberry, P.S., Hunt, J.R., McCown, R.L., Gartmann, A. *et al.* (2009) Re-inventing model-based decision support with Australian dryland farmers. 4. Yield Prophet helps farmers monitor and manage crops in a variable climate. *Crop and Pasture Science* 60, 1057–1070.

Hoogenboom, G., Jones, J.W., Wilkens, P.W., Porter, C.H., Batchelor, W.D., Hunt, L.A. *et al.* (2004) Decision Support system for agro-technology transfer, version 4.0 [CD-ROM]. University of Hawaii, Honolulu, Hawaii.

Hunt L.A. and Pararajasingham, S. (1995) CROPSIM-WHEAT – a model describing the growth and development of wheat. *Canadian Journal of Plant Science* 75, 619–632.

Jamieson, P.D., Porter, J.R. and Wilson, D.R. (1991) A test of the wheat simulation model ARCWHEAT1 on wheat crops grown in New Zealand. *Field Crops Research* 27, 337–350.

Jamieson, P.D., Porter, J.R., Goudriaan, J., Ritchie, J.T., van Keulen, H. and Stol, W. (1998a) A comparison of the models AFRCWHEAT2, CERES-Wheat, Sirius, SUCROS2 and SWHEAT with measurements from wheat grown under drought. *Field Crops Research* 55, 23–44.

Jamieson, P.D., Semenov, M.A., Brooking, I.R. and Francis, G.S. (1998b) Sirius: a mechanistic model of wheat response to environmental variation. *European Journal of Agronomy* 8, 161–179.

Janssen, S., Andersen, E., Athanasiadis, I. and Van Ittersum, M.K. (2009) A database for integrated assessment of European agricultural systems. *Environmental Science and Policy* 12, 573–587.

Jones, J.W., Hoogenboom, G., Porter, C.H., Boote, K.J., Batchelor, W.D., Hunt, L.A. *et al.* (2003) The DSSAT cropping system model. *European Journal of Agronomy* 18, 235–265.

Keating, B.A., Carberry, P.S., Hammer, G.L., Probert, M.E., Robertson, M.J., Holzworth, D. *et al.* (2003) An overview of APSIM, a model designed for farming systems simulation. *European Journal of Agronomy* 18, 267–288.

Li, A.-G., Trent, A., Wall, G.W., Kimball, B.A., Hou, Y.-S., Pinter Jr, P.J.R. *et al.* (1997) Free-air CO_2 enrichment effects on rate and duration of apical development of spring wheat. *Crop Science* 37, 789–796.

Li, A.-G., Hou, Y.-S., Wall, G.W., Trent, A., Kimball, B.A. and Pinter Jr, P.J. (2000) Free-air CO_2 enrichment and drought stress effects on grain filling rate and duration in spring wheat. *Crop Science* 40, 1263–1270.

Li, F.Y., Johnstone, P.R., Pearson, A., Fletcher, A., Jamieson, P.D., Brown, H.E. *et al.* (2009) AmaizeN: A decision support system for optimizing nitrogen management of maize. *NJAS – Wageningen Journal of Life Sciences* 57, 93–100.

Martre, P., Jamieson, P.D., Semenov, M.A., Porter, R.F., Zyskowski, J.R. and Triboi, E. (2006) Modelling protein content and composition in relation to crop nitrogen dynamics for wheat. *European Journal of Agronomy* 25, 138–154.

McCown, R.L., Carberry, P.S., Hochman, Z., Dalgliesh, N.P. and Foale, M.A. (2009) Re-inventing model-based decision support with Australian dryland farmers. 1. Changing intervention concepts during 17 years of action research. *Crop and Pasture Science* 60, 1017–1030.

McMaster, G.S. (1997) Phenology, development, and growth of the wheat (*Triticum aestivum* L.) shoot apex: a review. *Advances in Agronomy* 59, 63–118.

McMaster, G.S. (2005) Centenary review: phytomers, phyllochrons, phenology and temperate cereal development. *Journal of Agricultural Science (Cambridge)* 143, 137–150.

McMaster, G.S. (2009) Development of the wheat plant. In: Carver, B.F. (ed.) *Wheat: Science and Trade*. Wiley-Blackwell Publishing, Ames, Iowa, pp. 31–55.

McMaster, G.S. and Hargreaves, J.N.G. (2009) CANON in D(esign): Composing scales of plant canopies from phytomers to whole-plants using the composite design pattern. *NJAS – Wageningen Journal of Life Sciences* 57, 39–51.

McMaster, G.S. and Wilhelm, W.W. (2003) Phenological responses of wheat and barley to water and temperature: improving simulation models. *Journal of Agricultural Sciences (Cambridge)* 141, 129–147.

McMaster, G.S., Klepper, B., Rickman, R.W., Wilhelm, W.W. and Willis, W.O. (1991) Simulation of shoot vegetative development and growth of unstressed winter wheat. *Ecological Modelling* 53, 189–204.

McMaster, G.S., Morgan, J.A. and Wilhelm, W.W. (1992a) Simulating winter wheat spike development and growth. *Agricultural and Forest Meteorology* 60, 193–220.

McMaster, G.S., Wilhelm, W.W. and Morgan, J.A. (1992b) Simulating winter wheat shoot apex phenology. *Journal of Agricultural Science (Cambridge)* 119, 1–12.

McMaster, G.S., Ascough II, J.A., Dunn, G.A., Weltz, M.A., Shaffer, M., Palic, D. *et al.* (2002) Application and testing of GPFARM: A farm and ranch decision support system for evaluating economic and environmental sustainability of agricultural enterprises. *Acta Horticulturae* 593, 171–177.

McMaster, G.S., Ascough II, J.C., Shaffer, M.J., Deer-Ascough, L.A., Byrne, P.F., Nielson, D.C. *et al.* (2003) GPFARM plant model parameters: complications of varieties and the genotype x environment interaction in wheat. *Transactions of the American Society of Agricultural Engineers* 46, 1337–1346.

McMaster, G.S., White, J.W., Hunt, L.A., Jamieson, P.D., Dhillon, S.S. and Ortiz-Monasterio, J.J. (2008) Simulating the influence of vernalization, photoperiod and optimum temperature on wheat developmental rates. *Annals of Botany* 102, 561–569.

McMaster, G.S., White, J.W., Weiss, A., Baenziger, P.S., Wilhelm, W.W., Porter, J.R. *et al.* (2009) Simulating crop phenological responses to water deficits. In: Ahuja, L.R., Anapalli, S.A., Reddy, V.R. and Yu, Q. (eds) *Modeling the Response of Crops to Limited Water: Recent Advances in Understanding and Modeling Water Stress Effects on Plant Growth Processes*, Vol. 1, Advances in Agricultural Systems Modelling. ASA-SSSA-CSSA, Madison, Wisconsin, pp. 277–300.

Monteith, J.L. (1977) Climate and crop efficiency of crop production in Britain. *Philosophical Transactions of the Research Society London, Series B.* 281, 277–329.

Monteith, J.L. (1996) The quest for balance in crop modelling. *Agronomy Journal* 88, 695–697.

Norman, J.M. (1979) Modelling the complete crop canopy. In: Barfield, B.J. and Gerber, J.F. (eds) *Modification of the Aerial Environment of plants*. American Society of Agricultural Engineers, St. Joseph, Michigan, pp. 249–277.

Parker, P., Letcher, R., Jakeman, A., Beck, M.B., Harris, G., Argent, R.M. *et al.* (2002) Progress in integrated assessment and modelling. *Environmental Modelling and Software* 17, 209–217.

Passioura, J.B. (1996) Simulation models: science, snake oil, education, or engineering? *Agronomy Journal* 88, 690–694.

Porter, J.R. (1984) A model of canopy development in winter wheat. *Journal of Agricultural Science (Cambridge)* 102, 383–392.

Porter, J.R. (1993) AFRCWHEAT2: A model of the growth and development of wheat incorporating responses to water and nitrogen. *European Journal of Agronomy* 2, 69–82.

Porter, J.R., Jamieson, P.D. and Grace, P.R. (2007) Wheat production systems and global climate change. In: Canadell, J.G., Pataki, D. and Pitelka, L. (eds) *Terrestrial Ecosystems in a Changing World. The IGBP Series*. Springer-Verlag, Berlin and Heidelberg, pp. 195–209.

Prusinkiewicz, P. (1998) Modelling of spatial structure and development of plants: A review. *Scientia Horticulturae* 74, 113–149.

Richardson, C.W. and Wright, D.A. (1984) *WGEN: A Model for Generating Daily Weather Variables.* US Department of Agriculture, Agricultural Research Service, ARS-8, Washington, DC.

Rickman, R.W. and Klepper, B. (1995) The phyllochron: Where do we go in the future? *Crop Science* 35, 44–49.

Rickman, R.W., Waldman, S.E. and Klepper, B. (1996) MODWht3: A development-driven wheat growth simulation. *Agronomy Journal* 88, 176–185.

Ritchie, J.T. (1991) Wheat phasic development. In: Hanks, J. and Ritchie, J.T. (eds) *Modeling Plant and Soil Systems*. ASA-CSSA-SSSA, Madison, Wisconsin, pp. 31–54.

Ritchie, J.T. and Otter, S. (1985) Description and performance of CERES-Wheat: A user oriented wheat yield model. In: Willis, W.O. (ed.) *Wheat Yield Improvement*. USDA-ARS Publ. 38, National Technical Information Service, Springfield, Virginia, pp. 159–175.

Rivington, M., Matthews, K.B., Bellocchi, G., Buchan, K., Stockle, C.O. and Donatelli, M. (2007) An integrated assessment approach to conduct analyses of climate change impacts on whole-farm systems. *Environmental Modelling and Software* 22, 202–210.

Savabi, M.R. and Stockle, C.O. (2001) Modelling the possible impact of increased CO_2 and temperature on soil water balance, crop yield and soil erosion. *Environmental Modelling and Software* 16, 631–640.

Semenov, M.A., Martre, P. and Jamieson, P.D. (2009) Quantifying effects of simple wheat traits on yield in water-limited environments using a modelling approach. *Agricultural and Forest Meteorology.* 149, 1095–1104.

Sequeira, R.A., Sharpe, P.J.H., Stone, N.D., El-Zik, K.M. and Makela, M.E. (1991) Object-oriented simulation: Plant growth and discrete organ to organ interactions. *Ecological Modelling* 58, 55–89.

Sequeira, R.A., Olson, R.L. and McKinion, J.M. (1997) Implementing generic, object-oriented models in biology. *Ecological Modelling* 94, 17–31.

Shaffer, M.J., Bartling, P.N.S. and McMaster, G.S. (2004) GPFARM modelling of corn yield and residual soil nitrate-N. *Computers and Electronics in Agriculture* 43, 87–107.

Sinclair, T.R. and Seligman, N.G. (1996) Crop modelling: from infancy to maturity. *Agronomy Journal* 88, 698–704.

Steduto, P., Hsiao, T.C., Raes, D. and Fereres, E. (2009) Aquacrop – the FAO crop model to simulate yield response to water: I. concepts and underlying principles. *Agronomy Journal* 101, 426–437.

Steiner, J.L., Sadler, E.J., Wilson, G., Hatfield, J.L., James, D., Vandenberg, B. *et al.* (2009) STEWARDS Watershed Data System: system design and implementation. *Transactions of the American Society of Agricultural and Biological Engineers* 52, 1523–1533.

Stockle, C.O. (1992) A method for estimating the direct and climatic effects of rising atmospheric carbon dioxide on growth and yield of crops: Part I – Modification of the EPIC model for climate change analysis. *Agricultural Systems* 38, 225–238.

Stockle, C.O. and Debaeke, P. (1997) Modelling crop nitrogen requirements: a critical analysis. *European Journal of Agronomy* 7, 161–169.

Stockle, C.O., Donatelli, M. and Nelson, R. (2003) CropSyst, a cropping systems simulation model. *European Journal of Agronomy* 18, 289–307.

Tanner, C.B. and Sinclair, T.R. (1983) Efficient water use in crop production: research or re-search? In: Taylor, H.M., Jordan, W.R. and Sinclair, T.R. (eds) *Limitations to Efficient Water Use in Crop Production*. American Society of Agronomy, Madison, Wisconsin, pp. 1–27.

Tao, F., Yokozawa, M. and Zhang, Z. (2009a) Modelling the impacts of weather and climate variability on crop productivity over a large area: A new process-based model development, optimization, and uncertainties analysis. *Agricultural and Forest Meteorology* 149, 831–850.

Tao, F., Zhang, Z., Liu, J. and Yokozawa, M. (2009b) Modelling the impacts of weather and climate variability on crop productivity over a large area: A new super-ensemble-based probabilistic projection. *Agricultural and Forest Meteorology* 149, 1266–1278.

Tubiello, F.N., Rosenzweig, C., Kimball, B.A., Pinter Jr, P.J., Wall, G.W. Hunsaker, D.J. *et al.* (1999) Testing CERES-Wheat with free-air carbon dioxide enrichment (FACE) experiment data:

CO_2 and water interactions. *Agronomy Journal* 91, 247–255.

Van Ittersum, M.K., Ewert, F., Heckelei, T., Wery, J., Alkan Olsson, J., Andersen, E. *et al.* (2008) Integrated assessment of agricultural systems – a component based framework for the European Union (SEAMLESS). *Agricultural Systems* 96, 150–165.

van Keulen, H. and Seligman, N.G. (1987) *Simulation of Water Use, Nitrogen Nutrition and Growth of a Spring Wheat Crop*. Simulation Monographs, Pudoc, Wageningen, Netherlands.

van Laar, H.H., Goudriaan, J. and van Keulen, H. (1992) *Simulation of Crop Growth for Potential and Water Limited Production Situations (as Applied to Spring Wheat)*. Simulation Reports CABO-TT, 27. CABO-DLO/TPE-WAU, Wageningen, Netherlands.

Vos, J., Marcelis, L.F.M., de Visser, P.H.B., Struik, P.C. and Evers, J.B. (2007) *Functional-Structural Plant Modelling in Crop Production*. Springer Publishing, Dordrecht, Netherlands.

Wall, G.W., Garcia, R.L., Kimball, B.A., Hunsaker, D.J., Pinter Jr, P.J., Long, S.P. *et al.* (2006) Interactive effects of elevated carbon dioxide and drought on wheat. *Agronomy Journal* 98, 345–381.

Weir, A.H., Bragg, P.L., Porter, J.R. and Rayner, J.H. (1984) A winter wheat crop simulation model without water or nutrient limitations. *Journal of Agricultural Science (Cambridge)* 102, 371–382.

Weiss, A. and Moreno-Sotomayer, A. (2006) Modelling protein content and composition in relation to crop nitrogen dynamics for wheat. *European Journal of Agronomy* 25, 129–137.

Weiss, A., Hays, C.J. and Won, J. (2003) Assessing winter wheat responses to climate change scenarios: a simulation study in the U.S. Great Plains. *Climate Change* 58, 119–147.

Weiss, A., Baenziger, P.S., McMaster, G.S., Wilhelm, W.W. and Al Ajlouni, Z.I. (2009) Quantifying phenotypic plasticity using genetic information for simulating plant height in winter wheat. *NJAS – Wageningen Journal of Life Sciences* 57, 59–64.

Welch, S.M., Roe, J.L. and Dong, Z. (2003) A genetic neural network model of flowering time control in *Arabidopsis thaliana*. *Agronomy Journal* 95, 71–81.

Weyant, J., Davidson, H., Dowlatabadi, H., Edmonds, J., Grubb, M., Richels, R. *et al.* (1996) Integrated Assessment of climate change: an overview and comparison of approaches and results. In: Bruce, J.P., Lee, H. and Haites, E.F. (eds) *Climate Change 1995 – Economic and*

Social Dimensions. Cambridge University Press, Cambridge, UK, pp. 367–396.

White, J.W. (2006) From genome to wheat: Emerging opportunities for modelling wheat growth and development. *European Journal of Agronomy* 25, 79–88.

White, J.W. and Hoogenboom, G. (2003) Gene-based approaches to crop simulation: Past experiences and future opportunities. *Agronomy Journal* 95, 52–64.

White, J.W., McMaster, G.S. and Edmeades, G.O. (2004a) Physiology, genomics and crop response to global change. *Field Crops Research* 90, 1–3.

White, J.W., McMaster, G.S. and Edmeades, G.O. (2004b) Genomics, physiology, and global change: What have we learned? *Field Crops Research* 90, 165–169.

White, J.W., Herndl, M., Hunt, L.A., Payne, T.S. and Hoogenboom, G. (2008) Simulation-based analysis of effects of *Vrn* and *Ppd* loci on flowering in wheat. *Crop Science* 48, 678–687.

Wilhelm, W.W. and McMaster, G.S. (1995) The importance of the phyllochron in studying the development of grasses. *Crop Science* 35, 1–3.

Wilhelm, W.W., McMaster, G.S., Rickman, R.W. and Klepper, B. (1993) Above-ground vegetative development and growth as influenced by nitrogen and water availability. *Ecological Modelling* 68, 183–203.

Williams, J.R., Jones, C.A., Kiniry, J.R. and Spanel, D.A. (1989) The EPIC crop growth model. *Transactions of the American Society of Agricultural Engineers* 32, 497–511.

Yin, X. and Struik, P.C. (2009) C_3 and C_4 photosynthesis models: An overview from the perspective of crop modelling. *NJAS – Wageningen Journal of Life Sciences* 57, 27–38.

Yu, B. (2000) Improvement and evaluation of CLIGEN for storm generation. *Transactions of the American Society of Agricultural Engineers* 43, 301–307.

Zalud, Z., McMaster, G.S. and Wilhelm, W.W. (2003) Evaluating SHOOTGRO 4.0 as a potential winter wheat management tool in the Czech Republic. *European Journal of Agronomy* 19, 495–507.

10

The Way Ahead: From Science to Policy; Coordinating Efforts in a Global World

R. Ortiz

10.1 Green Revolution and Ecological Intensification

For more than half a century I have worked with the production of more and better wheat for feeding the hungry people, but wheat is merely a catalyst, a part of the picture. I am interested in the total development of human beings. Only by attacking the whole problem can we raise the standard of living for all people in all communities, so they will be able to live decent lives. This is something we want for all people on this planet.

Norman E. Borlaug

Crop yields in the developing world would have been 19.5–23.5% lower without the Green Revolution (Evenson and Gollin, 2003). Dramatic increases in crop yields have made food cheaper and more affordable since the 1950s. For example, equilibrium prices for all crops combined would have been from 18 to 21% higher than they were in 2000 without the research and innovations that were brought by the Green Revolution that was spearheaded by Norman E. Borlaug and co-workers in what later was known as the Consultative Group on International Agricultural Research (CGIAR). Without the bred-germplasm of main staples ensuing from the CGIAR, there would have been a drop of 4.5–5.0% in calorie consumption and 2.0–2.2% more malnourished children in the developing world.

As pointed out by Borlaug (2007), high-yield cultivars – mainly of rice and wheat, were the catalysts for the Green Revolution, but they needed fertilizers, timely weed control and optimum irrigation schedules to succeed and achieve impacts on yield potential. Indeed, the spectacular yield gains were also due to optimization of N-fertilizer used by semi-dwarf cultivars of rice and wheat (Trethowan et al., 2007). Dwarf genes (Rht) interfere with the action or production of gibberellin (GA); these genes encode growth repressors that are normally suppressed by GA (Hedden, 2003). About 1.1 billion ha of land was spared due to cereal crop yield gains in the remaining 660 million ha used to feed the world at the end of the 20th century (Borlaug, 2007). Selected maize, rice and spring wheat breeding research returns for the Green Revolution are given in Table 10.1.

Intensifying crop management, together with bred-seed and efficient use of inputs such as fertilizer and water, will be needed to provide enough food for the growing population of this 21st century using the same amount of land. Cassman (1999) indicated that further and sustainable gains at high crop yield levels need improvements in soil quality and precision management of all production factors in time and space. He concluded that global food security will depend on gaining more insights into the physiological basis of crop yield potential, analysing the processes involved in the relationship between soil quality and crop productivity and understanding the interacting environmental factors that affect crop yields. Ecological intensification will therefore rely on breakthroughs to be brought about by plant physiology, eco-physiology, agroecology and soil science.

In some rainfed environments, e.g. the savannahs of sub-Saharan Africa, soil

Table 10.1. Selected research returns of the Green Revolution: economic benefits of plant breeding from CGIAR and partners' research.

Crop	Annual benefits (US$ million)	Annual costs (US$ million)
Spring wheat	2,500	70
Rice (S-E Asia)	10,800	28
Maize (partial)	557–770	7–18

Source: CGIAR at www.cgiar.org

management combined with crop improvement and plant health research at farmers' level are addressing the intensification of cereal–grain–legume-based cropping systems (Sanginga et al., 2003). Similarly, in the better-endowed, irrigated, rice–wheat-cropping systems of the Indo-Ganges in South Asia, conservation agriculture practices led to diversification of such intensive agroecosystems (Jat et al., 2006), which are central to food security and reduction of poverty in Asia.

10.2 IAASTD

The International Assessment of Agricultural Knowledge, Science and Technology for Development (IAASTD) is a recent and unique global undertaking created to appraise the relevance, quality and effectiveness of agricultural knowledge, science and technology (AKST), and the effectiveness of public and private sector policies as well as institutional arrangements in relation to AKST (http://www.agassessment.org/). IAASTD was initiated by the World Bank in open partnership with a multi-stakeholder group of organizations, including UN agencies (FAO, GEF, UNDP, UNEP, WHO and UNESCO) and representatives of governments, civil society, private sector and scientific institutions from around the world. They used a strongly consultative bottom-up process that recognized the diverse needs of different regions and communities.

The Executive Summary of IAASTD indicates that AKST systems are needed to enhance sustainability while maintaining productivity in ways that protect the natural resource base and ecological provisioning of agricultural systems. Among the options to pursue are enhancing nutrient, energy, water and land use efficiency; improving the understanding of soil–plant–water dynamics; increasing farm diversification; supporting agroecological systems; and enhancing biodiversity conservation and use at both field and landscape scales. Likewise, IAASTD advocates promotion of the sustainable management of livestock, forests and fisheries; improving understanding of the agroecological functioning of mosaics of crop production areas and natural habitats; countering the effects of agriculture on climate change; and mitigating the negative impacts of climate change on agriculture.

In the view of some of its authors (Kiers et al., 2008), IAASTD claims that success is not based on technological performance in isolation, but rather on how technology builds knowledge, networks and capacity – i.e. innovation demands sophisticated integration with local partners. Moreover, the authors indicate that further advancements in science and technology need investments in rural infrastructure (physical, market and finance) and local governance for agriculture to succeed in the developing world, because those lagging behind in such investments cannot compete in domestic or international markets. Hence, investments that improve farmers' access to land and water resources are vital. Likewise, basic education investments, as they point out, are needed, e.g. farmers in the developing world who completed 4 years of elementary education had, on average, 8.7% higher productivity. They conclude by stating that most successful investments should 'increase the resilience of local and global food systems to environmental and economic shocks'.

10.3 GFAR

The Global Forum for Agricultural Research was founded in 1996 as a diverse community of regional and world organizations dedicated to harnessing agricultural research for sustainable development, a better environment and the alleviation of poverty (GFAR, 2004). This forum, better known by its English acronym GFAR, aims to identify research priorities and opportunities for the various stakeholder groups participating in agricultural research-for-development while regional or sub-regional groups do so at the regional or sub-regional level. GFAR draws on the stakeholder's complementary skills and strengths, and encourages inclusiveness while forging alliances and partnerships in the whole process: from setting the research agenda, throughout the implementing phase and measuring impacts, which should lead to a true ownership of the various actors engaged in finding research answers to the challenges being faced in attaining sustainable agriculture. GFAR offers the opportunity of creating an independent, unbiased agenda-defining and priority-setting process for the international agricultural research community.

Ortiz and Crouch (2007) suggested that an optimum model would dictate that this global agenda would feed into and align with the agendas of regional and sub-regional organizations and, in turn, into national, thematic and commodity-based groups. However, the extent to which individual institutions align themselves with this agenda depends increasingly upon the degree of reinforcement from major donors. Moreover, the relative success of this approach is also confounded by the impact of activities by major NGOs (non-governmental organizations) and the private commercial sector. Thus, one of the greatest challenges in this area is not in defining the agenda per se, but alignment with (as opposed to erosion by) the strategies of other major players in the same target domain. It is within this context that a systematic value chain approach becomes highly valuable, in order to define the major elements and linkages impacting the product development, delivery and impact following uptake of research outputs. In the past, it has been a great challenge to coordinate or converge the agendas of the many very diverse development investors, who each have minor positive or negative influence on these agendas.

GFAR is not itself an implementing agency and its operations rely on joint undertakings with its stakeholders, prominently the regional and sub-regional forums (Table 10.2). In addition to the

Table 10.2. Global, regional and sub-regional forums in agricultural research for development.

Acronym	Organization (website)
Global	
GFAR	Global Forum for Agricultural Research (http://www.egfar.org/home.shtml)
Regional	
AARINENA	Association of Agricultural Research Institutions in the Near East and North Africa (www.aarinena.org)
APAARI	Asia Pacific Association of Agricultural Research Institutions (http://www.apaari.org/)
EFARD	European Forum on Agricultural Research for Development (www.efard.org/)
FARA	Forum for Agricultural Research in Africa (www.fara-africa.org)
FORAGRO	Foro de las Américas para la Investigación y Desarrollo Tecnológico Agropecuario (http://www.iica.int/foragro/)
Sub-regional	
ASARECA	Association for Strengthening Agricultural Research in Eastern and Central Africa (http://www.asareca.org/)
CORAF/WECARD	West and Central African Council for Agricultural Research and Development (http://www.coraf.org/English/en.php)
SADC-FANR	Southern African Development Community – Food, Agriculture and Natural Resources Directorate (http://www.sadc.int/english/fanr/fanr_about/index.php)

setting of the research agenda at the global or regional level, national agricultural research systems (NARS), which should include or at least consult with commodity groups, farmer cooperatives and commercialization representatives, should also be active while setting national or local research agendas and focusing their roles according to respective comparative advantage.

In its last triennial conference, GFAR was asked to advocate for the need to change systems, institutions and technology-generation processes to become more pro-poor and biased towards satisfying the development needs of small landholders and the rural poor (GFAR, 2006). GFAR should also strengthen all stakeholders so they can contribute to agricultural research and innovation through an inclusive process, with the aim of alleviating poverty and eradicating extreme hunger. GFAR must therefore facilitate partnerships contributing to agricultural research and innovation that lead to elimination of hunger and reduction of poverty, as well as to mobilization and enabling of sharing and exchange of knowledge, skills and resources contributing to agricultural research and innovation at all levels: globally, regionally, nationally and locally.

10.4 CGIAR

The original CGIAR centres were conceived in the early 1960s (Ortiz et al., 2007a). The International Rice Research Institute (IRRI) was established in Los Baños (Philippines) in 1960 by the Ford and Rockefeller Foundations, in cooperation with the Government of the Philippines. IRRI remains the oldest and largest international agricultural research institute in Asia. The Office for Special Studies in Mexico – set up by the Government and the Rockefeller Foundation in the 1940s, evolved into the Centro Internacional de Mejoramiento de Maíz y Trigo (CIMMYT, Texcoco, Mexico), and later in the 1970s its potato programme was included in the newly launched Centro Internacional de la Papa (CIP, Lima, Peru). In 1967, two multi-crop centres, with like

names in English and Spanish but with clearly distinct agroecosystem geo-domains, were added. These centres were the Centro Internacional de Agricultura Tropical (CIAT) in Cali (Colombia) and the International Institute of Tropical Agriculture (IITA) in Ibadan (Nigeria). IRRI, CIMMYT, CIAT, IITA and CIP were the first international centres of the CGIAR network that today includes another ten institutes worldwide, addressing a broad range of issues in agriculture: fishery, forestry, livestock, water and other natural resources, plant genetic resources, research capacity building and food policy.

Due to the need for CGIAR to respond effectively to today's global agricultural challenges, the members (UN sponsors, governments, regional development banks and foundations) agreed to launch a Change Management process at the end of 2007 (http://www.cgiar.org/changemanagement/index.html). The most fundamental reason for this needed change in CGIAR was to strengthen its delivery of science in support of sustainable development. In this regard, the envisioning of CGIAR includes the following strategic objectives.

- Food for people: mobilize science and technology to accelerate sustainable increases in productivity and the production of healthy food, by and for the poor.
- Environment for people: mobilize science and technology to conserve, enhance and sustainably use natural resources and biodiversity to improve the livelihoods of the poor, and respond to climate change.
- Innovation for people: mobilize science and technology to stimulate institutional innovation and enabling policies for pro-poor agricultural growth and gender equity.

In the first strategic objective, creating and accelerating sustainable productivity and production increases of healthy food by and for the poor was given a high priority. Genetic improvement (including the use of molecular techniques and ensuing tools) to push the yield frontier and enhance its stability (especially in stressful environments) and sustainable intensification (through on-farm management, policy and institutional change) were indicated as key opportunities for increasing productivity.

Prior to the launching of the Change Management process, the Alliance of CGIAR Centers and the Generation Challenge Programme of CGIAR undertook a consultative process to address research on tolerance to selected abiotic stresses (Crouch and Ortiz, 2007). The participants indicated that the top priority for the CGIAR should be research and germplasm enhancement activities on improving abiotic stress tolerances for rainfed, drought-prone cropping systems and production systems with supplementary irrigation or prevalence of extreme high temperatures. In addition, it was concluded that a focus on soil nutrient deficiencies in Africa was also a very high priority. These are the highly complex, long-term recalcitrant problems that are key rate-limiting factors in large areas of production, with no sustainable alternative solutions other than genetic enhancement, and of greatest importance for resource-poor farmers. It was considered that where there were realistic options for addressing other abiotic stresses through improved agronomic practices then the CGIAR should only pursue genetic improvement activities that were completely embedded in natural resource management projects – for example, salinity tolerance in Asia. In addition, there may be justification for prioritizing tolerance to flooding in Asia and acidity tolerance in Latin America.

This consultative process also highlighted the urgent need for an increased emphasis on capacity building in phenotyping for physiological and genetic research, and for screening of priority abiotic stresses at the morphological, physiological and biochemical levels. In the participants' view, accurate plant phenotyping across a broad set of environments is the major bottleneck for mapping and breeding under abiotic stress conditions as field evaluation is confounded by environmental variation and adaptation of genotypes. Designing more effective physiological evaluation systems is a major constraint to better utilization of

genetic resources, development of marker-assisted selection (MAS) systems and the implementation of mainstream breeding for improved levels of tolerance. Precise site-characterization equipment is also critically important at these screening locations in order to quantify the spatial and temporal variation in a multitude of environmental factors that otherwise contribute to the error term of abiotic stress trials. The participants concluded that, although it may be convenient for conventional breeding programmes to simply select for yield under stress, such an approach does not appear to be appropriate for screening genetic resources, developing genomics tools or preliminary germplasm-enhancement undertakings. In these cases, the need remains to identify secondary traits that are closely correlated with the yield architecture under stress and that are related to underlying mechanisms and genetic controls. Likewise, their assessment indicated that global germplasm collections are under-characterized and in turn underutilized by plant breeders working on abiotic stress tolerance. They indicated that both areas would greatly benefit from a critical mass of researchers in CGIAR, strong national programmes and advanced laboratories working in interdisciplinary teams with physiologists and breeders to develop, validate, refine and implement new technologies for enhanced impact at both the phenotypic and the genotypic levels. They point out that these activities would be best organized within a global platform for accessing and evaluating genetic diversity of staple crops, which can then have spillovers into other crops and traits.

10.5 Global Challenges and Agriculture

No matter how excellent the research done in one scientific discipline is, its application in isolation will have little positive effect on crop production. What is needed are venturesome scientists who can work across disciplines to produce appropriate technologies and who have the courage to make their case with political leaders to bring these advances to fruition.

Norman E. Borlaug

The Millennium Development Goals (MDG) are the commonly accepted framework for measuring development progress in international research and development organizations. Any research and capacity-building agenda in agriculture must address them (Ortiz and Smale, 2007). For example, research that enhances sustainable crop yields may lead to improvement in rural and urban food security and raising of farmers' incomes, thereby contributing to *eradicating extreme poverty and hunger*. Moreover, an agenda targeting women and disadvantaged groups through community-based pro-duction systems by farmer associations and self-help groups, or using farmers' field schools for knowledge sharing, will *empower rural people*, thus promoting equity and *gender equality*. Furthermore, the diversifi-cation of the food basket through research and promotion of locally available crops with enhanced micronutrient content and protein quality will provide a means for healthy food options that may *reduce child mortality* and general malnutrition, which will allow better-fed children to achieve primary education. Likewise, active research on both conservation agriculture practices and preservation of genetic resources will foster *environmental sustainability*. Last but not least, any research and graduate training organizations should engage in *global partnerships for development* to acquire new methods and knowledge, leverage resources and ensure effective and efficient delivery of both technology and the associated 'know-how'.

Delivering public goods and building capacity to benefit resource-poor small to medium-size farmers, rural communities and local organizations – as well as commercial farmers and entrepreneurs – should be the aim of an international agenda for agriculture. However, the success in accomplishing that aim requires a fast and sustainable growth in the resource-poor areas of the world. Not surprisingly, most poverty alleviation approaches acknowledge the need to grow, and many of them should be improved by a better understanding of the sources of growth and of how this growth leads to reducing poverty. In this regard, 'international aid' organizations use the phrase 'pro-poor growth' to refer to the growth that benefits the poor in a location, country or region, where the incomes of the poor will rise faster than the overall average, thereby falling, inequality. Hence, an international agenda for agriculture should pursue a 'pro-poor growth'-driven mission that promotes both competitive (profitable) agriculture and sustainable management of natural resources while contributing towards social equity and betterment of life in target areas, where the rural sector remains a key contributor to economic and social development.

'Pro-poor growth' should be judged by how fast on average the incomes of the poor are rising, i.e. the speed at which absolute poverty reduces (DfID, 2004). A compre-hensive measure of 'pro-poor growth' considers therefore the non-income dimensions of poverty, such as malnutrition, access to education and child mortality (which are also at the core of MDG, as already indicated above). In this regard, giving the poor better access to assets and markets will lead to growth because it allows more of a country's resources – her people – to become highly productive. Hence, 'pro-poor growth' should enable poor people to participate in, as well as benefit from, the growth process. Why? High equality speeds growth – both by enhancing the ability of the poor to contribute to production and by reducing social and political tensions, which encourage investment. Without doubt, lowering the high levels of inequality reduces poverty and improves other aspects of social well-being.

10.6 Intensifying and Diversifying Agriculture through Eco-friendly Undertakings

Agriculture provides a means for ensuring food, earning incomes and improving livelihoods. None the less, farming needs to be in harmony with today's fragile environment, especially at a time of global climate change. Bio-energy, pro-active biodiversity conservation, transgenic technology, organic agriculture, food safety standards, carbon trade, corporate social responsibility, adding value through agribusinesses, global markets, land degradation and somewhat scanty natural resources are among the main topics to be addressed by an organization involved in the food value chain. Any research and capacity-building undertaking cannot therefore ignore such issues, and should direct its efforts towards adaptation to, and mitigation of, global warming that will significantly affect agriculture, especially in poor regions of the developing world such as sub-Saharan Africa, South Asia, Meso-America and the South American Andes. Likewise, research organizations need to provide technology and knowledge on healthy and nutritious food that requires safety measurements to avoid contamination, e.g. by pollutants such as chemical pesticides or fertilizers, or other food contaminants, or standards for delivering a produce that possesses appropriate feeding value. Such issues can be addressed by a client-oriented research agenda that will always bear in mind how to benefit the resource-poor.

In recent years, wealthy consumers in global, regional and even local markets are asking for better and safer food that should ensue from the use of sustainable and socially responsible practices. Hence, agriculture and environment should be regarded as complementary in today's world, and may be the two sides of the same coin in some areas of the world. For example, agroecosystems in densely populated rural areas, where cropping systems are intensive and complex, are driven by increasing demand for food crops and livestock, and the need for saving water and land resources. Farmers in such intensive agroecosystems need to sustain local communities and neighbouring cities, and are becoming more market oriented. Intensive agricultural systems are therefore a source of food and income security for rural and urban households in some areas of the developing world, e.g. in China, Latin America, North Africa or South Asia. These intensive systems are usually highly productive, feature multiple crops and are central to reducing poverty.

Sustainable agricultural intensification at a time of population growth, income improvement and limited natural resource base (land, water) will help to meet future food, feed, fibre and fuel needs. To address such a challenge, research should focus on farming practices that foster synergies, conserve water, enhance nutrient uptake efficiency, increase economic stability and promote equitable outcomes for small-scale farmers. In such agroecosystems, household strategies to improve livelihoods rely on the following.

- intensification of existing farm production patterns through enhanced use of quality inputs;
- diversification of production systems with emphasis on greater market orientation and value addition, and often through a shift to high-value products;
- enhancing off-farm income to supplement farming and provide financing for additional input use;
- withdrawal from agriculture, including migration from rural areas; and
- a better use of climate variations through optimal combinations of seasonal and perennial crops and trees.

The expected ecological impacts from doubling food production using past production strategies may result in production systems and associated ecological services becoming unsustainable. To avoid further expansion into natural ecosystems, agricultural systems must be intensified on existing land and with the available water resources, using more sustainable methods, and by changing current production systems towards more diverse and productive systems. Improved germplasm with enhanced efficiency for using added inputs, reduced nutrient losses from fertilizers and manures, increased water productivity, strengthened ecological resilience and reduced global warming potential are the most environmentally benign options that should translate into a high-quality produce output in intensive and diverse farming systems. Scientists should therefore conduct research on the basic underlying principles and approaches for developing sustainable agricultural intensification that will be complementary to integrated gene-natural resource management, and that will increase the productivity of existing land and water resources in the production of food, feed, cash crops (including fodder and biofuel feedstock), livestock and trees. In this regard, diversification, which should be understood as a change in current farming enterprise patterns to increase profitability or reduce risks, appears as an important option for sustainable intensification. Scientists should therefore engage in problem-solving research that requires inputs from the different parties across the entire value chain that brings their perspectives, and maybe change their views during a participatory consultative process in which stakeholders (including scientists) engage in practices of joint inquiry, collaborative and active learning and adaptive management (Ortiz and Crouch, 2007). A client-oriented agenda for intensifying and diversifying sustainably agricultural systems ensures that these key agricultural areas remain productive and ecologically sound into the future.

The main goal will therefore be to reduce poverty and conserve natural resources in densely populated areas, where intensive cropping systems underpin the livelihoods of the poor, by diversifying cropping systems, fostering expanded employment for the rural landless, improving food security for rural consumers and conserving water and land resources. By studying existing landscape mosaics of crops and tree systems, assessing their ecological sustainability and economic implications, any research organization should be able to select best-bets that can be tested in specific social, cultural and environmental conditions (e.g. see Ortiz (2008a) for boosting of crop yields in sub-Saharan Africa). The impacts of diversification on the environment, and the risk of unforeseen 'second-generation' management problems emerging in the future, will

be central to this research, whose agenda must ensure farmer participation and research-for-development approaches across the value chain.

10.7 A Global Research and Capacity-building Agenda

Addressing other issues affecting global development and agriculture today will be important for succeeding in the medium to long term. The four examples given below illustrate the design and implementation of a research and capacity-building agenda on such emerging issues affecting agriculture and the environment: (i) biodiversity and climate change; (ii) bio-energy; (iii) high-value crops and products; and (iv) food safety.

10.7.1 Biodiversity and climate change

The recent advances in agro-biotechnology (e.g. through genomics) offer a way towards better understanding of biodiversity at the species and gene levels that could lead to a more sustainable conservation of genetic resources through an appropriate use of such genetic endowments for plants, animals and other living organisms (Dwivedi et al., 2007). In this regard, adapting existing agrobiodiversity to biotic and abiotic stresses brought by climate change will be the main challenge to sustaining agro-ecosystems in the near future (Ortiz, 2008b). Indeed, rising temperatures and changes in water availability will lead to more stressful agroecosystems, exposing animals and plants to limiting factors such as heat, moisture extremes and evolving pest and pathogenic threats, which will further increase the vulnerability of rural populations, especially the resource-poor, to food insecurity, poverty and health risks. An eco-friendly, pro-poor research agenda should aim for the following.

• Define stresses affecting animals and plants by using knowledge-based scenario analysis, which combines most recent climate change and available agroecosystem information.

- Identify vulnerable agroecosystems (hot spots) that will be affected by climate change-induced stresses.
- Assess morphological and physiological changes in adaptation mechanisms that animals and plants will need to address climate variability.
- Utilize genetic enhancement and agroecosystem management (including integrated pest management) to provide technology options for farmers in 'climate change hot spots' by improving productivity, sustainability and reducing human health threats.
- Test best-bets options in hot spots through participatory technology exchange that will improve livelihoods and resilience to climate change while maintaining the resource base.
- Find solutions that address agroecosystem adaptation and maintain productivity while also contributing to mitigation in specific cases, e.g. greenhouse gas emissions, carbon storage.
- Preserve traditional ecological knowledge and strengthen household cultural bonds within communities.

This research agenda will test hypotheses in regard to the most important threats that climate variation brings to agroecosystems, and how much impact increased susceptibility of food security and livelihoods will have. Furthermore, this research will need to address questions regarding the genetic potential towards mitigation of the impacts of climate change-induced stresses and how traditional ecological knowledge can bring cultural benefits to households and communities. Finally, this research should be capable of assessing how a combination of genetic improvement and agroecosystem management will influence multiple ecosystem services, including water quality, energy conservation and human health, e.g. impact of pesticides on farm workers, carbon sequestration and greenhouse gas emissions. As a result of succeeding in the above undertakings, populations within climate change hot spots

may be able to benefit from the use of models and decision-making systems that will allow identification of agroecosystems prone to global warming, and to adopt technology options generated to address and counteract the negative impacts of climate change. More reliable, diverse, productive agroecosystems that ensure food supply sustainably in a changing world, with a basket of income options through better agrobiodiversity management, will be the major indicators of success. In this regard, Ortiz *et al.* (2008a) suggest that conservation agriculture practices and genetically enhanced technology that provide better ecosystem services and improve human health will be among the research outputs.

10.7.2 Bio-energy: growing energy on farms to generate income and protect the environment

With the recent policy developments regarding the use of alternative, renewable energy resources rather than fossil fuels, particularly in the industrialized world, the agriculture of the developing world will need to address the full integration of this emerging area as well as its impacts on food security, poverty alleviation, sustainable management of crop and natural resources, and the environment. Such a new challenge provides a means for a cross-cutting bio-energy research, which generates broad-based knowledge, ensuing technology and tools for assessment (Winslow and Ortiz, 2010). This engineering of new systems requires a holistic approach aiming at more efficient use of biomass by partitioning it between energy, feed, food and CO_2 fixation. The goal of this research should be, therefore, to provide more efficient and pro-poor farming systems using existing agricultural and other lands to exploit biodiversity and the new demands for energy, which will create new income options. Scientists dealing with international agriculture may play one or more of the following roles in bio-energy: 'developers' of analytical tools for energy-cropping options; policy 'analysts' and 'advocates' for energy, livelihoods and food security; genetic resources 'providers' of bio-energy plants or crops; trait and crop-resource management 'research catalysers'; proprietary technology 'brokers' to ensure energy at the village level; knowledge-sharing 'facilitators' throughout the bio-energy value chain; and knowledge 'integrators' for complex food–feed–fibre–fuel–environmental services systems (Iwanaga and Ortiz, 2007). A multidisciplinary research that balances agriculture and environment will address key issues, such as the following.

- possible food–feed–fibre versus fuel trade-offs: looking at the conditions under which the demand for biofuels (especially from food crop sources) could increase prices and have negative impacts on food and feed supply and food security;
- environmentally 'costing' of biofuels: the input–output energy balance, i.e. the energy output should be higher than that used for producing biofuels;
- less water-demanding biofuels than current alternatives such as sugarcane;
- environmental services: eco-friendly biofuels may reduce carbon emissions; mitigating climate change;
- opportunity 'windows' and risks from biofuels: an opportunity for tropical America's farmers and a possible risk to poor producers and consumers from this boom;
- policy-driven versus user demand: factoring in the role of governments and their expectations of unstable and perhaps rising future oil prices, as well as other motivations of political economy; and
- what kind of innovative research-for-development partnerships that combine food with eco-friendly energy production will meet the demand for both while expanding agriculture to 'marginal' (or waste-) land and that will increase incomes and provide new labour options for the poor.

Research organizations will also need to appraise their role in speeding up the adoption and development of 'second-generation' ligno-cellulosic biofuel technologies that will relieve pressures on food crops that are used in conventional conversion technologies (Ortiz *et al.*, 2006). Likewise, the agenda for biofuel crops will include areas dealing with increasing plant biomass, optimizing the chemical and physical attributes of biofuel sources, and traits for first- and second-generation biofuels. Scientists may consider undertaking frontier research in genetic resources by investigating the advantages of perennial biofuel plants and trees that can generate more annualized net photosynthesis; lower input costs (e.g. costs of tillage are eliminated after establishment); and longer life that leads to beneficial symbiotic interactions facilitating nutrient input and lower fertilizer runoff and where nutrients and organic matter can remain in the soil postharvest. Through alliances with the bio-energy sector, research organizations may also work both on eco-friendly industrial processes to adapt them to biomass sources, and on biomass to adapt it to promising, eco-friendly industrial processes.

10.7.3 High-value crops and products

The growth in high-value agriculture worldwide is partly driven by rising incomes, urbanization and perhaps changing preferences (World Bank, 2007). As income rises, the share of the food budget allocated to starchy staples declines relative to more expensive food items. High-value agricultural products (HVAP) with a high price per kilogram, per hectare or per calorie include fruits, vegetables, meat, eggs, milk, fish and non-timber forest products. With the knowledge available elsewhere on genetic resource enhancement and husbandry, as well as with overall assessments of cropping systems and value chains, the agenda for HVAP should be led by undertakings whose impacts will benefit the smallholders and poor consumers as well as their environments. Perhaps, the more interesting research hypothesis to pursue in this agenda should be in regard to income generation through organization of participatory value chains, promoting conservation through the use of the genetic resources endowment of the speciality traits for each HVAP and eco-friendly husbandry (Ortiz *et al.*, 2007b). The allocation of resources for research in HVAP may require consideration of the area under the target type(s) and their expected livelihoods impacts, as well as the potential changes in income and stability for these farmers due to the expected research outputs, and other benefits such as income gains for farm workers or gains in nutrition for poor consumers – especially if the lowering of prices enhances consumption of HAVP. Lastly, for HVAP to contribute to poverty reduction, the performance of value chains needs to be improved. An organizational and institutional analysis of the governance and coordination of these chains could provide policy and other solutions to improve benefits to farmers, without penalizing other actors. An analysis of HVAP chains may reveal where inefficiencies exist and, by bringing different stakeholders together, these value chains can be made to work more effectively and efficiently through participatory approaches, e.g. learning alliances aiming at linking small farmers successfully with markets. In some areas of tropical America, HVAP grown or bred sustainably, harvested as per international norms and meeting international food safety and trade market standards will be needed. However, in other areas it will be more important to learn how to produce and commercialize, in an eco-friendly way, some available HVAP rather than to commit research on what HVAP need to produce.

10.7.4 Food safety

Food is a necessary part of our daily lives, and the quality and safety of this food is a concern for many people, including international agencies such as the United Nations Food and Agriculture Organization (FAO) and the World Health Organization (WHO), who created the Codex Alimentarius as a food standard guideline. Many millions of people (both adults and children) suffer today from food-borne toxins and other

contaminants (arsenic, cadmium, pesticides). Staple crops or livestock products can also be the source of toxins, which are highly toxic metabolites produced at all stages of crop production: preharvest, harvest and storage, e.g. mycotoxins from fungi. Human exposure to levels of toxins and other food contaminants, from nanograms to micrograms per day, may occur through consumption of dietary staples in several tropical countries – e.g. see Williams *et al.* (2004) for an overview on aflatoxins. Because they are hazardous to health, toxins and other food contaminants are regulated through international markets and are considered non-tariff trade barriers. In the developed world, regulatory standards prevent exposure of humans and animals to dietary toxins and other food contaminants. These safety regulations reduce the risks of morbidity and mortality that are associated with the consumption of contaminated food.

In the developing world, however, monitoring and enforcement of standards are rare. Food products often fail to penetrate major markets due to the high quality standards set by importing countries. Costs to developing-world farmers include reduced income from outright food or feed losses and lower selling prices for contaminated commodities. The economic impact on livestock production includes mortality as well as reductions in productivity, weight gain, feed efficiency, fertility and ability to resist disease; both quantity and quality of meat, milk and egg production decrease. Any economic costs must be weighed against the costs of preventing toxins and other food contaminants through better production, harvesting and storage practices. Hence, Ortiz *et al.* (2008b) advocate that an international research agenda on food safety should consider integrated crop management and food processing packages; low-cost detection technology for rapid analysis to facilitate trade; an improved understanding of agroecosystems to provide guidance on toxin and other food contaminant management; and high-level panels with scientists, non-government organizations, farmers, traders, consumers, health officers and policy makers to monitor intervention strategies and organize awareness campaigns.

development agenda, scientists elsewhere will be serving the agroecosystems and people of this world.

10.8 Outlook: Mobilizing Resources for an Agenda on Agriculture and the Environment

Scientists need to tap significant resources to implement a research and capacity-building agenda that links agriculture and environment and addresses development issues of our time. There are multilateral, bilateral and local funds available from international aid, national governments, civil society and the private sector (that includes philanthropy) to support an agenda that includes some of the topics indicated above. Research organizations, therefore, should align their priority undertakings to addressing the interest of and capturing the investments of such funding sources whose main interest seems to be most often on global impacts. Succeeding in resource mobilization results from capturing specific investors' interests and, due to the acknowledgement by them of scientists' competitiveness and credentials at national, regional or global levels, where they will bid with others for their funds. Tapping resources from the private sector needs creative thinking for defining alliances or consortia with multinational or local enterprises in which the national public sector can also participate actively (Ortiz and Crouch, 2007). Such undertakings with the private sector should fit with the research organization's policy or guidelines on intellectual property and partnerships.

A research-for-development approach towards sustainable and profitable agro-ecosystems, bringing wealth and health to people and preserving the environment and its biodiversity, should be able to attract grants for large and medium-size undertakings if their ensuing international public goods (either technology or knowledge) show potential spillovers worldwide. Hence, by thinking and acting on global issues, and by implementing a regional research-for-

References

Borlaug, N.E. (2007) Sixty-two years of fighting hunger: personal recollections. *Euphytica* 157, 287–297.

Cassman, K.G. (1999) Ecological intensification of cereal production systems: Yield potential, soil quality, and precision agriculture. *Proceedings of the National Academy of Sciences (USA)* 96, 5952–5959.

Crouch, J.H. and Ortiz, R. (eds) (2007) *CGIAR System Priority 2B: Tolerance to Abiotic Stresses Framework Plan.* CGIAR Science Council, Rome.

DfID (2004) What is pro-poor growth and why do we need to know? *Pro-poor Growth Briefing* 1. Department for International Development, Policy Division, London, 4 pp.

Dwivedi, S.L., Crouch, J.H., Mackill, D., Xu, Y., Blair, M.W., Ragot, M. *et al.* (2007) Molecularization of public sector crop breeding: progress, problems and prospects. *Advances in Agronomy* 95, 163–318.

Evenson, R.E. and Gollin, D. (2003) Assessing the impact of the Green Revolution, 1960 to 2000. *Science* 300, 758–762.

GFAR (2004) *Annual Report 2003.* Global Forum on Agricultural Research, Rome, Italy.

GFAR (2006) *Reorienting Agricultural Research to Meet the Millennium Development Goals, Proceedings of the GFAR 2006 Triennial Conference,* New, Delhi, India, 8–11 November. GFAR Secretariat, Rome.

Hedden, P. (2003) The genes of the Green Revolution. *Trends in Genetics* 19, 5–9.

Iwanaga, M. and Ortiz, R. (2007) *Should Energy be a Product of 21st Century Agriculture in Developing Countries?* Centro Internacional de Mejoramiento de Maíz y Trigo, El Batán, Mexico, 18 pp. (http://www.cimmyt.org/english/docs/brochure/apaari2007.pdf).

Jat, M.L., Gupta, R.K., Erenstein, O. and Ortiz, R. (2006) Diversifying the intensive cereal cropping systems of the Indo-Ganges through horticulture. *Chronica Horticulturae* 46, 27–31.

Kiers, E.T., Leakey, R.P.B., Izac, A.-M., Heinemann, J.A., Rosenthal, E., Nathan, D *et al.* (2008) Agriculture at a crossroads. *Science* 320, 320–321.

Ortiz, R. (compiler) (2008a) *Alliance of CGIAR Centers Best Bets for Boosting Crop Yields in* sub-Saharan Africa. World Bank, Washington, DC, 147 pp. (http://www.worldagroforestrycentre.org/downloads/CGIAR_boosting_yields_ssa.pdf).

Ortiz, R. (2008b) Crop genetic engineering under global climate change. *Annals of Arid Zone* 47, 1–12.

Ortiz, R. and Crouch, J.H. (2007) Creating an effective process to define, approve and review the research agenda of institutions in the developing world. In: Loebenstein, G. and Thottappilly, G. (eds) *Agricultural Research Management*. Springer, Dordrecht, Netherlands, pp. 65–92.

Ortiz, R. and Smale, M. (2007) Transgenic crops: pro-poor or pro-rich? *Chronica Horticulturae* 47, 9–12.

Ortiz, R., Crouch, J.H., Iwanaga, M., Sayre, K., Warburton, M., Araus, J. *et al.* (2006) Bio-energy and agricultural research-for-development. In: *Vision 2020 for Food Agriculture and the Environment – Bioenergy and Agriculture: Promises and Challenges* 14(7). International Food Policy Research Institute, Washington, DC, 2 pp.

Ortiz, R., Mowbray, D., Dowswell, C. and Rajaram, S. (2007a) Norman E. Borlaug: The humanitarian plant scientist who changed the world. *Plant Breeding Reviews* 28, 1–37.

Ortiz, R., Pérez Fernandez, M., Dixon, M., Hellin, J. and Iwanaga, M. (2007b) Specialty maize: global horticultural crop. *Chronica Horticulturae* 47, 20–25.

Ortiz, R., Ban, T., Bandyopadhyay, R., Banziger, M., Bergvinson, D., Hell, K. *et al.* (2008a) CGIAR research-for-development program on mycotoxins. In: Leslie, J.F., Bandyopadhyay, R.

and Visconti, A. (eds) *Mycotoxins: Detection Methods, Management, Public Health and Agricultural Trade*. CABI Publishing, Wallingford, UK, pp. 415–424 (http://www.ifpri.org/2020/focus/focus14/focus14_07.pdf).

Ortiz, R., Sayre, K.D., Govaerts, B., Gupta, R., Subbarao, G.V., Ban, T. *et al.* (2008b) Climate change: can wheat beat the heat? *Agriculture, Ecosystems and Environment* 126, 45–58.

Sanginga, N., Dashiell, K.E., Diels, J., Vanlauwe, B., Lyasse, O., Carsky, R.J. *et al.* (2003) Sustainable resource management coupled to resilient germplasm to provide new intensive cereal-grain-legume-livestock systems in the dry savanna. *Agriculture Ecosystems and Environment* 100, 305–314.

Trethowan, R.M., Reynolds, M.P., Ortiz-Monasterio, I. and Ortiz, R. (2007) The genetic basis of the Green Revolution in wheat production. *Plant Breeding Reviews* 28, 39–58.

Williams, J.H., Philipps, T.D., Jolly, P.E., Stiles, J.K., Jolly, C.M. and Aggarwal, D. (2004) Human aflatoxicosis in developing countries: a review of toxicology, exposure, potential health consequences, and interventions. *American Journal of Clinical Nutrition* 80, 1106–1122.

Winslow, M.D. and Ortiz, R. (2010) Biofuels: Risks, opportunities and dilemmas in the context of international agriculture. In: Payne, W. and Ryan, J. (eds) *The International Dimension of the American Society of Agronomy: Historical Perspective, Issues, Activities and Challenges*. American Society of Agronomy, Madison, Wisconsin, pp. 99–106.

World Bank (2007) *Agriculture for Development: World Development Report* 2008. World Bank, Washington, DC, 365 pp.

Index